AF803387

Silk Roads

Silk Roads

From Local Realities to Global Narratives

edited by

Jeffrey D. Lerner and Yaohua Shi

OXBOW | books
Oxford & Philadelphia

Published in the United Kingdom in 2020 by
OXBOW BOOKS
The Old Music Hall, 106–108 Cowley Road, Oxford OX4 1JE

and in the United States by
OXBOW BOOKS
1950 Lawrence Road, Havertown, PA 19083

© Oxbow Books and the individual contributors 2020

Hardback Edition: ISBN 978-1-78925-470-9
Digital Edition: ISBN 978-1-78925-471-6 (ePub)

A CIP record for this book is available from the British Library

Library of Congress Control Number: 2020938425

All rights reserved. No part of this book may be reproduced or transmitted in any form or by any means, electronic or mechanical including photocopying, recording or by any information storage and retrieval system, without permission from the publisher in writing.

Printed in Malta by Melita Press

Typeset by Versatile PreMedia Services (P) Ltd

For a complete list of Oxbow titles, please contact:

UNITED KINGDOM
Oxbow Books
Telephone (01865) 241249
Email: oxbow@oxbowbooks.com
www.oxbowbooks.com

UNITED STATES OF AMERICA
Oxbow Books
Telephone (610) 853-9131, Fax (610) 853-9146
Email: queries@casemateacademic.com
www.casemateacademic.com/oxbow

Oxbow Books is part of the Casemate Group

Front cover: "Interior of Tilya-Kori Madrasah on Registan Square in Samarkand, Uzbekistan." Photo: J.D. Lerner.

Contents

Preface vii
Acknowledgements ix
Contributors xi

Introduction 1
Jeffrey D. Lerner and Yaohua Shi

Section One: Acculturation and Hybridization

1. The 'Birth' of the Silk Road Between Ecological Frontiers and Military Innovation 11
   Nicola Di Cosmo

2. Sogdians in Shanxi (386 CE–618 CE): Literary and Archeological Evidence 21
   Xiaoyan Qi

3. From Exotic Toys to Objects of Scientific Inquiry: A Special Way of Transmitting European Optical Knowledge in the Qing Dynasty 31
   Yunli Shi

4. The Karakorum Highway: Gateway of Empires, Religions, and Commerce 49
   Saba Samee

Section Two: Understanding Spice Through Interdisciplinarity

5. A TRP Along the Silk Roads: How and Why We Detect and Use Spices 65
   Wayne L. Silver and Cecil J. Saunders

6. Silk Road Pharmacy: Debating Theriac and Defining the Natural World 81
   Monique O'Connell

7. Spice and Taste in the Culinary World of the Early Modern Mediterranean 97
   Eric Dursteler

Section Three: Tradition as Continuity and Change
8. Devotional Prints and Practice: Woodcuts from the Library Cave at Dunhuang 113
Bernadine Barnes

9. Dome of Heaven: From the Lantern Ceiling to the Chinese Wooden Dome 131
Di Luo

10. 'Malacca' – From Fabled Port to Muddy Lagoon: A Cautionary Tale of Ecological Disaster 161
Margaret Sarkissian

11. 21st-century Trading Routes in Mongolia: Changing Pastoral Soundscapes and Lifeways 177
Jennifer C. Post

12. Erasing the Local, Celebrating the Local: Tracing the Contradictions of the Silk Road in Pakistan 197
Chad Haines

Section Four: Cultural Transactions
13. Arsacid Economic Activity on the Silk Road 215
Touraj Daryaee

14. Pearls and Power: Chōla's Tribute Mission to the Northern Song Court within the Maritime Silk Road Trade Network 223
James A. Anderson

15. 'Flying Cash': Credit Instruments on the Silk Roads 237
Dan Du

Section Five: Long-Distance Commodity Trade
16. The Case for Shipwrecked Indians in Germany 267
Jeffrey D. Lerner

17. Samuel Shaw's 'Maritime Silk Road' from American Independence towards Monopoly, 1784–1794 285
John A. Ruddiman

Preface

In the Fall 2016 semester an interdisciplinary group of Wake Forest University faculty came together to form Silk Roads Winston-Salem (SRWS) with the intent of encouraging teaching and scholarship related to the social, economic, religious, and artistic exchanges along the historical Silk Roads. From this group, the faculty formed an ad hoc Steering Committee to organize events and programming. The six members of the Steering Committee were Stew Carter (Music), Elizabeth Clendinning (Music), Andrew Gurstelle (Anthropology), Jeffrey D. Lerner (History), Yaohua Shi (East Asian Languages and Cultures), Qiong Zhang (History), and Monique O'Connell (History). In the Spring 2017 semester, the Steering Committee announced the SRWS project to the general faculty at the College Faculty Meeting on 13 March 2017. Events that had already been planned for the semester were brought together with SRWS support and promotion. Following success with these early events, the Steering Committee decided to focus its efforts over the course of two years, spanning the 2017–2018 and 2018–2019 academic years. The planned events culminated in a conference, *Silk Roads: From Local Realities to Global Narratives*, that brought leading scholars of Silk Roads studies to the Wake Forest campus.

A multiple-day, interdisciplinary conference was held 27–30 March 2019. Instead of adhering to a geographic or temporal focus, the theme of the conference emphasized exchange and transformation along the Silk Roads – moments of acculturation or hybridization that contributed to novel syncretic forms.

Acknowledgements

This book is the product of a closed conference held on 27–30 March 2019 hosted by Silk Roads Winston-Salem on the Wake Forest University campus at Reynolda House Museum of American Art in Winston-Salem, North Carolina.

Above all, the editors wish to extend their deep appreciation to the contributors of this volume for their responsiveness to and observance of the deadlines and instructions throughout the different phases of the project. Despite occasionally difficult circumstances and busy personal schedules, deadlines were met. The authors' commitment to the project is commendable and deeply appreciated.

As this book goes to the printer, we would like to extend our appreciation to those who made this book possible by offering their goodwill and support. We need first to thank J. Kline Harrison, Associate Vice President and Associate Provost for Global Affairs, for his invaluable input, support and inspiration throughout this project. We also want to thank our colleagues in the Department of History who have supported us throughout this endeavor. Finally, we owe our special gratitude to the Office of the Dean.

There are also two individuals without whose assistance this book would not have come together. Our special thanks goes out to Julie Pechanek for being so supportive and diligent in her work on this project, particularly as she kept us on task. We want also to thank our Technology Specialist Tyler Pruitt who was always in the right place with the right fix when we needed it most.

Jeff Lerner is most grateful to his wife, Colleen, for inspiring him to complete the volume and for her continued support, although it meant that he had to devote the majority of his time to the project.

Contributors

James A. Anderson is Associate Professor of History at the University of North Carolina at Greensboro. He has written *The Rebel Den of Nùng Trí Cao: Loyalty and Identity along the Sino-Vietnamese Frontier* (2007). He co-edited *The Tongking Gulf Through History* (2011) with Nola Cooke and Li Tana as well as *China's Encounters on the South and Southwest: Reforging the Fiery Frontier Over Two Millennia* (2015) with John Whitmore. He is currently finishing a new book on the Southwestern Silk Road, 9th–13th centuries.

Bernadine Barnes is Professor of Art History at Wake Forest University. She is a specialist in early modern art and the history of prints. Her most recent research is on the uses of devotional prints in the 16th century.

Touraj Daryaee holds the Maseeh Chair in Persian Studies and is the Director of the Dr Samuel M. Jordan Center for Persian Studies and Culture at the University of California, Irvine. He is a historian of the ancient Iranian World, as well as Zoroastrian religion and Iranian languages and cultures. His latest book is *Sasanian Iran in the Context of Late Antiquity: The Bahari Papers at Oxford University*, published in 2018.

Nicola Di Cosmo is Henry Luce Professor of East Asian Studies at the Institute for Advanced Study. He is a historian of China and Inner Asia, from the ancient to the early modern period. He has written on Mongol and Manchu history, and, more recently, on climatic issues in the history of nomadic empires.

Dan Du is Assistant Professor of History at the University of North Carolina – Charlotte. She completed her PhD at the University of Georgia in 2017. Her current book project is titled, *This World in a Teacup: Chinese-American Tea Trade in the Nineteenth Century.* Her research explores US tea consumption after the American Revolution as well as China's banking and credit culture.

Eric Dursteler is Professor and former Chair of the Department of History at Brigham Young University. He completed his PhD at Brown University in 2000. His research focuses on gender, religious identity, language, and food in the early modern Mediterranean. He is the author of several books, including most recently *In the Sultan's Realm: Two Venetian Ambassadorial Reports on the Early Modern Ottoman Empire* (2018). He is currently completing a book on food and identity in the early modern Mediterranean.

Chad Haines is Cultural Anthropologist and Assistant Professor of Religious Studies at Arizona State University. His research focuses on everyday peace and lived ethics with a focus on Pakistan/South Asia, Islam, urbanism, and cosmopolitanism. He is the author of *Nation, Territory, and Globalization in Pakistan* and the forthcoming *Muslim Pathways: Religion in the Alleys of Cairo, Islamabad, and Dubai* along with being co-editor of *Women and Peace in the Islamic World* and *People's Peace: Prospects for a Human Future.*

Jeffrey D. Lerner is Professor at Wake Forest University. He is a historian of Ancient History specializing in Central Asia from the Achaemenids to the Kushans. He has written on art and archaeology, epigraphy, history, and numismatics.

Di Luo is Chu-Niblack Assistant Professor of Art History and Architectural Studies at Connecticut College. Her research includes Chinese wooden architecture, the building traditions of Asia, and the application of digital technologies in art history, urban studies, and cultural heritage preservation.

Monique O'Connell is Professor of History and Department Chair at Wake Forest University, where she holds the James P. Barefield Endowed Faculty Fellowship. She has two main areas of research interest: the political and social history of Renaissance Venice and the history of connection and exchange in the medieval and early modern Mediterranean. Her publications include her book *The Mediterranean World: From the Fall of Rome to the Rise of Napoleon*, co-authored with Eric Dursteler (Baltimore 2016).

Jennifer Post is Senior Lecturer in ethnomusicology at the University of Arizona. Her most recent work focuses on music and sonic practices of Kazakh pastoralists in Mongolia. She is currently co-editing a volume titled *Mongolian Sound Worlds.*

Xiaoyan Qi is Associate Professor at Changzhi University, China. She completed her PhD at Nankai University, China, in 2017. Her main research interest concerns the study of Sogdians and Sogdiana in light of East–West interaction. Her current book project involves a study on the spread of Hellenism into China by the Sogdians.

John Ruddiman is Associate Professor of History at Wake Forest University. He completed his PhD at Yale University in 2010. His 2014 book, *Becoming Men of Some Consequence: Youth and Military Service in the Revolutionary War*, explores the lives and choices of young men in the Continental army.

Saba Samee is Associate Professor and Coordinator in the School of Culture and Language at the Institute for Art and Culture, Pakistan. She completed her Masters in Archaeology of Buildings from the University of York, UK, in 2008. Her research interest focuses on the relationship of people with their places. The northern provinces of Khyber Pakhtunkhwa and Punjab are the regions where she bases her research. She has published numerous papers and book chapters on the Built Heritage of Pakistan and the values people associate with these historic structures.

Margaret Sarkissian is Professor of Music at Smith College in Northampton, Massachusetts. She completed her PhD at the University of Illinois, Champaign-Urbana and is author of *D'Albuquerque's Children: Performing Tradition in Malaysia's Portuguese Settlement* (2000) and, with Ted Solís, *Living Ethnomusicology: Paths and Practices* (2019). Her fieldwork and research interests lie in the city of Melaka, Malaysia, where she works primarily with the Portuguese and Straits-Chinese (Baba Nyonya) communities. When in Melaka, she plays accordion with the 1511 O Maliao Maliao Cultural Troupe.

Cecil J. Saunders is Research Assistant Professor in the Biology Department at Wake Forest University. He completed his PhD in Neuroscience at the University of Colorado Medical Camus in 2014. His research questions are focused on learning how evolution has shaped and retooled the cellular and molecular mechanisms of chemosensory systems across different organisms.

Yunli Shi is Professor of History of Science at the University of Science and Technology of China. His main research interests are the history of astronomy and the dissemination of Western sciences in China before the 19th century.

Wayne L. Silver is Professor in the Biology Department and Co-Director of the Undergraduate Neuroscience Program at Wake Forest University. His main research interest focuses on chemesthesis, the chemical stimulation of the somatosensory system, which is largely responsible for our detection of spices.

Yaohua Shi is Associate Professor of East Asian Languages and Cultures at Wake Forest University. His research interests include pre-modern Chinese vernacular fiction and East–West cultural relations.

Introduction

Jeffrey D. Lerner and Yaohua Shi

In recent decades there has been a new surge of interest in the history and legacies of the Silk Roads both within academic and public discourses. A field of Silk Roads Studies has come into its own. Consciously mirroring the temperament of its subject, the field has moved out of the narrow niches of particular disciplines to become a truly interdisciplinary endeavor. New research findings about the historical operations of the Silk Roads and interpretations of their legacies for the modern and contemporary world have broken down geographical and temporal divides that once demarcated the Silk Roads as primarily pre-modern and Old World-centered conduits of globalization. In light of these developments, the time is ripe to begin formulating a new definition of the contour of Silk Roads Studies and laying a new foundation for further work in this field.

To that end, *Silk Roads: From Local Realities to Global Narratives* stems from a conference of the same name that occurred in a multiple-day, interdisciplinary meeting at Wake Forest University on 27–30 March 2019. The volume consists of an array of approaches by 18 leading scholars and researchers from the United States and abroad.

Rather than adhering to a specific geographic or temporal focus, the theme of the volume as with the conference emphasizes exchange and transformation along the Silk Roads. *Silk Roads: From Local Realities to Global Narratives* contrasts with other works on the subject that are restrictive in terms of the temporal and spatial context they cover as well as the disciplinary approach to the subject. By contrast, this volume differs in that it brings together contributions intended to facilitate the sharing and critiquing of multi-disciplinary perspectives. This holistic approach to understanding globalization, exchanges, transformations, and movements rooted in the past, and their continued relevance in the present, is in line with contemporary academic trends toward interdisciplinarity. Indeed, the Silk Roads is such an expansive topic that many approaches to its study must be included in order to do justice to representing its many facets.

Silk Roads: From Local Realities to Global Narratives takes a comprehensive approach to defining the field of Silk Road studies. It does so in three ways:

1. It highlights the multiplicity of networks that constituted the Silk Roads. Crucially, this volume will incorporate both land and maritime routes into a single focus.
2. It approaches the Silk Roads from a plethora of time periods spanning from antiquity to China's One Belt One Road Initiative.
3. It includes the works of 18 scholars from many disciplines, rather than catering to a single academic specialty.

The intent of this volume is to present a format that will lead to a new definition of the contours of Silk Roads Studies. The aim is to establish an interdisciplinary approach to current research. The civilizations and stateless societies that formed this exceptionally creative intersection of peoples, goods and ideas serve a particularly rewarding theme for comparative analysis. From our vantage point, we are in a position to appreciate the Silk Roads as systems of networks[1] in both the physical sense of place and object and as a series of social networks connoting power and commerce. The consequences both envisioned and accidental led to a series of interconnected trade routes formed over long distances in which people traveled through an array of landscapes that served as a medium of cultural and personal identity construction. In this regard, trade networks acted as places of interaction. We employ the term Silk Roads as partly a recognizable representation to explain past events, and partly an analytical instrument to describe past actions.

We have chosen to refer to the Silk Roads in the plural, because as a term it best represents the multiplicity of land and sea trading routes that initially linked East Asia with the Afro-Eurasian continent in antiquity, but subsequently came to embrace the Americas in the early modern period serving as the basis of what we today recognize as Globalism. The idea of the 'Silk Road' was coined in the 19th century by the German geographer Ferdinand von Richtofen.[2] For some the term has outlived its usefulness in light of new approaches gleaned from new evidence. We, however, chose to employ the term precisely because it is recognizable and thereby has advantages over other less understandable and more restrictive foci. Until a better term – conceptually and methodologically – can be found to replace it, we are constrained to retain the use of 'Silk Roads' from a matter of convenience.

The book is intended to be useful not only to those engaged in Silk Road Studies, but to scholars involved in comparative cultures and religions, as well as for socio-economic and intellectual historians interested in the comparative analysis of states and stateless societies. It is also designed to help link the fields of European, African, Asian, and American studies to wider interdisciplinary debates within the humanities and social sciences.

The interdisciplinary team of historians, archaeologists, ethno-musicologists, art historians, biologists, political scientists, and anthropologists whose contributions make up this volume provide a variety of different perspectives of artistic, scientific, social, intellectual, and economic development. The book is divided into five sections consisting of 17 articles.

Section One: Acculturation and Hybridization concerns the circuitous processes in which different social and economic relationships are continually in a state of flux, and through which entirely new social and material conditions emerge. Acculturation processes may vary with respect to levels of technology, circumstances of distance, power dynamics, and the individual motivations that drive movement, contact and exchange. In contact situations, uncertainty and equivocation may inexplicably assist hybridized cultures to become more cogent, because the cryptic meanings and understandings that emerge in establishing identity can serve to link cultural differences and solidify relationships.[3] Nicola Di Cosmo focuses on the 5th–3rd century BCE, a period marked by increasing militarization in China ('Warring States') and the social and economic changes of the societies situated in the steppe region along China's borderlands. In rejecting the accretional expansion of cultural networks dating back to the 3rd millennium BCE (or even earlier), and the sudden-burst explanation of a relatively abrupt event following the expansionist rush into Central Asia of a newly minted imperial China, Di Cosmo offers a different perspective. He examines how frontier dynamics may have affected existing patterns of exchange and created the conditions for the later expansion of this region as an overland series of transregional networks and trade routes. Xiaoyan Qi analyzes the participation of Sogdians in the East–West economic, political, and cultural exchanges. Although initially they participated on a small scale, this changed in the early 3rd to the 7th century CE when they became the most active traders along the Silk Roads, many of whom came to inner China by way of Xinjiang. Within this broad context, a great number of Sogdians settled in Jin (modern Shanxi) where they established political, economic, and cultural contacts between Sogdiana and Shanxi. It is within this context that she attempts a reconstruction of the route that the Sogdians took to enter Shanxi, the places where they settled, and their interaction with the elites. Yunli Shi argues that clocks, medicines, musical, and scientific instruments are no less important as books in long-distance transmission of knowledge. However, compared with knowledge disseminated through linear language, knowledge transmitted through material objects involves a far more complex and subtle process of acculturation and adaptation. Focusing on optical instruments, Yunli Shi shows how European missionaries, particularly the Jesuits, exhibited them as exotica to convince the skeptical Chinese of the superiority of European knowledge and religion. Tracing part of the ancient Maritime Silk Road, Jesuit fathers arrived in China in the 17th and 18th centuries. Optical instruments and their knowledge of optics won them access to the imperial court. Shi shows how over time Chinese craftspeople and scientists decoded and appropriated European optical instruments for local uses. Saba Samee surveys the legacy of Pakistan as an important crossing point for trade and commerce testified by the remains of diverse historic monuments. Buddhist relics, medieval fortresses, and an array of other archaeological remains bear witness as cultural expressions of these historic trade routes that led travelers into surrounding regions. While the Silk Roads brought economic prosperity and socio-cultural exchanges, they also brought conflict. She looks at China's 'One Belt One Road Initiative' (*i.e.* 'the China-Pakistan Economic Corridor') as a new challenge confronting Pakistan today from the perspective of whether this economic belt will surpass the role played by the ancient Silk Roads, or if it will create its own series of mixed horrors and benefits.

Section Two: Understanding Spice Through Interdisciplinarity reflects a practice that is increasingly becoming normative scholarship: the engagement in research and teaching that cuts across multiple academic disciplines in the course of researching a subject. This requires a scholarship that necessitates learning the languages, cultures, methods, and knowledge of our colleagues in different disciplines.[4] In this section, the study of spice exemplifies the intersection of different disciplines to address new ways of approaching a subject. Wayne L. Silver and Cecil J. Saunders begin this section by examining spices from a biological perspective. They note that humans have used spices for millennia without knowing how they work. Spices are plant-derived, containing chemicals, which plants use primarily for defense – repelling herbivorous predators. The active ingredients in most are repellant because they stimulate pain receptors, yet humans seek out and intentionally consume these substances. They are the only animals to do so. The receptors for many of the irritants in spices are Transient Receptor Potential (TRP) channels found in the mouth, nose, and eyes. While the molecular and cellular mechanisms by which active ingredients in spices stimulate nociceptors are understood, their paper seeks to explain why humans enjoy the sensations associated with these compounds. Monique O'Connell investigates the late 15th century marking Venice's Golden Age of Spices: in 1496, for example, four galleys from Alexandria arrived in the city carrying approximately 4 million pounds of spices, mostly pepper. Her paper looks at spices that were sold in Venice as simple remedies or compound medicines in Venetian pharmacies. Venice had more than 50 apothecary shops in the early modern era. Although they used spices in medicines, these shops also acted as centers of knowledge, museums of exotic and strange material from around the world, offering medical advice, theological speculation, and healthy doses of gossip and news. Venetian pharmacies were microcosms of Silk Road trading zones, mixing the exotic and the familiar for the medical benefit and the entertainment of the city. Eric Dursteler notes that the prevailing narrative holds that beginning in the 16th century the 'spicy' Middle Ages were supplanted by a shift toward more subtle flavorings produced by indigenous European herbs, rather than exotic eastern spices. He argues that the movement away from spices was gradual and not universal. An examination of cuisine in the early modern Mediterranean provides ample evidence of the persistence of spice usage, even as foodways in this period experienced important changes as a result of the introduction of new foods and flavors from both the new and the old worlds.

Section Three: Tradition as Continuity and Change examines how technology, art and architecture, and ethnomusicology can be used as vehicles for understanding how heritage defines a culture. Peoples' interactions are not limited by clearly defined borders in social structure, space, or time, because tradition is transferred diachronically or synchronically, even when it is not fully appreciated or recognized.[5] Bernadine Barnes focuses on the woodblock prints found in the so-called 'Library Cave' at Dunhuang. These are the earliest surviving examples of woodblock prints in the world, some dating from as early as the 7th century. She examines the function of the prints, as icons, relics, and elements of religious practice, and considers how these prints were distributed or preserved. The survival of the prints in Cave 17 is remarkable, but it is an indication of more widespread visual culture. In their technique and uses, the prints create an interesting comparison to 15th-century woodcuts in Europe. She questions whether there was a connection between the works made hundreds of years apart. Di Luo draws on the earlier work of Alexander Soper to examine a type of ceiling dome popular in Asian architecture from India to Korea: the 'diamond in the square'. By analyzing their differences in building material (wood or stone), use in religious function and ritual practice, ornamentation associated with a celestial realm, she is able to demonstrate how certain formal features combine to create a symbolic and ritual space. Although generally adopting the same formal pattern albeit with variations, she argues that the significance and popularity of this architectural construction was purposefully selected for its structural intricacy and ingenuity. She explains how they were able to transform the interior of architecture into a ritual space and to connect it with an imaginary portal to heaven. Margaret Sarkissian explores the fluctuating fortunes of Malacca and its transnational communities, primarily through the lens of a small coastal fishing community known as the Portuguese Settlement as an example of hybridization and acculturation. Its residents, who identify themselves as 'Portuguese descendants' (though some have Dutch or British family names), continue to speak a unique Portuguese creole language, profess the Catholic faith, and maintain a diasporic Portuguese performing tradition. Having successfully navigated life in post-independence Malaysia by positioning themselves as a distinct and colorful cultural minority, this community is now challenged by a new threat: China's new 'Silk Road Initiative'. The Chinese-funded 'Melaka Gateway', a massive development project that is building a chain of islands and a new city in the sea, has the potential to erase historical Malacca entirely and destroy the fragile environment inhabited by the Portuguese community. Finally, in her analysis of new trade routes and other pathways related to economic, ecological, and cultural exchange now crisscrossing pastoral lands in Asia, Jennifer Post notes how these offer expanded opportunities for travel for local residents and outsiders. Yet they have also had a significant impact on local lifeways, including soundways. Paths used by herders as they support local lands using socially and ecologically established patterns, combined with routes that are the result of new forms of mobility, contribute in different ways to changes in soundscapes, sound practices, and the land itself. Most of the movement of people, goods, and ideas threaten sounds that have traditionally reinforced the human–animal–human–plant relationships that herders have maintained for generations to access critical information on ecosystem health. By drawing on music, ecology, geography, and ethnobiology to consider how the new trade networks in Mongolia, and in other pastoralist sites, create new cultural scenes, she is able to explain how changes in the economic patterns and ecological soundscapes affect pastoralists' engagement in sustainable land management as well as local human and environmental well-being. The section concludes with Chad Haines re-examining the contradictory imagining of the Silk Roads as a historical narrative of cosmopolitanism through interconnectedness, cultural exchange, and human creativity, while simultaneously used to legitimize nationalist and racist discourses, provides avenues of conquest and subjugation, and commodifies cultural differences. Focusing on the conjuring of the Silk Road in Pakistan in historiographical studies, political rhetoric, and development discourses, he analyzes how this tract of the Silk Roads myth is forged in identity politics, naturalizing the nation-state idea of bounded belonging rather than providing an avenue of transcending its artificial borders. He proposes that the Silk Roads as manifested in Pakistan reflects how mythologizing is employed to advance particularistic and bounded identities. The Silk Roads myth, therefore, is produced and circulated globally, nationally, and locally, taking on differing meanings, often in conflict with one another, but ultimately reinforcing nationalist interests.

Section Four: Cultural Transactions underscores the inherent tension between the notion of recognition in relation to identities of culture and the redistribution of power and resources. It tackles the question of how culture can be reified through material. The notion considers how cultures can be seen to reproduce their values through institutions and practices even at the expense of another's cultural capital.[6] While the interest in late antique studies has largely concentrated on Roman-Sasanian interactions, Touraj Daryaee examines the large output of Sasanian silver coins, primarily drachms, used as military pay and economic exchange found in hoards mainly to the east of the Empire. By mapping out some of these hoards on land and the coast, it becomes clear that the economic interest of the Persians was towards the East, and not the West. These coin hoards also demonstrate the economic reach of the Persians, not as a military power, but as traders on the Silk Roads. Furthermore, the non-Sasanian coins struck between Persia and China, which imitate the Sasanian drachms, suggest the prestige of an Empire, which continued beyond the 7th century CE. James A. Anderson investigates an 11th-century anecdote about an emissary from the South Indian kingdom of Chōla recorded by the Chinese scholar, Peng Cheng, of the Northern Song dynasty. Anderson uses this account and draws from others to understand the nature of Song China's relations with the maritime kingdoms beyond its borders. From these records came an official understanding of 'barbarian decorum', against which the Chinese state continued to measure its own cultural norms. A study of China's relations with its southern neighbors reveals the many levels of images employed by the Song as expressions of the power and prestige of the Chinese cultural empire in encounters with powerful but distant courts along the Maritime Silk Roads. Dan Du considers the role played by currency along the Silk Roads, principally evidence of pre-modern, non-European credit instruments. Although most research on the Silk Roads tends to highlight commodity monies, such as spice and silk, or hard money, like gold and silver, the cost and risk of transporting cash long distances led to the establishment of cashless payments. She explores well-documented credit instruments in China, India, and the Islamic Empire, the three major powers on the Silk Roads, and revisits the financial structure of pre-modern world trade. The credit instruments employed as fiscal and commercial devices in China, India, and the Islamic Empire highlight the sophistication of credit techniques in the East. The preoccupation with commodity and hard monies, however, sets the nature of world trade during the Silk Roads period apart from that in the European-dominated era, which featured a credit economy.

Section Five: Long-Distance Commodity Trade examines how long-distance commodity trade remained even in the early modern period a relatively rare phenomenon as travel was slow and vehicles lacked high freight capacity. Instead the majority of people engaged in trade and commerce closer to home. The two articles that compose this section concern seaborne trade by means of inland waterways and transoceanic voyages. Jeffrey D. Lerner examines three ancient versions derived from the lost work of Cornelius Nepos (*c.* 110–24 BCE), who wrote that in 62 BCE a group of shipwrecked Indians in Germany were presented as a gift to the Proconsul Northern Italy, Quintus Caecilius Metellus Celer. This episode was preserved as proof that a continuous river called Okeanos ('Ωκεανός) flows around the earth, and that it is possible to circumnavigate it through a northwest or northeast passage. Although recent scholarship is dismissive of the incident, Lerner attempts to show that the episode is entirely plausible: a group of Indians could have made their way to Germany using a system of inland waterways that formed part of the northern route of the Silk Roads linking Afghanistan to the Black Sea where they subsequently found themselves shipwrecked on the banks of the Danube. John Ruddiman probes Samuel Shaw's service as the supercargo in 1784 on the Empress of China, the first American flagged vessel to sail in the 'China Trade'. Having served eight years in patriot forces in the Revolutionary War, and unable to support himself in peacetime, he sailed to 'the other side of the world' in pursuit of wealth and fame. His journals and correspondence in the 1770s and 1780s place his two voyages to China into a broader context with his military experiences, his earlier travels in North America, his financial ambitions, and his personal connections with the political elite of the new nation. In doing so, Ruddiman ascertains how Shaw's military experiences, particularly as a staff officer, shaped his subsequent economic and diplomatic endeavors.

Notes

1 The field of social network theory and analysis is a small but growing discipline. For an introduction to the terminology of social network analysis, see *e.g.* Scott 2000; Prell 2015. On network analysis, see Carrington *et al.* 2009; Fu *et al.* 2017. For the applicability of network science in archaeology and history among other disciplines, see Knappett 2013; Seland and Teigen 2017.
2 For an overview of the concept in English, see Waugh 2007.
3 Friedman 1997, 82, 88.
4 Froderman 2017, 6–8.
5 Cf. AlSayyad 2004, 7–8.
6 Hall 2000, 210.

Bibliography

AlSayyad, N. 2004. The End of Tradition, or the Tradition of Endings?, in: AlSayyad, N. (ed.), *The End of Tradition?*, London, 1–27.

Carrington, P.J., Scott, J. and Wasserman, S. (eds) 2009. *Models and Methods in Social Network Analysis*, Cambridge.

Friedman, J. 1997. Global Crises, the Struggle for Cultural Identity and Intellectual Porkbarreling: Cosmopolitans versus Locals, Ethnics and Nationals in an Era of De-hegemonisation, in: Werbner, P. and Modood, T. (eds), *Debating Cultural Hybridity*, London, 70–89.

Froderman, R. 2017. The Future of Interdisciplinarity: An Introduction to the 2nd Edition, in: Froderman, R, Thomspon Klein, J. and Dos Santos Pacheco, R.C. (eds), *The Oxford Handbook of Interdisciplinarity*, 2nd edn, Oxford, 3–8.

Fu, X., Luo, J.-D. and Boos, M. (eds) 2017. *Social Network Analysis: Interdisciplinary Approaches and Case Studies*, Boca Raton.

Hall, S. 2010. Conclusion: The Multi-cultural Question, in: Hesse, B (ed.), *Unhettled Multiculturalisms: Diasporas, Entanglements, Transruptions*, London, 209–241.

Knappett, C. 2013. *Network Analysis in Archaeology: New Approaches to Regional Interaction*, Oxford.

Prell, C. 2015. *Social Network Analysis: History, Theory & Methodology*, Los Angeles.

Scott, J. 2000. *Social Network Analysis: A Handbook*, 2nd edn, London.

Seland, E.H. and Teigen, H.F. (eds) 2017. *Sinews of Empire: Networks in the Roman Near East and Beyond*, Oxford.

Waugh, D. 2007. Richtofen's 'Silk Roads': Toward the Archaeology of a Concept, *The Silk Road* 5.1, 1–10.

Section One

Acculturation and Hybridization

Chapter 1

The 'Birth' of the Silk Road Between Ecological Frontiers and Military Innovation

Nicola Di Cosmo

In a nutshell, what goes under the name of Silk Road usually describes a system of communication, made of trails and routes, towns and cities, peoples and goods, that connected distant places. Such a system provided the skeletal structure of commercial exchange, population movements, and transmission of knowledge across Eurasia, both by land and by sea. This essay tries to answer a limited but necessary question: How and when was the Silk Road 'born'? That is to say: did it come into existence through a sudden, epochal event, or rather through a slow accretional and evolutional process? The question is whether we think about the Silk Road as an historical, rather than purely cultural, construct. As such, it implicitly requires the identification of changes and transformations of that system, which in turn connect the Silk Road with macro-events that today we recognize as the 'matter' of global history. Therefore, investigating the 'birth' of the Silk Road is not about 'origins' but rather about the establishment of contexts fit to provide an explanation about what changed and why. In other words, it is about focusing not on the general reconstruction of 'connectivity' but on the particular events that, deliberately or accidentally, allowed conditions that favored connectivity to emerge. In the same line of reasoning, it is essential to take a closer look at the interaction between natural and human factors, which can point to new ways to interpret the evolution of the Silk Road.

When we speak of natural-human interaction in the continental zone of Eurasia, one element appears to be especially significant: climate. The long-term impact of climatic variability on the expansion of nomadic cultures may have contributed, in view of the latter's mobility range and migratory patterns, to the creation of interconnected exchange networks across Asia. In a nutshell, the creation of climatic and environmental conditions favorable to the spread of pastoral societies may have also favored the spread of a material culture that expressed itself through a common idiom of objects, themes, and artistic forms, even though some local features remained distinctive. The Scythian, or Scytho-Siberian civilization of the Eurasian nomads usually dated from the first centuries of the 1st millennium BCE, has long been known as a cultural complex that exhibits numerous shared elements across different times and regions, especially at the level of elite culture. The nomads have been regarded as prime actors in the creation of a 'geo-economic' environment favorable to cultural conductivity.

The spread of a nomadic culture characterized by comparable cultural traits in the Bronze Age, however, does not in itself explain the connection between East Asia, and in particular China, and the rest of Asia through a system of long-distance communication based on commercial and diplomatic exchanges. Surely, exchanges between two materially very different worlds existed, and have been documented extensively, but these exchanges, prior to the second half of the 1st century BCE, appear as sporadic, occasional, and isolated phenomena that did not amount to a 'system'. There is a world of difference between objects that circulate by happenstance, through a whole host of possible means of transfer, and the deliberate export and import of goods within a design of economic relations emanating from, and consistent with, the social and political lives of distant and separate communities. Registering that difference provides the basis for the possible evolution from a mode of material diffusion that is documented in the evidential record, but cannot be seen as connected with the workings of a society as a whole, and one that is intentionally directed and operated by forces located within a society, and expressed coherently with that society's level of development.

If we accept these premises, the beginning of the Silk Road can be located at the intersection of two historical moments, one longer, the other quite short. The first moment can be identified in the expansion of nomadic cultures whose elites developed a taste for mutually recognizable luxuries, expressed in their own distinctive forms. Typically, these consisted of wearable or portable objects made of precious metals such as gold and silver, often with animal motifs, ornamental plaques and appliques in bronze, as well as tools and weapons. The second historical moment concerns the Late Warring States period in China (475–221 BCE), during which commercial relations between China and the steppe world were subject to a steep and sudden acceleration.

These two historical contexts converged to make the late Warring States period, that is, the 4th and 3rd centuries BCE, the historical moment in which the Silk Road came into existence as a historical artefact. Traditional chronologies date its beginning to the journey to Central Asia by Zhang Qian, the envoy sent in 138 BCE by the Han emperor Wudi to Central Asia with the mission to establish relations with nomadic peoples, in the context of the Han-Xiongnu wars, followed by the Han expansion into the Western Regions (today's Xinjiang). In partial contrast with this narrative, what appears to be more significant is an earlier moment of expansion of contacts and trade that is intimately connected with the deep transformation of the Chinese political landscape that preceded the Qin unification. While the Han expansion of the 1st century BCE is of course of great historical importance, and brought some states of Central Asia within the sphere of Han political influence and control, the foundation of the historical Silk Road has to be searched for in the Chinese economic and political transformations that preceded the imperial unification of 221 BCE. More specifically, it can be traced back to the heightened climate of war and competition that enveloped the Central Plains towards the end of the Warring States period. It is at this time that a strategic relationship between China and the steppe developed, and arguably created the basic interdependency and commercial integration that provided the rationale for long-distance exchange. Shifting our attention from the Han dynasty to the end of the Warring States is significant first of all because it changes the terms upon which the beginning of the Silk Road is predicated. It replaces the geopolitical narrative of Han imperial expansion into the Western Regions with a hypothesis of radical rebalancing of commercial relations between Chinese states and nomadic elites generated by China's dependency upon pastoral goods, and influx of Chinese goods into the nomadic exchange networks.

In order to explore in greater detail such a hypothesis, we need to take a step back. The role of nomads in the formation of arteries of communication across Central Asia has been pointed out by David Christian, Elena Kuzmina, and more recently Jessica Rawson among others.[1] Usually the arguments for a pre-history of the Silk Road focus on the transfer of various civilizational components – for lack of a better word – from domesticated crops to horse riding, chariots, and metal technology between Western Inner Asia (the Black Sea, the Ukraine and the Altai) and East Asia (Mongolia and China). This is what Andrew Sherratt called 'Trans-Eurasian Exchange' and others have called Bronze Age Eurasian Network.[2] While it is important to be conscious of how such elements were transmitted and received, how non-local phenomena were localized, and what impact they had in various parts of continental Eurasia, their connections with the later, historical Silk Road are dim at best, since 'transmission' is a concept that does not explain the 'leap' from the random diffusion of material goods documented across regions and cultures from the earliest times, to the systemic connectivity that we associate with the Silk Road. In other words, the transformative aspect represented by the coming into being of the Silk Road should be based not just on the documentation of contacts but rather on an analysis of the underlying historical contexts and causes leading to a structural transformation of international trade. From a methodological standpoint, the idea that the Silk Road came into existence as an ever-growing web of networks and ever-growing volume of exchange, detached from the societies it is supposed to connect, is surely flawed, in that it represents a shortcut that overlooks the historical circumstances that materially generate change. The theory that international communication networks are simply the product of incremental exchange and connections, a theory that is functional to the general paradigm of globalization studies, of which the Silk Road is seen as an important component, cannot be taken as a general assumption. Rather, specific historical contexts need to be explored to identify the transformative circumstances that made distant societies closer and more connected. In order to focus on those changes that, as mentioned above, converged to produce the historical context that favored the creation of the Silk Road, I will focus on two aspects: a cursory view of the environmental reasons for the expansion of nomadic cultures in Eurasia, and the relationship between nomads and Chinese states in the late Warring States period.

The 'nomadic pastoral complex'

David Christian's description of the proto-Silk Road as 'trans-ecological exchange', that is, exchange between societies occupying different ecological zones, is critical to grasping the importance of the environmental aspect in the evolution of nomadic cultures. Ecology is transformed as the environment changes due to climatic variability.[3] According to climatologists, a change that may have had momentous consequences for the history of the Silk Road occurred in the 9th or 8th centuries BCE. At this time, climatic changes in northern Asia enabled advances in pastoral nomadism as a specific mode of production, and, in turn, the emergence of aristocratic elites that commanded great political and military power, and shared a rich material inventory. This phenomenon has been identified archaeologically with the birth and expansion of the Scythian cultural complex. According to the authors,

> In south-central Siberia archaeological evidence suggests an acceleration of cultural development and an increase in the density of nomadic populations around 850 BC. We hypothesize a relationship with an abrupt climatic shift towards increased humidity caused by a decline of solar activity. Areas that initially may have been hostile semi-deserts changed into attractive steppe landscapes with a high biomass production and high carrying capacity. Newly available steppe areas could be invaded by herbivores, making them attractive for nomadic tribes. The central Asian horse-riding Scythian culture expanded, and an increased population density was a stimulus for westward migration towards southeastern Europe.[4]

Based on a close collaboration between climatologists and archaeologists, this study found that both the development of the Tagar culture in the Minusinsk region (South Siberia) from the early 1st millennium BCE – a nomadic culture closely connected with the later Scythian phenomenon – and the movement of nomadic peoples into the Tuvan region coincided with an increase in humidity due to diminished solar activity. These findings seemed to support the hypothesis that a climatic impulse might have favored the expansion of Scythian cultures across South Siberia.

This study was received with some skepticism and taken to task for not considering the anthropogenic impact that should have accompanied an increase in human and animal occupation,[5] but the authors responded to such criticism by citing as evidence the large increase of archaeological sites:

> [i]t is clear from the combined archaeological, paleoenvironmental and radiocarbon evidence that there was a sudden human occupation of a hitherto almost 'empty' Tuva.[6]

Their conclusions were based, 'in the first place on the observed high number of sites from the Scythian period compared to the evidence from the Late Bronze Age.'

The possibility of an expansion of nomadism in the same areas is based also on other climatic reconstructions, albeit based on different premises and results. According to Blyakharchuk *et al.*, the reconstructed climate for the mid-Holocene in Tuva and Siberia indicates a turn towards aridization around the 2nd and 1st millennium BCE (that is 4000–3000 BP) that transformed forested areas into a grassland environment. As a result, high-elevation areas, given the reduction of forest coverage, became suitable for mountain pastures, and for pastoral occupation. While this type of climatic change began probably 4000 years ago, it was accelerated 3000 years ago:

> In high-mountain areas of southeastern Altai and southwestern Tuva forest vegetation disappeared and high mountain tundra and tundra– steppe developed. This climatic cooling along with increased continentality and aridity became especially pronounced after 3500 14 C yr BP.[7]

A similar climatic pattern, which favored a reduction of forested areas and increase in grasses was found approximately at the same time (2000 BCE), in western Mongolia, where steppe vegetation and grasslands expanded.[8]

Whether such paleoclimatic studies attribute the changes in climatic conditions to greater humidity, which increased the grassland biomass, or to a more arid climate, replacing high-elevation forests with pastures, these changes seem to have favored the formation of abundant pasturelands in the regions of the Altai, Tuva, and Minusinsk basin. If climatic studies suggest a transformation of the environment that favored the spread of mounted pastoral nomadism and made it the dominant economic mode of production across the Inner Asian steppes, archaeological evidence suggests that after the mid-9th century BCE the nomadic societies become more complex, richer, and more connected.[9]

Such a nomadic expansion may have also affected the 'northern zone' of China proximate to the Shang and Zhou domains, not in the sense that we find here a distinct and identifiable Scythian culture, but rather in the sense that ecological frontiers became more clearly demarcated by the formation of a complex of nomadic societies that emerged in today's Mongolia, Inner Mongolia, the Ordos desert, and parts of Qinghai, Gansu and Ningxia. Here too, the affirmation of pastoral nomadism as a distinct economic specialization became in some areas predominant over other economic forms, and contacts among the elites evolved through the circulation and production of objects inspired by a broadly unified taste and artistic idiom. Artwork and luxury objects of this nomadic complex, at the same time, remained separate and distinct from China, where wealth and status were displayed on the basis of an entirely different cultural canon.

The funerary record of nomadic cultures in the 'Northern Region' is especially rich in luxury goods, and shows connections with the Altai, Tuva, Siberia, and further West. For instance, the small gold appliques that appear in late Warring States China in sites such as Aluchaideng, Majiayuan, and Xinzhuangtou, in the shape of various animals (rams, horses, wild pigs) have their clear counterpart in the Altai, in sites such as Pokrovka in the southern Urals and Arzhan II in Tuva.[10] Similar appliqués are found in mound 15 of the Pazyryk-type burials at Khankarinski Dol in the Northeastern Altai (Kazakhstan, dated to the 4th century BCE).[11]

At the same time, in the last phase of the Pazyryk culture (4th to 2nd century BCE) we have the earliest findings of Chinese silk in a nomadic context (in kurgans no. 3 and 5) as well as in others kurgan complexes (Katanda and Bashadar) also belonging to the Pazyryk culture, and dated to the 3rd century.[12] The silk certainly came from China, and was probably manufactured in the southern Zhou state of Chu. Particularly impressive is a long silk shirt found in the burial mound of a noble woman in the Altai republic of Russia, dated to the 3rd–2nd century BCE, which indicates a particular type of high-end export.[13]

What scholars have also pointed out is that some of the precious ornaments in the 'nomadic style' show that they were 'made in China' and produced specifically for the nomadic market. According to Yan Liu,

> Craftsmen in the northern Chinese states played a critical role in cultural transmission by making prestige goods in the style favoured by peoples of the steppes. This is also indicated by bronze plaques and ceramic models decorated with animal motifs recovered from the tomb of the Qin state metalworker in Shaanxi.[14]

But what is the historical background of this exchange? The evidence found in the material record is, in my view, closely connected with the growth of the commercial traffic between Chinese states and nomadic peoples, and the reasons for such a qualitatively and quantitatively significant increase have to be sought in the historical developments taking place in China during the late Warring States period.

The Warring States Crucible

The 4th and 3rd centuries BCE are a time when the few Chinese states that emerged from a century-long process of consolidation of power by a handful of ruling families were going through a process of increasing militarization, as a consequence of a constant state of war and evermore violent internecine struggle. Yet, at the same time, several textual references point to increasingly frequent contacts with the steppe region.

First of all, the reform of cavalry by King Wuling of the state of Zhao in the 4th century BCE points not just to cultural contacts but also to the need to import horses. This debate specifically refers to nomadic (*hu*) attire as a necessary import meant to facilitate riding horses. Adopting 'nomadic cavalry' meant not just getting more horses (surely from pastoral 'producers') but also, presumably, weapons and tactics, which implies the use of grooms and trainers from the steppes. All of this suggests a substantial increase in contacts with nomads as a consequence of Chinese military developments.[15]

The nomads could also provide goods needed to equip increasingly large infantry armies, which comprised hundreds of thousands of soldiers. Such needs included pack animals for transport, and hides and leather for armor. Armies on the move required a large number of pack animals for logistic operations. Moreover, Chinese soldiers at the time, as we can see plainly from the terracotta army, used leather armor for bodily protection. In addition, sheep and cattle could be used to feed the troops on campaigns. For all these needs the closest, most abundant, and easily reachable 'market' was the economy of the pastoral nomads.

One key sign of the importance attributed to the grassland regions by the northern states was the massive construction of so-called 'long walls'. In the 4th and 3rd century BCE, the states of Qin, Zhao, and Yan began expanding into, and protecting with lines of fortifications, areas occupied by nomadic peoples with the goal, presumably, of preserving their own access to these economic zones. Far from being a defense against bellicose barbarians, the walls, extending deeply into the grassland, were meant to fence those areas and thereby secure the supply chain of products necessary to keep their armies outfitted.[16]

In terms of textual evidence, we can point to at least three instances of commercial exchanges with the nomads that took place prior to the unification of China. In the *Zhanguoce* (*Intrigues of the Warring States*), a text of the 3rd century BCE, there are occasional, yet clear indications that the Zhou states were importing horses and furs from the nomads.[17] The *Mu Tianzi zhuan* (*The Biography of Son-of-Heaven Mu*), a text dated around the 4th or 3rd century BCE, though largely fictional, mentions information that must have originated in actual practices and customs. In the course of his travels, Emperor Mu is said to have exchanged gifts with foreign chiefs, probably northern nomads. In one exchange with what was presumably a tribal king or chieftains, known as Ruonu, he received goat and sheep milk, 300 horses, 5000 heads of cattle and sheep, 1000 wagons of grain and 30 wagons of 'desert millet', as well as jade artifacts.[18] The gifts presented by King Mu in return were primarily precious artifacts, such as deer made of silver or gold, a silver bird, necklaces of gold or precious stone beads, pearls, gold bullion, belts adorned with precious shells. One might wonder whether these gifts find their material counterparts in the recumbent bronze deer and golden animal-style plaques so common in the late Warring State nomadic burials. Finally, trade contacts are mentioned in the *Shiji* (*The Historian's Memoirs*, ca. 100 BCE), by Sima Qian. Here we find that the land 'north of Longmen and Jieshi is rich in horses, cattle, sheep, felt, furs, tendons and horns'.[19] The commercial value of this land was not lost on the merchant Wuzhi Luo, who became rich in the state of Qin. He traded silk for domestic animals with the king of a barbarian (Rong) tribe, thereby increasing enormously the size of his own herds of horses and cattle, whose number 'could only be estimated by the valleyful'.[20]

What these texts arguably point to is a commercial exchange between nomadic elites and Chinese states in which the latter exported luxury products in order to procure pastoral products. Key to this trade's success was that the Chinese states had to find a 'currency' that enabled them to pay for what they needed. Initially, the most common kind of payment must have been luxury objects similar to those already circulating among nomads, such as bronze and golden plaques and especially gold and silver ornaments. At the same time, uniquely Chinese products found their way into the nomadic lands, and especially silk. Silk was highly valued by the nomads and abundant in China, thus making it the perfect monetary instrument to purchase large quantities of horses, cattle, and other goods. Its light weight and durability made it easily transportable (unlike coins) and tradeable across long distances.

Conclusion

Based on this extremely cursory analysis of the archaeological record (as fuzzy as it is), and of the textual evidence (as sparse as it is), we can begin to put together a hypothesis concerning the context in which the 'Silk Road' was formed that this evidence suggests. To do so, we have identified the late Warring States as the period in which China became more fully integrated with the nomadic economic sphere. In particular, I have argued that a possible reconstruction of the historical opening of regular, sustained, and large-volume trade between China and the steppes, leading to an acceleration in both commercialization and militarization of the 'northern zone', was directly connected to the expansion of military activities that took place at that time. The pastoral economy that had developed along the 'ecological frontier' with China, whose first impulse can be seen in the centuries-long expansion of grasslands and steppe vegetation in eastern Inner Asia, were behind the qualitative transformation of Chinese warfare, to which it contributed the reform of cavalry, and military supplies that allowed the movement and equipment of large armies. China paid not just by bartering luxury objects consistent with the nomads' value system, but also by exporting authentically Chinese products, first of all silk.

How Chinese silk reached Pazyryk and other nomadic sites in the 3rd century BCE has been so far just a matter of conjecture.[21] Surely this transfer cannot be simply attributed to nomadic 'mobility' or migrations. The hypothesis presented here, instead, provides at least a partial explanation, namely, that silk entered the nomadic economy as an integral part of an exchange system that developed along the frontier when the Chinese states became dependent on the pastoral economy, predominantly for their military needs. It is thanks to such economic mechanisms that China became truly and organically integrated with the wider Inner Asian and Central Asian world. Arguably, the beginning of the historical Silk Road is a result of such integration, linked to new conditions arising in both societies, which favored their 'trans-ecological exchange'.

A few additional points should be made. In the first place, if the establishment of this exchange mechanism can be dated to the 4th and 3rd centuries BCE, and depended largely on the evolution of the political situation in China as well as on the northern frontier of China, this new phase cannot be conceived as a simple evolution of pre-existing Bronze Age networks. In fact, specific historical events played an accelerating role on previous trends, and actually transformed them in the process. If previous exchanges within the nomadic world may have included objects coming directly from China (and this point is moot) they could not be sourced to a specific commercial development such as that emerging in the late Warring States. Likewise, the received narrative that dates the 'opening' of the Silk Road to the Han dynasty, and in particular to the expedition of Zhang Qian to the Western Regions should also be seen as an event generated by an entirely new political situation. In other words, each period (Bronze Age, Warring States, or Han dynasty) presents specific characteristics and shows how the Silk Road 'trans-Eurasian exchange' was repeatedly re-configured.

Secondly, much of what happened beyond the horizon of the written sources requires careful interpretation. Even a cautious approach, however, should be open to piecing together different classes of evidence. In this short essay we have seen how paleoclimatic studies, material artefacts, and textual sources can be related to a common picture of change and transformation that includes both long and short-term trends. Examining the interactions between phenomena with different temporalities is key to understanding the evolution of cross-cultural contacts. In our case, we find evidence in Chinese sources of at least three ways in which Chinese goods entered the 'trans-Eurasian' networks: tribute and diplomacy, private trade, and state trade. Likewise, we know that foreign products, especially luxuries, came into China through analogous routes. These forms of exchange cannot be regarded as variables independent from the historical circumstances that, at different points in time, favored one or the other form of contact. Each opening or closure of what we call the Silk Road responded to specific pulls and pushes generated from within the societies that participated in the exchange networks.

Finally, something needs to be said in regard to environmental changes. The ecology of nodal places through which people moved was modified by climatic fluctuations. These had at times a transformative impact on local societies as well as on long-distance communication. Climatic reconstructions should enter more organically the study of the Silk Road, not to reduce its history to a series of climatic pulsations that closed or opened routes of communication, but to understand the effects of environmental changes and the options available to people.

Notes

1 See Christian 1998; Kuzmina 2008; Rawson 2017, 375–388.
2 Sherratt 2006, 30–61; Jaang 2015, 179–213.
3 Christian 2000, 1–26.
4 van Geel *et al.* 2004, 1735–1742. See also Bokovenko 2006, 21–33.
5 Riehl and Pustovoytov 2006, 143–144.
6 van Geel *et al.* 2006, 145–148.
7 Blyakharchuk *et al.* 2007, 518–534.
8 Rudaya *et al.* 2008, 2–14.
9 Davis-Kimball *et al.* 1995.
10 Yablonsky and Malashev 2004, 259–292.
11 Dashkovskiy and Usova 2011, 78–84.
12 Chang *et al.* 2003, 298–312.
13 Polosmak 1999, 125–163.
14 Liu 2017, 1596.
15 On the adoption of cavalry by King Wuling of Zhao see Di Cosmo 2002, 134–138.
16 Di Cosmo 2002, 138–155.
17 Crump 1964, 55, 324.
18 Frühauf 2004, 173–200.
19 *Shiji* 129, 3254; Watson 1993, 2, 434.
20 *Shiji* 129, 3260; Watson 1993, 2, 440.
21 Good 1995, 959–968.

Bibliography

Blyakharchuk, T.A., Wright, H.E., Borodavko, P.S., van der Knaap, W.P. and Ammann, B. 2007. Late Glacial and Holocene Vegetational History of the Altai Mountains (Southwestern Tuva Republic, Siberia), *Palaeogeography, Palaeoclimatology, Palaeoecology* 245.3–4, 518–534.

Bokovenko, N.A. 2006. Migrations of Early Nomads of the Eurasian Steppe in a Context of Climatic Changes, in: Scott, E.M., Alekseev, A.Y. and Zaitseva, G. (eds), *Impact of the Environment on Human Migration in Eurasia*, Dordrecht, 21–33.

Chang, C., Benecke, N., Grigoriev, F.P., Rosen, A.M. and Tourtellotte, P.A. 2003. Iron Age Society and Chronology in South-east Kazakhstan, *Antiquity* 77.296, 298–312.

Christian, D. 1998. *Inner Eurasia from Prehistory to the Mongol Empire*, Oxford.

Christian, D. 2000. Silk Roads or Steppe Roads? The Silk Roads in World History, *Journal of World History* 11.1, 1–26.

Crump, J.I. 1964. *Intrigues: Studies of the Chan-kuo ts'e*, Ann Arbor.

Dashkovskiy, P.K. and Usova, I.A. 2011. Pazyryk Burial at Khankarinsky Dol in the Northwestern Altai, *Archaeology, Ethnology and Anthropology of Eurasia* 39.3, 78–84.

Davis-Kimball, J., Bashilov, V.A. and Yablonsky, L.T. (eds) 1995. *Nomads of the Eurasian Steppes in the Early Iron Age*, Berkeley.

Di Cosmo, N. 2002. *Ancient China and Its Enemies: The Rise of Nomadic Power in East Asian History*, Cambridge.

Frühauf, M.W. 2004. Tribut und Geschenke im Mu Tianzi Zhuan, *Zeitschrift der Deutschen Morgenländischen Gesellschaft* 154.1, 173 200.

Good, I. 1995. On the Question of Silk in Pre-Han Eurasia, *Antiquity* 69.266, 959–968.

Kuzmina, E.E. 2008. *The Prehistory of the Silk Road*, V.H. Mair (ed.), Philadelphia.

Jaang, L. 2015. The Landscape of China's Participation in the Bronze Age Eurasian Network, *Journal of World Prehistory* 28.3, 179–213.

Liu, Y. 2017. Exotica as Prestige Technology: The Production of Luxury Gold in Western Han Society, *Antiquity* 91.360, 1588–1602.

Polosmak, N.V. 1999. The Burial of a Noble Pazyryk Woman, *Ancient Civilizations from Scythia to Siberia* 5.3, 125–163.

Rawson, J. 2017. China and the Steppe: Reception and Resistance, *Antiquity* 91.356, 375–388.

Riehl, S. and Pustovoytov, K. 2006. Comment on van Geel *et al.*, Journal of Archaeological Science 31 (2004) 'Climate change and the expansion of the Scythian culture after 850 BC: a hypothesis', *Journal of Archaeological Science* 1.33, 143–144.

Rudaya, N.A., Tarasov, P., Dorofeyuk, N.I., Kalugin, I., Andreev, A., Diekmann, B. and Darin A. 2008. Environmental Changes in the Mongolian Altai During the Holocene, *Archaeology, Ethnology and Anthropology of Eurasia* 36.4, 2–14.

Sherratt, A. 2006. The Trans-Eurasian Exchange: The Prehistory of Chinese Relations with the West, in: Mair, V.H. (ed.), *Contact and Exchange in the Ancient World*, Honolulu, 30–61.

Shiji 1972. By Sima Qian. 10 vols. Beijing.

van Geel, B. 2006. Reply to S. Riehl and K. Pustovoytov (*Journal of Archaeological Science* 33 (2006), 143–144), *Journal of Archaeological Science* 33.1, 145–148.

van Geel, B., Bokovenko, N.A., Burova, N.D., Chugunov, K.V., Dergachev, V.A., Dirksen, V.G., Kulkova, M., Nagler, A., Parzinger, H., van der Plicht, J., Vasiliev, S.S. and Zaitseva, G.I. 2004. Climate Change and the Expansion of the Scythian Culture after 850 BC: A Hypothesis, *Journal of Archaeological Science* 31.12, 1735–1742.

Watson, B. 1993 [1961]. Transl. *Records of the Grand Historian by Sima Qian*, 2 vols. New York.

Yablonsky, L. and Malashev, V. 2004. Early Nomads in the Southern Foothills of the Urals Based on Materials from the Pokrovka Burial-Ground, *Ancient Civilizations from Scythia to Siberia* 10.3–4, 259–292.

Chapter 2

Sogdians in Shanxi (386 CE–618 CE): Literary and Archeological Evidence[1]

Xiaoyan Qi

The period from 386 CE to 618 CE spanned the co-existent Northern dynasties and Southern dynasties as well as the Sui dynasty. The dynasties in North China consisted of the Northern Wei, Eastern Wei, Western Wei, Northern Qi and Northern Zhou, which are collectively called the Northern dynasties so as to distinguish them from the dynastics in South China. The Northern dynasties, along with the Sui dynasty, saw the acceleration of fusion between the Han Chinese and other ethnic groups as well as the development and prosperity of the overland Silk Roads.

Shanxi is located on a plateau made up of higher ground in the east and the west, the Taihang Mountains and the Lüliang Mountains respectively, as well as a series of valleys in the center. The Fen River runs through Shanxi and the Yellow River forms its western border. Shanxi borders Hebei to the east, Henan to the south, Shaanxi to the west, and Inner Mongolia to the north. Cities like Pingcheng (modern Datong), Jinyang (modern Taiyuan), and their surrounding areas played an important role in the Silk Roads, especially during the Northern dynasties. Pingcheng once served as the capital of the Northern Wei dynasty, and Jinyang was the auxiliary capital of the Northern Qi dynasty. A palace was also established in Jinyang during the Sui dynasty.

Many archeological discoveries – tombs, murals, and burial goods – suggest the extent of Sogdian influence in Shanxi. In addition, Chinese written sources, such as *The Book of Wei* (*Weishu* 魏書), *The History of the Northern Dynasties* (*Beishi* 北史) and *The Book of Northern Qi* (*Beiqishu* 北齊書), also provide rich information for reconstructing the Sogdian experience in Shanxi. Therefore, the combination of archaeological evidence and written documentation provides a wealth of knowledge for the study of Sogdians in Shanxi. However, compared with the study of Sogdians in Henan, Shaanxi and Inner Mongolia, the study of Sogdians in Shanxi is not proportionate with their frequent activities. In this paper, I will mainly discuss the route that the Sogdians took to enter Shanxi, the places where they settled, their interaction with the local elites and their cultural contributions to Shanxi life.

The term Sogdiana and *Sute* (粟特)

It is generally thought that the term Sogdiana corresponds to Sute in Chinese. This identification went through a long process. *The Records of the Grand Historian* (*Shiji* 史記) locates Sogdiana in a broad region between Ferghana (Dayuan 大宛) and Parthia (Anxi 安息): 'Although the states from Dayuan west to Anxi speak rather different languages, their customs are generally similar and the languages are mutually intelligible. The people all have deep-set eyes, profuse moustaches and beards. They are skilled in commerce and will haggle over every penny.'[2]

Sogdiana is named Liyi 栗弋 in *The Book of the Later Han* (*Houhanshu* 後漢書). This is thought to be the first reference to Sogdiana. However, it is not commonly used to refer to Sogdiana by ancient or modern scholars. According to 'Account of the Western Region' in *The Book of the Later Han*, 'Liyi belongs to Kangju 康居, which has fine horses, cattle and sheep, grapes and many fruits. Its soil and water are excellent, so their grape wine is well known.'[3]

The name Sute first appeared in *The Book of Wei*. Sute, along with the other western states, paid tribute to the court of the Northern Wei, and the distance between Sogdiana and Dai (another name for Pingcheng) was 16,000 *li.*[4] *This description laid the foundation for later records of Sogdiana and suggested that the Silk Roads extended eastward to Jin.*

Purposes of the Sogdians in Shanxi

Based on Chinese written sources, Sogdiana sent 17 missions to North China during the Northern dynasties, including 15 missions to Pingcheng, Shanxi, and two to Chang'an. The Sogdians sent one mission to Jiankang, capital of the Southern dynasties, and many missions to Chang'an during the Daye reign of the Sui dynasty. Two activities stood out from the missions to Shanxi. In 419 the Sogdians participated in the Northern Wei sacrifices to heaven. In 457 they sent envoys to redeem the Sogdians who had been captured and taken to Pingcheng from the capital of the Northern Liang in 437.

The Sogdians paid tribute to the northern Chinese courts in Shanxi in order to expand the silk trade. According to *The Book of Wei,*

> From the time Taizu stabilized the Central Plains the middle and lower reaches of the Yellow River to the time Shizu suppressed the battles, the courts of the Northern dynasties gained numerous treasures. Government coffers accumulated surpluses ... In the winter of the second year of Heping, the imperial court decreed that 200,000 lengths of silk and cotton be made available from the treasury to court officials as prizes for archery competitions.[5]
>
> In the fourth month of the first year of the Taihe reign, the envoy Biba arrived to pay tribute and brought fine horses and marten coats. Biba and his entourage said, 'We hear that

> the celestial court has an abundance of resplendent treasures and respectfully request to view them.' Therefore, the emperor ordered the officials to take out gold and jade, embroideries, utensils, horses in royal stables, exotic birds and rare animals and every convenience in the world and display them in stores in the capital for the delegation to view. When Biba and his entourage saw the treasures, they exclaimed to one another, 'The wealth and splendor of a great country is something that we have never witnessed before.'[6]

It can be concluded from the records that Shanxi was rich in silk and silk products. Therefore, one reason that the Sogdians came to Shanxi where the capital of Northern Wei was located was the silk trade. But political reasons should not be ignored. The Sogdians hoped that the court of the Northern Wei would help them get rid of the control of the Hephthalites. In addition, in 437 the court of the Northern Wei conquered the Northern Liang. Consequently, a large number of Sogdians, along with the locals, were captured and moved to Pingcheng. Twenty years later, the king of Sogdiana sent envoys to redeem the hostages. From the Chinese perspective, it is interesting that as *Sanqi shilang* 散騎侍郎 (assistant ministers), Dong Wan and Gao Ming's mission to Sogdiana and the Northern Wei conquest of the Northern Liang happened in the same year. This is similar to the purpose and result of the missions of Zhang Qian to the Western Regions. His main purpose was to form an alliance with Yuezhi against Xiongnu, but his missions officially opened the Silk Roads. The Northern Wei conquest of the Northern Liang meant the unification of North China, so the court of the Northern Wei hoped to get help from the Western Regions to strengthen its control and sovereignty. In order to show their friendship, the Sogdians sent envoys to Pingcheng. Therefore, it cannot be denied that the mission of Dong Wan and Gao Ming played an important role in the prosperity and eastward extension of the Silk Roads.

Based on the discussion above, the Sogdians came not only from Sogdiana but also its colonies in the Northern Liang. Furthermore, their influence in ancient Shanxi was not only economic but also political.

Roles of the Sogdians in Shanxi

After they reached Shanxi, the Sogdians participated in local affairs and played different roles. First was to be traders. According to *The Book of Wei*,

> The State of Kang [Samarkand] is successor to Kangju [the Talas basin, Tashkent and Sogdiana]. Its people are nomads and have no fixed abode. Since the Han dynasty, the State of Kang has continued from one generation to the next. Its royal family is surnamed Wen and is of Yuezhi descent. The country used to occupy Zhaowu north of the Qilian Mountains. After the Xiongnu conquered the city, its people crossed the Congling (Pamir Mountains) and reestablished the state [in Samarkand] and separated into different kingdoms. The confederation under Kang inherited the clan names of Zhaowu so as not to forget its roots. ... Mi, Shi, Cao, He, An, Xiaoan, Nasebo, Wunahe and Mu are all subsidiary states of Kang.[7]

The Sogdians often appeared as *Hu* 胡 or barbarian traders in Chinese historical sources. Sometimes they were also called the Nine Barbarian Clans (*Jiuxinghu* 九姓胡). According to *The Book of Northern Qi*, 'twelve palaces were built in Jinyang, and the magnificence of Jinyang was beyond that of Ye [modern Linzhang county in Hebei province]. Li Longzhi was King of the Plains (*Pingyuan wang* 平原王) and Assistant Grand Tutor (*Taibao* 太保). His neighbor Yang Yin often met the rich barbarian merchants in Li Longzhi's house.'[8] The last sentence suggests the presence of a local community of Central Asian traders. What is more, many images of 'barbarians' have been found in Shanxi. A significant number are of Central Asians, for example, murals that depict the selling and buying of horses in the tomb of Sima Jinlong in Jinyang[9] as well as the mural of horse-trading of Jiuyuangang in Xinzhou.[10] These important archaeological finds suggest the activities of Central Asian merchants in ancient Shanxi in the Northern Wei and Northern Qi periods.

The second role of the Sogdians was as officials. An Tong, Kang Da and Cao Yi are typical examples. The surnames An, Kang and Cao represented the people originating from Bukhara, Samarqand and Ustrushana in Sogdiana.

An Tong received the titles of General of Martial Glory (*Guangwu jiangjun* 光武將軍 and General of Border Peace (*Anyuan jiangjun* 安遠將軍) during the Northern Wei because of his merits. According to *The Book of Wei*:

> In the early years of the Dengguo reign ... Emperor Taizu awarded officials based on merits. An Tong was awarded a wife and several concubines, 30 households of slaves, 2 horses, 50 sheep, and titled *Guangwu jiangjun*.
>
> [...] An Tong was sent to Jiang to supervise levies, but found a gulley east of the Fen River. It ran more than 300 *li* east to west with any crossings. Yao Xing and his troops invariably took the commanding position from west of the Fen and marched all the way to Chaibi. Thus, enemy forces converged from within and without to besiege the country and were impossible to defeat. An built a pontoon over the Fen and a solid wall on the west bank, which left the bandits with no recourse ... Because of the success of his strategy, An Tong received the rank Marquis of Beixin and the title General of Border Peace (*Anyuan jiangjun*).[11]

The epitaph on Kang Da's tomb excavated in Luoyang, Henan province, states, 'His ancestor [Kang] Ling once administered Yanmen county and took the position of *Shangyitong* 上儀同 (Grand Commandant-Equivalent) during the Northern Qi.'[12] The epitaph of Cao Yi was excavated in Fenyang, Shanxi. It reads, 'His ancestor [Kang] Gui took the title of General of Martial Prowess (*Zhuangwu jiangjun* 壯武將軍) during the Northern Qi. His father [Kang] Zun was Metropolitan Chariot Officer (*Cheji qiduwei* 車騎騎都尉) and *sabao* of Jiezhou Prefecture (*Jiezhou saobao fu* 介州薩寶府) during the Sui dynasty.'[13]

General of Martial Glory, General of Border Peace, Grand Commandant-Equivalent, and Metropolitan Chariot Officer were military titles. The term *sabao* appeared first in the Northern Qi, but until the Sui dynasty *sabao* designated an administrator of Sogdian affairs. Therefore, I propose that the Sogdians' official positions were most likely related to the political climates of the Northern and Sui dynasties. According to the above-mentioned records, the Sogdians participated in the unification battles of North China against the invasion from the surrounding nomadic tribes and from the courts of the Southern dynasties.

The third role of the Sogdians was as entertainers. The ancient Chinese term *yue* 樂 refers to music and dance, which experienced a gradual development from *Qiyue* 七樂 (Seven Music and Dance Arts), *Jiuyue* (九樂) (Nine Music and Dance Arts) to *Shiyue* 十樂 (Ten Music and Dance Arts). They include music and dances from China, the Western Regions and other foreign states. *The Old Book of Tang* details some information about the *yue* of An and Kang, including their ensemble instruments and their dance called barbarian whirling.

The *yue* of Kang and the *yue* of An are described in the following manner:

> Musicians from the State of Kang wear a black silk cap and a crimson silk gown with a brocade collar. A pair of dancers wear a crimson coat with a brocade collar and brocade sleeves, green twill breeches and red leather boots and carry a white pouch. They whirl rapidly like the wind, giving rise to the nickname 'the barbarian dance'. The orchestra employs two flutes, a first and a second drum and a pair of bronze cymbals.
>
> Musicians from the State of An wear a black silk cap and a purple-sleeved vest with a brocade-bordered collar. A pair of dancers wear a purple coat and red boots and carry a white pouch. The orchestra employs [four] and five-stringed lutes, vertical harps, vertical and transverse flutes, bagpipes, first and second drums, cymbals and [horizontal] harps. Five-stringed lutes no longer survive.[14]

Abundant archeological evidence testifies to the presence of 'barbarian' entertainers. On the stone engraving unearthed from the ruins of the Northern Wei to the south of Datong, one man is raising his head and dancing with a bent leg. Another is sitting cross-legged while playing a lute.[15] The figures of entertainers unearthed in the tomb of the Northern Wei at Yanbei Normal University in Datong have been identified as acrobats.[16]

Route of the Sogdians into Shanxi

It is generally accepted that the route from Sogdiana to the Hexi Corridor was relatively smooth after the opening of the Silk Roads. When the Sogdians arrived at Dunhuang, they would continue to migrate eastward to Jiuquan, Zhangye and Wuwei. According to the research of Rong Xinjiang, when the Sogdians arrived at Wuwei, their way to Shanxi went northeast to Lingzhou, Taiyuan, Datong, then Beijing.[17] However, I think that this route mainly reflected the migration from Samarqand to Shanxi during the Sui and Tang dynasties. How the Sogdians migrated into Shanxi during the Northern dynasties requires further study.

First, it should be noted that it was not a good choice for the Sogdians to reach Shanxi via Chang'an because the Northern and Southern dynasties were fighting over Chang'an and the road was unsafe. According to *The Book of Wei*, the Sogdians once tried to pay tribute to the Southern dynasties but ended in failure, which reflected not only the small number of activities of the Sogdians in South China, but also the difficulty of entering Shanxi through Chang'an.

The historical events surrounding the unification of the Northern Wei are helpful in outlining the route from Wuwei to Pingcheng. From 438 to 439 CE, Emperor Taiwu led his army to conquer the court of the Northern Liang. The campaign started from Pingcheng along the southern edge of Mongolia, continued across the Yellow River by way of the Junzi Ford through the town of Tongwan, and finally arrived at Wuwei. After being defeated by the troops of the Northern Wei, the local Chinese and many Sogdian inhabitants were captured and taken to Pingcheng, thus suggesting the viability of the route along southern Mongolia.

After the Sogdians entered Shanxi, they migrated southward. In fact, the route from Jinyang to Pingcheng had become safe. According to *Universal Geography of the Taiping Era* (*Taiping huanyu ji* 太平寰宇記), there were three routes from Hebei to the frontier:

> From Zhou, Qin, Han and Wei on, there were only three routes for the northern military expeditions. Among them the middle route started due north from Taiyuan, passing through Yanmen, Mayi and Yunzhong, exiting Wuyuan Fortress and extending directly to Dragon City where the Xiongnu leader worshipped heaven during the grand ceremony conducted every year during the tenth month.[18]

In addition, some epitaphs demonstrate that the Sogdians in Luoyang came from Shanxi. According to the inscription on the tomb of a woman surnamed He, the deceased's family was originally from Taiyuan. One of her ancestors became an official in Luoyang.[19] The epitaph on Zhai Tusuo' tomb reads in part 'Tusuo, style name Bohebiduo (薄賀比多) was from Taiyuan. His father Suomoke, style name Bohebiduo (博賀比多) was a grand *Sabao*. Tusuo fell ill on the 18th day of the first month in the 11th year of the Daye reign and died at 70 in Jiashanli, Chongyexiang, Luoyang county.'[20] Likewise, the inscription on Kang Wutong's tomb records that Kang Wutong from Qi county in Taiyuan rose to the position of general in Yangcheng county and died at his residence in Zhongshan Ward in Luoyang in 629,[21] so it is possible that the Sogdians in Henan came from Shanxi with the change of the capital of the Northern Wei from Shanxi to Henan.

Therefore, it can be concluded that during the Northern dynasties, the Sogdians arrived at Dunhuang, continued to migrate through Jiuquan, Zhangye, Wuwei, Tongwan, and along the southern edge of Mongolia, before finally arriving at Pingcheng. After they entered Pingcheng, the Sogdians migrated southward to Taiyuan and Luoyang. This route reflects the history of the Sogdian migration and colonies in North China during the Northern dynasties. When the Sui dynasty unified China, Chang'an again became the capital, so it became feasible to go to Shanxi through Chang'an. Based on the discussion above, I propose that it is necessary to add the north route from Sogdiana to Shanxi to show the whole history of Sogdian migration in China.

The Impact of the Sogdians on Shanxi

The Sogdians made a great contribution to the spread of Buddhism and Zoroastrianism as well as the eastward spread of grapes in Shanxi. Scholars have confirmed the role of the Sogdians in both. In this part, I will mainly discuss the spread of sorghum and the opening of the Tea-and-Horse Road in Shanxi by the Sogdians.

Dahe 大禾 (sorghum) is also called *gaoliang* 高粱 in Chinese. *Da* means 'big' or 'high'; *he* 'grain'. The name *dahe* is also used to distinguish it from *xiaohe* 小禾 (millet). Sorghum was introduced to Shanxi by the Sogdians during the Northern Wei period. Chinese sources such as *Techniques to Harness the People's Livelihood* (*Qimin yaoshu* 齊民要術), *Comprehensive Statutes* (*Tongdian* 通典) and *Imperial Reader of the Taiping Era* (*Taiping yulan* 太平御覽) are important references for the study of sorghum in China.

Techniques to Harness the People's Livelihood is the most completely preserved of the ancient Chinese agricultural texts. It was written by Jia Sixie during the Northern Wei period. According to Jia, *dahe* from Sogdiana is more than 10 Chinese feet and its seedlings are like small beans.[22]

Comprehensive Statutes written by Du You of the Tang dynasty covers fields such as economics, politics, government, rites and ceremonies, music, and geography. According to Du:

> Liyi lying in the west of Congling is a great state and has made contact with China from the Wei period on. It is also called Sute or Tejumeng (特拘夢). It has fine horses, cattle, sheep, grapes and many kinds of fruit. Its soil and water are excellent, so their grape wine is well known. Dahe is one native grain with a height of more than ten Chinese feet, and its seedlings are like hudou (胡豆 barbarian beans). It is five thousand li north of Parthia and has more than four hundred affiliated states.[23]

Imperial Reader of the Taiping Era is a Chinese *leishu* 類書 (lit., 'category books') encyclopedia edited by a number of officials under Li Fang in the Song dynasty (960–1276). It records, '*Huohe* 火禾 came from Sogdiana. Its height is more than 1 *zhang* 丈 and its seedlings are like small beans.'[24]

These sources state that *dahe* came from Sogdiana. One *zhang* is about 3.33 m, and this is the main reason why it was called *dahe*. *Huo* means 'fire', here referring to the color. After grinding, *dahe* was called 'red flour' by the locals. *Dahe* contributed significantly to the life of Shanxi, especially for places such as Xinzhou, Wutai and Datong. Even now, *dahe* is a popular staple food in these places. Another Sogdian contribution was to Fen liquor. There are many records of wine-making techniques during the Spring–Autumn and Warring States period, but Fen wine was highly praised and treated as a royal wine by Emperor Wucheng of the Northern Qi. One of the three raw materials for making Fen liquor is *dahe*. Nowadays, Fen liquor is one of the most famous liquor brands in China. Therefore, we cannot overlook the role of the Sogdians in the popularity of Fen liquor.

The Tea-Horse Road started from the later period of the Ming dynasty and arrived at its peak in the middle of the Qing dynasty, lasting around 300 years. It connected

Mongolia, Russia, Europe, and the Central Asian states. Shanxi lies between Luoyang and Inner Mongolia. Shanxi merchants (*Jinshang* 晉商) became some of the most prominent in the tea-horse trade, buying and selling salt, iron, cotton, fur, silk, and tea. Compared with the ancient Silk Roads, this route had many similarities with the Sogdian migration route from Pingcheng to Luoyang. To some extent, the Sogdian migration route laid a foundation for the opening of the Tea-Horse Road. The Sogdians not only greatly contributed to the development of the ancient Silk Roads eastward to Shanxi, but also left many legacies there.

Today Shanxi plays an important role in the construction of the One Belt One Road Initiative. On 25 February 2019, Shanxi was chosen as the first province to be introduced to the world by the Ministry of Foreign Affairs of the PRC. In this context, the study of the Sogdians and their legacies in Shanxi acquires a contemporary significance.

Notes

1 This paper is based on two research projects, 'Hellenistic Civilization in the Far East and the Silk Roads: Historical Research on Sogdiana' and 'Special Topics in Shanxi Mercantile Culture: Data Collection, Collation, and Research on Ze-Lu Business Groups during the Ming and Qing Dynasties', respectively sponsored by the Young Scholars Fund of the Chinese Ministry of Education (Project No. 18YJC770027) and the Shanxi Mercantile Culture Foundation.
2 Sima Qian 1959, 3174.
3 Fa Ye 1965, 2912–2913. Unless otherwise noted, all translations are mine.
4 During Zhou, Qin, and Han, 1 Li equals 415.8 m.
5 Wei Shou 1974, 2851.
6 Wei Shou 1974, 2296.
7 Wei Shou 1974, 2281.
8 Li Baiyao 1972, 457.
9 Zhang Qingjie 2010, 147.
10 Bai Shuzhang *et al.* 2016, 28–34.
11 Wei Shou 1974, 712.
12 Rong Xinjiang and Zhang Zhiqing 2004, 116.
13 Li Rui *et al.* 2014, 28–32.
14 Liu Xu 1975, 1071.
15 Wang Yanqing 2011, 567–577.
16 Wang Yanqing 2011, 567–577.
17 Rong Xinjiang and Zhang Zhiqing 2004, 4–5.
18 Yue Shi 2007, 1036.
19 Zhou Shaoliang 1992, 585.
20 Rong Xinjiang and Zhang Zhiqing 2004, 144.
21 Rong Xinjiang 2001, 84–89.
22 Jia Sixie 2015, 1203
23 Du You 1988, 5273–5274.
24 Li Fang 1960, 3749.

Bibliography

Bai Shuzhan 白曙璋 *et al.* 2016. Shanxi Xinzhou Jiuyuangang Beichao bihuamu de fajue 山西忻州九原崗北朝壁畫墓的發掘, *Dazhong kaogu* 大眾考古 5, 28–34.

Du You 杜佑. 1988. *Tongdian* 通典, Beijing.

Fa Ye 範曄. 1965. *Houhanshu* 後漢書, Beijing.

Jia Sixie 賈思勰. 2015. *Qinmin yaoshu* 齊民要術, translated and annotated by Shi Shenghan 石聲漢, Beijing.

Li Baiyao 李百葯. 1972. *Beiqishu* 北齊書, Beijing.

Li Fang 李昉. 1960. *Taiping yulan* 太平御覽, Beijing.

Li Rui 李瑞 *et al.* 2014. Shanxi Fenyang Tang Cai Yi mu fajue jianbao 山西汾陽唐曹怡墓發掘簡報, *Wenwu* 文物 11, 28–32.

Liu Xu 劉昫. 1975. *Jiutangshu* 舊唐書, Beijing.

Pulleyblank, E.G. 1952. A Sogdian Colony in Inner Mongolia, *T'oung Pao* 41, 317–356.

Rong Xinjiang 榮新江. 2001. Sui ji Tangchu Bingzhou de Saobaofu yu Sute juluo 隋及唐初并州的薩保府與粟特聚落, *Wenwu* 文物 4, 84–89.

Rong Xinjiang 荣新江 *et al.* 2004. *Cong Sama'erhan dao Chang'an – Suteren zai Zhongguo de wenhua yiji* 從撒馬尔罕到長安 – 粟特人在中國的文化遺跡, Beijing.

Sima Qian 司馬遷. 1959. *Shiji* 史记, Beijing.

Wang Yanqin 王雁卿. 2011. Beiwei Pingcheng Huren de kaoguxue guancha 北魏平城胡人的考古學觀察, *Zhongguo Weijin Nanbeichao shi xuehui di shi jie nianhui ji guoji xueshu lunwenji* 中國魏晉南北朝史學會第十屆年會暨國 际學術論文集, Shanxi, 567–577.

Wei Shou 魏收. 1974. *Weishu* 魏書, Beijing.

Yue Shi 樂史. 2007. *Taiping huanyuji* 太平寰宇記, Beijing.

Zhang Qingjie 張慶捷. 2010. Beichao Tangdai de Hushang yong Hushang tu yu Hushang wenshu 北朝唐代的胡商佣、胡商圖与胡商文書, in: Zhang Qingjie, *Minzu huiju yu wenming hudong – Beichao shehui de kaoguxe guancha* 民族匯聚与文明互动北朝社會的考古學觀察, Beijing, 141–191.

Zhou Shaoliang 周紹良. 1992. *Tangdai muzhi huibian* 唐代墓誌彙編, Shanghai.

Chapter 3

From Exotic Toys to Objects of Scientific Inquiry: A Special Way of Transmitting European Optical Knowledge in the Qing Dynasty

Yunli Shi

Objects often played an important role in the long-distance transmission of knowledge in a transcultural context like the Silk Roads, because to most people they were more concrete and attractive than abstract knowledge. In many cases, however, the knowledge transmitted through objects was complicated, because it often involved a far more subtle and complex process of understanding compared to the information transmitted by linear language. A good example is the transmission of optical knowledge from Europe to China through a number of optical toys in the 17th–18th centuries primarily by Jesuit missionaries.

In early modern Europe, a number of optical toys were invented and circulated for both social and intellectual play. On the one hand, the mysterious, miraculous visual tricks and illusions they produced became a source of amusement. On the other, these tricks and illusions were meant to be a way to grasp the preternatural, a realm 'of those wonders or marvelous events and objects that fell outside the ordinary course of nature, but of which the cause was nevertheless not supernatural', because with the new mathematical and experimental method mathematicians were able to expose the underlying mathematical rules and principles of these toys.[1] Eventually, this type of toy went by such designations as the 'Instrument for Recreation and Amusement',[2] or simply 'philosophical toys'.[3]

Great attention was paid to these toys by the Jesuits in Europe, who attached religious meanings to the study and play of light, as seen in the writing of the *Ars Magna Lucis et Umbrae* (*The Great Art of Light and Shadow*) by the famous Jesuit polymath Athanasius Kircher (1601–1682), who tried to reveal the nature of light and the principles for its applications, including a number of optical toys that were popular in his day.[4] For the Jesuits coming to China, these toys were important for a different reason. Not only did the Jesuits use these toys to surprise, amuse and thus to attract the attention of the Chinese, who had inexhaustible curiosity over exotica, but they used them to demonstrate the soundness and superiority of European knowledge and arts, and thereby to prove 'visually' the credibility of their European religion. As a result, a number of these toys were introduced into the palace and displayed in Jesuit churches. Together with such optical devices as eyeglasses and telescopes, they left discernible imprints on Chinese culture and even led to the rise of a new profession of opticians in China.

In the meantime, the Jesuit missionaries did not publish any up-to-date explanations about these optical devices and toys. The only Chinese treatise about the lens and telescope, the key components of these devices and toys, was *An Explanation of Far-Seeing Glasses* (*Yuanjing shuo* 遠鏡說) published by the German Jesuit Johann Adam Schall von Bell (Tang Ruowang 湯若望, 1592–1666) in 1626.[5] Unfortunately, the work was unusable owing to its simplicity and obsolete ideas that prevented any real understanding of the underlying principles of these devices and toys, because it was written long before the first publication of the law of refraction by René Descartes (1596–1650) in 1637, and the theory of the lens by Bonaventura Cavalleri (1598–1647) in 1647.[6] Under these circumstances, few Chinese were aware of the scientific, or 'philosophical', meaning that these toys held for Europeans, although they might have had some idea that these toys were somehow connected with European arts and geometry. It was not until the early 19th century that they finally captured the attention of a Chinese mathematician named Zheng Fuguang 鄭復光 (1780–*c*. 1853). He was an expert in European mathematics and natural philosophy that the Jesuit missionaries introduced into the country in the late Ming dynasty. Drawing on what he had learnt from Chinese books on 'western learning', he carried out a systematic study on these toys in an attempt to reveal the underlying 'scientific' principles embodied in them. The effort ended up with a book entitled *Blazoning Infatuation in Front of Mirrors* (*Jingjing lingchi* 鏡鏡詅癡) in which he established a special system of optics of his own.

Toys in the Palace

From the very beginning of their arrival in China, Jesuit missionaries realized the strong attraction that exotica held for the Chinese regardless of class. Therefore, European books, paintings, automatic striking clocks, musical organs, and astronomical instruments became the most common gifts that Matteo Ricci (1552–1610) presented to Ming officials and the emperor for official permission for the Jesuits to stay in Beijing and other cities.[7] The Jesuits employed the same strategy as noted by Jean-Baptiste Du Halde (1674–1743):

> They did not afterwards fail to comply with the Emperor's Taste, for great Quantities of curious things were sent out of Europe by Christian Princes, who had the Conversion of this great Empire at Heart, insomuch that the Emperor's cabinet was soon fill'd with various Rarities, especially Clocks of the newest Invention, and most curious Workmanship.[8]

Emperor Kangxi of the Qing dynasty presented the Jesuits in Beijing with a different challenge, which they used as a new opportunity. He was not interested in these kinds of objects from pure curiosity, because he took a personal interest in studying Western science and technology. Once again, Du Halde provides us with insight into his situation:

> The late Emperor *Cang hi* [*i.e.* Kangxi], whose chief Delight was to acquire Knowledge, was never weary of seeing or hearing them: On the other hand the Jesuites, perceiving how necessary the Protection of this great Prince was to the Progress of the Gospel, omitted nothing that might excite his Curiosity, and satisfy this natural relish for the Sciences.[9]

Needless to say, objects that embodied the elements of both scientific novelty and curious amusement were most suitable for this niche. So, after the restoration of European astronomy at the Bureau of Astronomy in 1668, Ferdinand Verbiest (Nan Huairen 南懷仁, 1623–1688) initiated a process to introduce to Emperor Kangxi European sciences by making objects useful, curious, or amusing. Except for the large cannons that he manufactured to rearm the imperial troops and the six new astronomical instruments that he constructed to re-equip the imperial observatory, Verbiest constructed for the emperor a number of objects of wonder imbued with European knowledge in gnomonics, mechanics, hydragogics, optics, dioptrics,[10] perspective, statics, hydrostatics, hydraulics, pneumatics, horology, metrology, and the like.[11] Under the categories of optics and dioptrics, there were, in addition to 'all sorts of Telescopes, [...] and likewise Glasses for diminishing, magnifying and multiplying',[12] a number of optical toys popular in contemporary Europe.

The first optical toy mentioned by Verbiest is an optical camera obscura:[13]

> In the course of the preceding years, I presented to the Emperor a semi-cylinder of the proper size, made of boards of a fairly light kind of wood. In the centre of the axis of the semi-cylinder I put a proportioned glass, and I projected into the hollow and dark room inside the cylinder pictures of all the things which that glass beheld; it was just like the eye in the forehead of a giant, because this instrument could be turned to all direction.[14]

This portable camera obscura is similar to the 'picture box' invented and presented to the Royal Society by Robert Hook (1635–1703) in 1694.[15] Greatly amused by this device, Emperor Kangxi ordered Verbiest 'to construct a similar instrument in his gardens, close to the Palace, so that he could observe everything that passed outside on the public road by means of inward projection, without himself being seen by anyone.'[16] Using 'a glass with the greatest possible diameter' and a window in the wall of the palace facing a busy public road, Verbiest constructed a room-sized camera obscura, through which the Emperor and his consorts could watch the crowds passing by.[17] Apparently, this camera obscura is similar to the type depicted by A. Kircher in his *Ars Magna*.[18]

Another type of optical toy that the Jesuit presented to Emperor Kangxi for his amusement was an anamorphosis, a visual trick popular in Europe since the

Renaissance.[19] Basically, there are two types of anamorphic techniques: images with a distorted perspective (perspective anamorphosis) or projection (mirror anamorphosis), respectively. The former can be restored to its natural form view with the naked eye from a particular point or angle, while the latter is reflected through a curved mirror. In the *Ars Magna*, Kircher discussed both types of the art.[20]

According to Verbiest, the Italian Jesuit Philippo Claudio Gramildi (1638–1717) presented a number of anamorphic pictures to Emperor Kangxi:

> The garden of our residence in Peking is enclosed by four walls, and on each of them Father Filippo Grimaldi has drawn, according to the optical laws, a picture which, in line with the wall, extends over about 50 feet. When viewed from the front, it shows only mountains, woods, hunting-parties, etc., but, if one looks at it from a special angle, it conjures up a very accurately delineated picture of ... a man of at least of a living thing. The day that the Emperor visited our residence he looked at these pictures, one by one, long and thoroughly, [...] Still other things related to optics remain, for instance all those pictures that are sketched on a panel and that look deformed and without any proportion between their parts if viewed by the naked eye as if they are thrown down and distorted, but if they are set up and restore to their beauty by a reflecting cylinder or a reflecting cone, they please the eye.[21]

It is clear from this description that Filippo Grimaldi constructed both the perspective anamorphosis and mirror anamorphosis in Beijing.

Under the category of dioptrics, Verbiest singled out three 'tubes' that he thought 'most amazing'. 'The first was the "displaying" tube which, (inserted) in a horizontal octagonal column, showed with its eight facets eight different manifestations of the objects, and represented everything true and real in accordance with the parastatic magic.'[22] The main component of this toy is an octagonal prism, while the 'tube' itself seems to be a precursor of the kaleidoscope invented by David Brewster (1781–1868).[23] The major difference is that the eight rectangular surfaces inside the prism reflect eight different images from outside the tube, while the kaleidoscope proper uses several rectangular mirrors that are attached lengthwise to reflect the loose, colored objects inside the tube.

The second tube is also very interesting:

> The second tube was equipped with a polygonal glass. It transferred the components of a picture from right to left, from top to bottom, and vice versa, and it gathered and ordered the pieces of pictures, which were mutually sawn up and separated, to one perfect image. Though the picture, to the naked eye, only represented woods, fields and cattle, this tube clearly revealed a human face and several other things that the eye could not see without the tube![24]

Apparently, this 'tube' is a kind of faceted lens that was popular in 17th-century Europe, of which the French mathematician Jean-François Niceron (1613–1646) provides a detailed discussion in his *La Perspective Curieuse*.[25]

Verbiest's description of the third tube is all too brief:

> Finally, the third was a nocturnal tube, which we may justly call the 'wonder-performing one' and others call the 'magic' one. For, at night or in a dark room, depending on whether it is closer to or further opposite wall pictures of things through the light of a lantern, which is closed off on all sides, and it represents the figure, however little it may be, very clearly and distinctly.[26]

Fortunately, Du Halde provides a clearer depiction:

> There was also another Machine which contained a lighted Lamp, the Light of which came through a Tube, at the end whereof was a Convex-Glass, near which several small Pieces of Glasses painted with divers Figures were made to slide: These Figures were seen upon the opposite Wall of a Bigness in Proportion to the Distance of the Wall; this Spectacle in the Nigh-time, or in a very dark Place, frightened those who were ignorant of the Artifice, as much as it pleased those who were acquainted with it: On this Account they have given it the Name of the Magick-Lanthorn.[27]

In fact, this is the so-called 'magic lantern', an image projector believed to have been invented around the mid-17th century in Europe, of which Kircher gave the first description in his *Ars Magna*,[28] published in 1646 and 1671, respectively.

Interestingly, the Jesuits in Beijing even combined a magic lantern with a clock to create a noiseless time-projecting device for use at night so as not to disturb anyone's slumber:

> A wheel-clock, accurately regulating the hours with its pendulum, projects through such a nocturnal-tube each hour and its subdivisions on the wall of a dark room, by the light of a night-lamp. The eye, in case it would be on the watch, could in this way immediately read the hour and the quarter on the opposite wall at any moment, without any striking of bells, which interrupt the sleep, while a dark pointer runs around an illuminated clock-face, just like the gilded pointer does on the clocks of tower in Europe.[29]

Here we find an example of applying the magic lantern for a practical purpose.

In order to demonstrate to the Emperor their theory on the causes of such atmospheric phenomena as parhelia, rainbows, and halos, the Jesuits at court even built an optical machine:

> It was a Drum made very close, and whitened on the inside; the inward Surface represented the Heavens; the Light of the Sun entering through a little Hole pass'd through a triangular Prism of Glass, and fell upon a polish'd cylinder; from this cylinder it was reflected upon the Concavity of the Drum, and exactly painted the Colour of the Rainbow. From a part of the Cylinder, a little flattened, was reflected the Image of the Sun; and by other Refractions and Reflections were shewn the Haloes about the Sun and Moon, and all the rest of the Phaenomena relating to Celestial Colours, according as the Prism was more or less inclined towards the Cylinder.[30]

From the description this machine is similar to a device described by Kircher in his *Ars Magna*.[31]

Just as Verbiest repeatedly noted in the *Astronomia Europaea*, these optical toys caught the interest and attention of Emperor Kaingxi. For example, after his description of the octagonal prism 'tube', Verbiest added: 'Because of the pleasant images of the objects it showed, this tube kept the eyes of the Emperor firmly in its grips for a long time.'[32]

Toys in Churches and Their Imitations

For the Jesuits in China, churches were an important stage for them to reach a wider audience, for which their European objects of wonder turned out to be indispensable props. Beginning with Matteo Ricci, Jesuits paid a great deal of attention to decorating their churches with all kinds of objects attractive to the public. In discussing the large clock constructed by Thomas Pereira (1645–1702) in his church at Beijing,[33] Verbiest attached great importance to this device of wonder for the Jesuit mission in China:

> Because this clock echoes far and wide, and makes resound the name of our church all through the city, a big crowd flocks to view our church. [...] Although the greater part consists of heathens, they honour the status of our Saviour with genuflections and by repeatedly bowing their heads to the ground. I hope that, thanks to the soon-to-be completed organ, on which Father Tomé (Pereyra) is working at the moment, as I mentioned above, still more people will flock to the new *theatre* and that, in this way, every soul will praise GOD on drums, organ and harmonious cymbals![34]

In such a context, optical toys were employed and brought about similar success. For example, the huge anamorphic pictures that Grimaldi made in the church not only grasped Emperor Kangxi's attention, but also attracted many other elites and officials, so that 'up to the present day dignitaries and the most prominent mandarins continue to frequent our residences daily, to look at these and other similar (specimens). There are none that do not admire this art.'[35]

The use of optical toys in churches left deep impressions on Chinese men of letters, as seen in Qu Dajun's屈大均 (1630–1696) work. For example, he mentions in about 50 Chinese characters the different kinds of clocks, world maps, astronomical instruments, and a large organ in the churches at Macao, but he devotes 132 characters to the optical glasses housed in one church:

> There is a thousand people glass (*boli qianrenjing* 玻瓈千人鏡), which, while hung up, can behold the [images of] everything. There is a multiple treasure glass (*duobao jing* 多寶鏡), which can produce millions of the images of one person standing afar, so that the golden god in the church appears to really have trillions of avatars. There is a thousand-mile-seeing glass (*qianli jing* 千里鏡), in which the tip and the bell-strings of a pagoda 30 miles distant appear as clear as the level and erect strokes in a Chinese character without any confusion, while the interior of the moon looks like a calyx of water with black paper dregs floating on the surface[...]. There is a magnifying glass (*xianwei jing* 顯微鏡), through which one can see the tiny worms on a flower's pistils carrying their seeds on their backs.[36]

Wang Shizhen 王士禛 (1634–1711) also mentioned the names of these optical devices of wonder in his discussion of the Catholic churches in Macao.[37] Later, Yin Guangren 印光任 (1691–1758) and Zhang Rulin 張汝霖 (*fl.* 1735) mentioned these devices in their description of Macao as well.[38]

As the church was a meeting place between the Jesuit missionaries and the wider Chinese public, it is reasonable to regard it as a site where knowledge was disseminated, especially European knowledge about science and technology. A good example is the case of Huang Lüzhuang 黃履莊 (1656–?),[39] a mysterious figure who was famous for his expertise in making various kinds of ingenious machines and devices, or *Qiqi* 奇器 (devices of marvels or wonders) as his contemporaries called them. Apparently, he acquired this knowledge from the missionaries themselves. According to *A Short Biography of Huang Lüzhuang* (*Huang Lüzhuang xiaozhuan* 黃履莊小傳)[40] by his cousin Dai Rong 戴榕 (1656–), his devices of marvel included a mini automaton 3-*cun* high that could walk, a two-wheeled vehicle 3-*chi* long that was able to carry a rider without any outside power, a wooden dog whose bark sounded like a real dog, a wooden bird in a cage that could flip and twitter, and a mini fountain that was able to produce a stream of water 5-*chi* high. In addition to these items, Dai attached 'A Brief Catalogue of Devices of Marvels' (*Qiqi mulue* 奇器目略) to *A Short Biography of Huang Lüzhuang* wherein we find a number of glasses and anamorphic pictures:

2. Various Glasses (*zhu jing* 諸鏡): Whether your virtue is thick or thin only a friend can see; whether you appear ugly or pretty, only a glass can tell. The use of glasses is nothing more than both self-watching and object-seeing, and therefore different types are made to extend the application.

2.1 Thousand-mile-seeing Glass (*qianli jing* 千里鏡), in various sizes
2.2 Fire-fetching Glass (*quhuo jing* 取火鏡), for catching fire from the sun
2.3 Figure-diminishing Glass (*suorong jing* 縮容鏡)
2.4 Picture-copying Glass (*linhua jing* 臨畫鏡)
2.5 Water-fetching Glass (*qushui jing* 取水鏡), for collecting water from the moon
2.6 Magnifying Glass (*xianwei jing* 顯微鏡)
2.7 Multiple-object Glass (*duowu jing* 多物鏡)
2.8 Auspicious-light Glass (*ruiguang jing* 瑞光鏡), in different sizes, the biggest one 5 to 6 *chi* in diameter. Illuminated with a lantern, its light can be projected as far as several *li*. Sitting in its light during the winter, you can feel as warm as sitting in the sunshine

3. Different Pictures (*zhuhua* 諸畫): Pictures are used for virtual sights, either turning a one-dimensional scene into one with depth, or turning a single surface into multiple surfaces. All are variations of pictures.[41]

3.1 Far-sighting Picture (*yuanshi hua* 遠視畫)
3.2 Side-sighting Picture (*Pangshi hua* 旁視畫)
3.3 Picture in the Glass (*jingzhong hua* 鏡中畫)
3.4 Glass-tube-sighting Picture (*guankuijing hua* 管窺鏡畫). What looks totally unlike a picture will vividly resemble something real when viewed through the tube
3.5 Up-and-down Picture (*shangxia hua* 上下畫). One picture becomes two while simultaneously seen separately up and down

3.6 Triple-faceted Picture (*sanmian hua* 三面畫). One picture with three facets appears as three pictures when seen [from different angles][42]

We do not know whether this catalogue is a list of devices that Huang Lüzhuang actually made, or if it is just a list of devices of which he knew. It is quite clear, however, that what we have here is basically a list of objects of European origin. From Dai Rong's description of Huang's life, we can figure out the source of his knowledge about these devices:

> Mr. Huang Lüzhuang is one of my cousins. As a little boy, he was very smart and was capable of reciting anything that he had read just a couple of times. He particularly liked to make various ingenious devices that sprang from new ideas ... When he was ten, my uncle passed away. He moved to Yangzhou to live with me and therefore began learning about Western knowledge, such as geometry and mechanics. The level of his craftsmanship became more sophisticated ... Those who are curious about his marvelous talents suspect that he must have had some esoteric books or an esoteric apprenticeship. As someone who has had the longest and most intimate association with him, I never find him [to possess] these kinds of books. According to my investigation about the sources of his knowledge, he really had no teacher at all. To my inquiry he just replied: 'how can I be anything marvelous myself? Heaven, earth and man are all marvelous devices. Moving like heaven, stationary like the earth, intelligent like the man, and abstruse like myriads of things, nothing is not marvelous, but none of their marvels are of their own account. There has to be something that is the source and the governor of their marvels, something that is ultimately marvelous but is not marvelous to itself, just as paintings have to have a painter and buildings have to have an architect. That is the very thing which is the supreme marvel.[43]

Although Dai Rong does not make clear how, or from whom, Huang Lüzhuang knew Western geometry and mechanics, his references to Huang Lüzhuang's 'supreme marvel' and his 'source and governor' of 'all marvelous devices' remind us of none other than the Christian God. Considering that Yangzhou was such an important center of the Jesuit mission in China in the late Ming and throughout the Qing period,[44] it is not unreasonable to consider that Huang Lüzhuang might have had some connection with the Jesuits and their church in the city, or he might even have been a Catholic convert.

Since Dai Rong made it clear that Huang Lüzhuang was born in 1656 and was 28 years old when he wrote his biography, we can approximate that he completed his catalogue in ca. 1684. This is also the period when Verbiest and his Jesuit brothers in Beijing were busy constructing all kinds of devices of marvels for Emperor Kangxi. Just three years earlier, another list of 'optical devices of marvel' was published, called *A History of Glasses* (*Jingshi* 鏡史) by Sun Yunqiu 孫云球 (*c.*1650–?), a famous optician based in Suzhou.[45] It turns out that *A History of Glasses* is not a book about the manufacturing of optical glasses as we might expect. Instead, it looks more like a brochure advertising the author's optical products, including a list of eyeglasses and a number of optical toys. With the exception of 24 glasses for hypermetropia, 24 for myopia, and 24 for 'young eyes', the list includes far-seeing glasses (*yuanjing* 遠鏡), *i.e.* telescopes, fire glasses (*huojing* 火鏡), face-checking glasses (*duanrong* *jing* 端容鏡), incense-burning glasses (*fenxiang jing* 焚香鏡), light-absorbing glasses (*sheguang jing* 攝光鏡), *i.e.* a camera obscura, sunglasses (*xiyang jing* 夕陽鏡), magnifying glasses (*xianwei jing* 顯微鏡), and myriad flower glasses (*wanhua jing* 萬花鏡), which showed 10 images of an object. In the brief introduction to each of these glasses, Sun Yunqiu also included a number of 'glasses that men do not often use, or use purely for entertainment': Mandarin duck glasses (*yuanjing* 鴛鏡), semi glasses (*banjing* 半鏡), multi-faced glasses (*duomian jing* 多面鏡), illusive-image glasseses (*huanrong jing* 幻容镜), magnifying glasses (*chawei jing* 察微鏡), celestial-observing glasses (*guanxiang jing* 觀象鏡), gunnery glasses (*zuopao jing* 佐炮鏡), light-emitting glasses (*faguang jing* 發光鏡), one-thread-of-sky glasses (*Yixiantian* 一線天), and one-beam-of-light glasses (*yixianguang* 一線光).[46]

From the prefaces and colophons of *A History of Glasses*, it is clear that Sun Yunqiu had learned the technique for making optical devices from a group of opticians living in and around Hangzhou and Wuxian, two other centers of the Jesuit mission in China. Although trade with European merchants could have been a channel for the importation of optical instruments and toys into China,[47] this does not mean that the technical know-how would have necessarily derived from these items. As a result, it is reasonable to consider that this group of Chinese opticians would have had some direct connection with the Jesuit missionaries in these cities.

Because of the emergence of these opticians, eye-glass making became a new business in China, while Suzhou became a center of this trade because of men like Sun Yunqiu, while optical toys were also important products of local eyeglass workshops. In his report of the handicraft industry in Suzhou, Gu Lu 顧祿 (1793–1843) notes the production of optical toys in the Huqiu region, observing that 'the methods of "image play" (*yingxi* 影戲) and "western pictures" (*yanghua* 洋畫) were transmitted from European counties in the Western Ocean, whereas people today in Huqiu are able to produce all of them.' Moreover, Gu Lu's report provides a detailed description of structure and use of both 'image play' or 'image play with a lantern' (*dengying zhixi* 燈影之戲), and 'western pictures'. In other words, the former is actually the magic lantern, while the latter is the peepshow box. In addition, he mentions the names of other optical toys that were produced in Huqiu, including 'tubes of myriad flowers' (*wanhua tong* 萬花筒), 'hexagonal western glasses' (*liujiao xiyangjing* 六角西洋鏡), and 'heaven eye glasses' (*tianmu jing* 天目鏡). Gu Lu attributes the origin of this local tradition to Sun Yunqiu and lists a number of glasses that Sun Yunqiu made.[48] According to his report in another book, Jiangning was also famous for the production of 'western glasses' (*xiyang jing* 西洋鏡) and 'western pictures', which are apparently the peepshow boxes and zograscope glasses respectively as seen from Gu Lu's description.[49]

From Devices of Wonder to Objects of Scientific Inquiry

Living in the 21st century, we may not be able to realize the degree to which optical toys from Europe were popular in the Qing dynasty before the Opium

War in 1840, and to what degree they influenced the daily life of the Chinese. We do, however, have one pertinent example. According to the famous playwright Kong Shangren 孔尚任 (1648–1718), author of *The Peach-Blossom Fan* (*Taohua shan* 桃花扇) and a contemporary of Filippo Claudio Grimaldi, Huang Lüzhuang, and Sun Yunqiu, mirrors and glasses played important roles in his hometown of Qufu in Shangdong during various festivals throughout the year. Thus on New Year's Day, mirror cases in every household should be opened to 'discharge brightness' (*kai guangming* 開光明), people should adorn themselves with a small mirror, called a 'brightness-discharging mirror', while at midday, one should fetch fire from a fire glass, called the genuine fire of the sun. On the first day of the 8th month, the elderly should put on eyeglasses. The 15th day of the same month, corresponding to the Mid-Autumn Festival, was the day to grind mirrors and to see the 'precious glasses' (*kan boli bojing* 看玻璃寶鏡), including 'facial glasses' (*mianjing* 面鏡), 'eyeglasses' (*yanjing* 眼鏡), magnifying glasses, 'far-seeing glasses' (*yuan jing* 遠鏡), 'glasses of multiple glasses' (*duojing jing* 多鏡鏡), 'fire glasses' (*huojing* 火鏡) and 'dyeing glasses' (*ranjing* 染鏡). On the ninth day of the 9th month, marking the Double Ninth Festival, one should 'climb to the top of a mountain or the town's gate tower and look into the distance with a telescope'.[50] It is not difficult to find that a number of optical devices and toys of European origin was used during these festivals. The implication is that these devices and toys had become part of the local customs.

More interestingly, besides hundreds of poems on eyeglasses, we can also find many poems about optical toys from Europe: 'Six Poems on the Peepshow Box from the West' (*Xiyang jingxiang shi liushou* 西洋鏡箱詩六首) by Xu Qianxue 徐乾學 (1631–1694),[51] 'Poems on Six Glasses in Response to Jiang Qingshan' (*Liujing shi he Jiang Qingshan* 六鏡詩和蔣靜山) by Yuan You 袁佑 (1634–1699),[52] 'Poems on Four Glasses from the West' (*Xiyang sijiang shi* 西洋四鏡詩) by Honglan Zhuren 紅蘭主人 (1671–1704),[53] 'Poems on Six Glasses Composed for Honglan Zhuren' (*Liujing shi wei Honglan Zhuren fu* 六鏡詩為紅蘭主人賦) by Tao Xuan 陶煊, who flourished during the reign of Emperor Kangxi,[54] and 'A Song of the Little View through the Western Magnifying Glass' (*Xiyang xianwei jing xiaojing ge* 西洋顯微鏡小景歌) by Jiang Yü 江昱 (1706–1775)[55] to name but a few.

In the jottings of the literati from the Qing dynasty, we also find numerous descriptions of optical toys. For example, Wang Shizhen recorded his experience with the anamorphic arts:

> Western paintings and drawings: Glass devices made by Westerners are full of wonders. I once saw the portrait of a person that they drew. At first I was not able to make clear the distinction between the head, eyes, hands and feet. But when a mirror was applied, the countenance of [the figure in the portrait] turned out to be very nice and beautiful. The mirror is long and conical, looking just like the conical head of a writing brush. They also drew gloriettes, pavilions, palaces and houses, and displayed the picture on a wall. Viewed from more than ten paces, it revealed layers of gates and doors leading to infinite series of

complexes of buildings, as deep as a princely palace. Watching closely, one could only discern hundreds of levels and erect strokes in the shape of a chess board.[56]

From his words, we can feel his amazement at these exotic tricks.

By the late 18th and early 19th centuries, the peep show and magic lantern show had become customary attractions of the New Year celebration in Nanjing and Suzhou, as Gu Lu 顧祿 recorded in two of his books.[57] In Yangzhou, there was even a Western style building where a mechanical clock was combined with optical devices to amaze visitors:

By the hill to the left was a large house with a front portico built in imitation of the architecture of Westerners. Viewed from the outside, the house seemed to have hundreds of layers of rooms, with perplexing turning points and myriad passage ways that made one confused and frightened. The only way to get through it was to follow the signal of the clock and move in accordance with it, because a clock chamber was set up in the middle of the house, and its triggering mechanism was attached to every corner in the house so that one signal corresponded to one turn. The interior wall was decorated with pictures of landscapes and seascapes. Opposite the clock chamber was an image projecting lantern [*i.e.* a magic lantern] that was used to project the image painted inside the house. On top of the house there was a dormer window of more than one *chi* wide, which let in the light of the sky and the shade of clouds and mixed them together, adding to them sunshine and moonlight and making the image so pure and bright that nothing was comparable to it.[58]

Apparently, this hall is a very complicated mirror maze equipped with a magic lantern and a kind of camera obscura.

These kinds of optical toys in places such as Yangzhou excited the imagination of young people like Zheng Fuguang, who lived in Shexian County of Huizhou Prefecture in the south of the present Anhui Province. This was one of the wealthiest Chinese areas in the Ming and Qing period, as well as one of the most important Chinese centers of knowledge, especially in textual scholarship and exegetics. Thanks to the enduring influence of Mei Wending (梅文鼎, 1633–1721), the region was also the heart of mathematical sciences. Zheng Fuguang was educated in the Confucian tradition, but he took a special interest in mathematics. He passed the provincial examination and matriculated at the Imperial College in Beijing. Without further success in civil examinations, he primarily made his living as a tutor and an aide in the houses of high officials and men of wealth.

Apparently, Zheng Fuguang invested much of his spare time in mathematical studies and wrote several books on mathematics, *An Illustrated Explanation of the General Procedure of the Six Formulae for the Regular Arc* (*Zhenghu liushu tongfa tujie* 正弧六術通法圖解), *A Brief Explanation of Written Calculations* (*Bisuan shuolue* 筆算說略), *A Brief Explanation of Calculation with Napier's Rods* (*Chousuan shuoleu* 籌算說略),[59] *Illustrated Explanations of and Supplement to the Collective Methods of Geometrical Construction* (*Jihe qiuzuo huibu tujie* 幾何求作匯補圖解) contained in *Posthumous Manuscripts of Zheng Fuguang* (*Zheng Huangxiang Yigao* 鄭浣香遺稿),[60] and *A Simplified Commentary on the*

Zhou Dynasty Canon of Gnomon Computation (*Zhoubi suanjing qianzhu* 周髀算經淺注).[61] Another book on mathematics is his *Table of the Cyclotomic Arc* (*Geyuan huji biao* 割圓弧積表), which was almost lost.

In addition to mathematics, Zheng Fuguang also wrote on physics (*wuli* 物理, literally, 'principles of things'). It seems that he was influenced to some extent by the earlier publication of Fang Yizhi 方以智 (1611–1671), a scholar from southern Anhui of the late Ming and early Qing period. His masterpiece, *Miscellaneous Notes on the Principles of Things* (*Wuli xiaozhi* 物理小識), ushered in a new trend on how to study nature by integrating the 'western knowledge' of Aristotelian natural philosophy with Neo-Confucian tenets of natural law. Zheng Fuguang's *Notes on the Profound and Subtle Principles of Things* (*Feiyin yuzhi lu* 費隱與知錄), is an examination of the underlying causes and principles of natural and preternatural phenomena in the same vein as Fang Yizhi's *Miscellaneous Notes on the Principles of Things*, though on a smaller scale. The book was published in 1843, but Zheng Fuguang dated its conception to 1816, when he sojourned in Yangzhou with his younger cousin Zheng Beihua 鄭北華 (fl. 18 16–1846), who shared similar interests. His *Miscellaneous Notes of Zheng Fuguang* (*Zheng Yuanfu zhaji* 鄭元甫札記) is a collection of notes taken of his readings on natural studies, containing excerpts from books by scholars like Ge Hong 葛洪 (284–364), Shen Kuo 沈括 (1031–1095), Bo Ming 博明, and Jiang Yong 江永 (1681–1762) on nature-related topics.[62]

In late 1819, Zheng Fuguang visited Yangzhou for the second time and was likewise accompanied by Zheng Beihua. This time they were both fascinated by the performance of the magic lantern and paid close attention to optical phenomena, even conducting a series of experiments on grinding ice into a fire-lens.[63] Once back home, they embarked on a joint research project designed to unlock the principles of the magic lantern and other optical toys and instruments in China, which eventually gave rise to the first draft of *Blazoning Infatuation in Front of Mirrors* more than 10 years later. They continued revising the draft until the work's publication in 1846.

The work consists of four volumes. Volume 1 *An Exposition of the Essentials* (*Mingyuan* 明原) is devoted to the most essential properties of color, light, shadow, light-rays, eyes, and glasses, as well as the rules of their behavior. Volumes 2–3 *An Explication of the Circular [Lens and Mirror]* (*Shiyuan* 釋圓) are devoted to the rules governing the feature and behavior of lenses and mirrors, both convex and concave. Volume 4 *Instructions for Work* (*Shuzuo* 述作) is devoted to the structure, function, design and manufacture of 17 types of optical devices, including prisms, lenses, eyeglasses, telescopes, the zograscope (magnifying lens), peep-show boxes, multifaceted mirrors, the optical camera obscura, magic lanterns, the sunglass, the telescopic sextant, the search light, kaleidoscopes, and the like.

In early modern Europe, these objects were so inextricably connected to science that their working principles were systematically studied and encoded into theories. They arrived in China, however, not as a branch of science, but as exotic phenomena and wondrous marvels.[64] As more and more Chinese craftsmen began to understand how these devices operated, Zheng Fuguang became the first who took it as his responsibility as a Confucian scholar to create a theoretical guide. In this way, he turned these instruments and toys into objects for scientific inquiry. To achieve his goal, he spent nearly 30 years collecting raw data on a variety of devices, models, books, and technical diagrams, drawing on every piece of Chinese and European knowledge that he could find. His systematic approach to investigating these devices was unprecedented in China. His efforts led him to establish a theory largely modeled on the axiomatic system of Euclidean geometry, applicable not only qualitatively but also quantitatively.[65] Taken as a whole, his system is completely unlike anything developed by contemporary optical physicists in Europe. His was a hybrid distillation of knowledge that was cobbled together using all the available knowledge he had at his disposal, regardless if the source of that knowledge was textual or physical, foreign or domestic.

Concluding Remarks

In discussing the first significant wave of the transmission of European science and technology to China before the Opium War in 1840, it is very easy to overlook the role played by objects other than books, such as clocks, musical organs, medicine, and scientific instruments. These objects were no less the bearers of knowledge than were books, not just in terms of the technical knowledge for how to produce them and how they functioned, but also the underlying principles and theories about how they operated. An abundance of these objects found their way into the inner court of the early and mid-Qing dynasty through Jesuit missionaries, emissaries, and merchants from Europe and doubtlessly contributed to the dissemination of European science and technology in China. Most of these objects are still preserved in the Palace Museum in Beijing.[66] Among these objects are those of modern science, such as heliocentric orreries, refracting and reflecting telescopes, pendulums, Napier's rods, and Pascalines. It is clear that these Chinese replicas are the result of knowledge that was transmitted to the Chinese craftsmen at the Palace Workshop (*Zaoban chu* 造辦處) of the Qing dynasty. The textual accounts of many of these instruments, including two heliocentric orreries,[67] were incorporated in such books as *The Illustrated Norms of the Ceremonial Instruments of the Present Dynasty* (*Huangchao liqi tushi* 皇朝禮器圖式),[68] which helped to expose them to a wider public beyond the Forbidden City.

In this paper, we have examined a category of objects that consist of optical toys. They were introduced to China by Jesuit missionaries mostly as exotica in order to show the superiority of European science and arts. In the eyes of the ordinary Chinese literati and public, these objects were so strange and marvelous that they became subjects of literature as well as playthings during festivals and ceremonies. In the hands of Chinese craftsmen and mathematicians, however, the 'object-borne knowledge' in these 'Western Wonders' meant more than objects of purely entertainment value.

They were studied, deciphered, copied, and then adapted for local consumption. At the same time, this understanding was combined with a 'text-borne knowledge', which began a fascinating process of hybridization that led to Zheng Fuguang's *Blazoning Infatuation in Front of Mirrors*. In this long chain of events, we see how these objects changed depending on who was handling them at any given time within different socio-cultural contexts. It was in this milieu that the Jesuit missionaries made a special contribution to the development of optics and the optical industry in China in the 17th and 18th centuries.

Notes

1 Dupre 2008, 219–244.

2 For example, in Jones 1795 there is a category under this name with the note: 'As the study of natural philosophy is often excited in young minds by amusing and entertaining recreations and experiments; the following are a select variety, made intentionally with that view, and are presumed to be the most instructive of the kind.' See also Stafford and Terpak 2001, 176.

3 This term was first used by Charles Wheatstone in 1827. In the beginning of his paper on the kaleidoscope (Wheatstone 1827), he gives the following note: 'The application of the principles of science to ornamental and amusing purposes contributes, in a greater degree, to render them extensively popular; for the exhibition of striking experiments induces the observer to investigate their causes with additional interest, and enables him more permanently to remember their effects. I shall not, therefore, need an apology for presenting the tyro in science with another combination of philosophy with amusement, in addition to those already extant.'

4 The book was first published in 1646, and then republished in 1671 as an enlarged edition. For the religious meanings attributed to light and optical toys by A. Kircher, see Vermeir 2005.

5 For an account of the contents of the book, see D'Elia 1960, 35–41; Hashimoto 1988, 173–175.

6 Crew 1928, 144–152.

7 Interestingly, in the gift-list of Matteo Ricci on these occasions, there were already optical items, namely prisms. See Liu Dong 2001.

8 Du Halde 1741, 77.

9 Du Halde 1741, 74.

10 Verbiest mistook this for 'Catoptrics', but the objects he described under this title belong to the realm of refractive, rather than reflective, optics.

11 Verbiest provides a detailed introduction to these works in his *Astronomia Europeae*. See Golvers 1993, 102–129. For a much abridged introduction of these activities, see Du Halde 1741, 75–76.

12 Du Halde 1741, 74.

13 A pinhole camera obscura was nothing new in both Europe and China at the time, but it was not until the end of the 16th century that the pinhole camera obscura was equipped with lenses and mirror and, therefore, transformed into the early modern optical camera obscura, which forms sharper and clearer images at long distances. On the invention of the optical camera obscura, see Lefèvre 2007, 5–30. Before the Kangxi period, the optical camera obscura had been introduced into China as an auxiliary tool for painting by Adam Schall von Bell in his *Yuanjing shuo* 遠鏡說 (Tang Ruowang 1626, 379), and used for observing the solar eclipse during the calendar reform of the late Ming Dynasty; see Wang Guangchao *et al.* 2008.

14 Golvers 1993, 114. A similar story is retold by Du Halde 1741, 72, in which Verbiest's role is not mentioned and 'a convex-glass' is used instead of 'a proportioned glass'.

15 Lefèvre 2007, 15.
16 Golvers 1993, 114.
17 Golvers 1993, 114–115; Du Halde 1741, 72–73.
18 Kircher 1646, figs 807; 1671, 709. For a recent discussion, see Lefèvre 2007, 14.
19 For a detailed introduction of different forms of anamorphic arts in Europe in the 16th–18th century, see Stafford and Terpak 2001, 235–248.
20 Kircher 1646, 177–187; 1671, 124–136.
21 Du Halde 1741, 73–74; Golvers 1993, 115.
22 Du Halde 1741, 74; Golvers 1993, 116.
23 Brewster 1858.
24 Du Halde 1741, 74; Golvers 1993, 116.
25 For a detailed introduction to the history of faceted lenses in Europe, see Stafford and Terpak 2001, 184–191.
26 Golvers 1993, 116.
27 Du Halde 1741, 74–75.
28 Kircher 1646, 887; 1671, 768–769. For the history of the magic lantern in the Europe, see Mannoni 2000, 3–74.
29 Golvers 1993, 116.
30 Du Halde 1741, 78.
31 Kircher 1646, 818 with the three figures that follow; 1671, 718.
32 Du Halde 1741, 74; Golvers 1993, 116.
33 Pereira equipped the clock with a set of bells in musical scale that could announce the time by automatically playing different pieces of Chinese and European music. See Du Halde 1741, 74; Golvers 1993, 127.
34 Golvers 1993, 127–128.
35 Golvers 1993, 115.
36 Qu Dajun 1985, 37.
37 Wang Shizhen 1982, 517.
38 They added two kinds of glasses that Qu Dajun did not mention: Houzi jing 火字鏡 (Character-firing glass) and Zhaozi jing 照字鏡 (Character-illuminating glass). In addition, they also indicated that 'Duobao jing consisted of a mass of small glasses and could, therefore, turn one person into millions of himself' (Yin Guangren and Zhang Ruling 1988, 78).
39 For a brief discussion of Huang Lüzhuang in English, see Needham 1956, 516–517.
40 Zhang Chao 1985, 113–116.
41 From the description, the items under this category are anamorphic pictures.
42 Zhang Chao 1985, 114–116 (I have added numbers for sake of clarity).
43 Zhang Chao 1985, 113.
44 Jiangsusheng zhengxie wenshi ziliao weiyuanhui *et al.* 1999, 37–48.
45 For the most recent study of Sun Yunqiu on the basis of the discovery of his *Jingshi*, see Sun Chengsheng 2007.
46 Sun Yunqiu 1680, main part, 1a–12b.
47 For instance, in *Changshui zeli* 常稅則例 and *Yue haiguan zhi* 粵海關志, we do find the rules for taxing optical goods like glass mirrors, spectacles (glass or crystalline), and *Xiyang jing* 西洋景, see Anonymous 1772–1735, 432–433, 483; Ling 1838, 613.
48 Gu Lu 1999, 156–157.
49 Gu Lu 1980, 13.
50 Kong Shangren 2016, 805, 846, and 851.
51 Xu Qianxue 1631–1694, 421–422.
52 Yuan You 1717, 44–45. The six glasses are the panel mirror, eyeglasses, fire glass, telescope, multi-eyed glass, and magnifying glass.

53 Zha Weiren 1741, 134–135. The four glasses are the telescope, fire glass, multi-treasures glass, and magnifying glass.
54 Tao Xuan 1662–1722, 367–368.
55 Ruan Yuan 1798, 182. Here the 'Little View of Western Magnifying Glass' refers to the zograscope.
56 Wang Shizhen 1982, 631–632.
57 Gu Lu 1980, 157; Gu Lu 1999, 14.
58 Li Dou 1960, 270. The building is inside *Garden of Pure Aroma (Jingxiang yuan* 淨香園).
59 The *Bisuan shuolue* and *Chousuan shuolue* are housed in the library of the Institute for the History of Natural Sciences, Chinese Academy of Sciences, but they are reprinted in the *Xuxiu siku quanshu* 續修四庫全書.
60 Housed in the Anhui Provincial Museum.
61 Conserved in the Shandong Provincial Library.
62 The autographed manuscript of the notebook is now preserved in the Anhui Provincial Museum, while a late Qing copy of it is located in the library of the Institute for the History of Natural Sciences of the Chinese Academy of Science and Technology.
63 This experiment was so important to Zheng Fuguang that he mentioned it in both *Notes on the Profound and Subtle Principles of Things* and *Blazoning Infatuation in Front of Mirrors* with its exact date, the process used, and the results obtained. See Zheng Fuguang 1842, 33ab; 1846, 60.
64 Although Chinese astronomers also used the telescope in their observations, it was still regarded as a wonderful optical toy in the eyes of the ordinary public, as seen in one of Li Yu's (1999 [1611–1680]) stories in *Shi'er lou* 十二楼. In this case, he describes how a playful young voyeur won the love of a beautiful girl by observing her through a telescope. In his comments on the story, Li Yu lists a number of other optical devices, which he notes 'can generate a great many fantastical things'. See Li Yu 1999, 285–292.
65 For the most recent analysis of Zheng Fuguang's special system of optics, see Qian Changyan 2002.
66 An exquisitely illustrated introduction to these holdings, including Chinese replicas, can be found in Liu Lu 1999.
67 For a detailed discussion of these orreries, see Sivin 1973; Xi Zezong *et al.* 1973.
68 Yun Lu 1759, j. 3, 1a–82a.

Bibliography

Anonymous. 1772–1735. Changshui zeli 常稅則例, *Xuxiu siku quanshu* 續修四庫全書 j. 0834, Shanghai.

Brewster, D. 1858. *The Kaleidoscope: Its History, Theory, and Construction with its Application to the Fine and Useful Arts*, London.

Crew, H. 1928. *The Rise of Modern Physics*, Baltimore.

D'Elia, P. 1960. *Galileo in China: Relations Through the Roman College Between Galileo and the Jesuit Scientist-Missionaries 1610-1640*, Cambridge, MA.

Du Halde, J.-B. 1741. *The General History of China: Containing a Geographical, Historical, Chronological, Political and Physical Description of the Empire of China, Chinese-Tartary, Corea and Thibet: Including an Exact and Particular Account of their Customs, Manners, Ceremonies, Religion, Arts and Sciences* vol. 3, London.

Dupre, S. 2008. Inside the Camera Obscura: Kepler's Experiment and Theory of Optical Imagery, *Early Science and Medicine* 13, 219–244.

Golvers, N. 1993. *The Astronomia Europaea of F. Verbiest, S.J. (Dillingen, 1687): Text, Translation, Notes and Commentaries*, Nettetal.

Gu Lu 顧祿. 1980 (1842). *Tongqiao yizhao lu* 桐橋倚棹錄, Shanghai.
Gu Lu 顧祿. 1999 (1830). *Qing jia lu* 清嘉錄, Nanjing.
Hashimoto, K. 1988. *Hsü Kuang-Ch'i and Astronomical Reform: The Process of the Chinese Acceptance of Western Astronomy 1629–1635*, Kansai.
Jiangsu sheng zhengxie wenshi ziliao weiyuanhui 江蘇省政協文史資料委員會 *et al.* 1999. Yangzhou zongjiao 揚州宗教, *Jiangsu wenshi ziliao* 江蘇文史資料 115, Nanjing.
Jones, W. and Jones, S. 1795. *A Catalogue of Optical, Mathematical and Philosophical Instruments Made and Sold by William and Samuel*, London.
Kircher, A. 1646. *Ars Magna Lucis et Umbrae*, Rome.
Kircher, A. 1671. *Ars Magna Lucis et Umbrae*, Amsterdam.
Kong Shangren 孔尚任. 2016 (1648–1718). *Jiexu tongfenglu* 節序同風錄, Ma Siding 馬斯定 (ed.), Hangzhou.
Lefèvre, W. (ed.) 2007. *Inside the Camera Obscura - Optics and Art under the Spell of the Projected Image*. Max Planck Institute for the History of Science Reprint no. 333, Berlin.
Li Dou 李斗. 1960 (1795). *Yangzhou huafang lu* 揚州畫舫錄, Beijing.
Li Yu 李漁. 1999 (1651–1661). *Shier lou* 十二樓, Beijing.
Liang Tingyue 梁廷玥. 1838. Yue haiguan zhi 粵海關志, *Xuxiu siku quanshu* 續修四庫全書 j. 0835, Shanghai.
Liu Dong 劉東. 2001. Li Madou de sanlengjing 利瑪竇的三棱鏡, *Dongfang wenhua* 東方文化 6, 51–52.
Liu Lu 劉潞. 1999. *Qinggong xiyang yiqi* 清宮西洋儀器, Shanghai.
Mannoni, L. 2000. *The Great Art of Light and Shadow: Archaeology of the Cinema*, Exeter.
Needham, J. 1956. *Science and Civilisation in China*, vol. 2. Cambridge.
Qian Changyan 錢長炎. 2002. Guanyu *Jingjing lingchi* zhong toujing chengxiang wenti de zai tantao 關於《鏡鏡詅癡》中透鏡成像問題的再探討, *Ziran kexueshi yanjiu* 自然科學史研究 21.2, 135–145.
Qu Dajun 屈大均. 1985 (ca. 1700). *Guangdong xinyu* 廣東新語, Beijing.
Ruan Yuan 阮元. 1798. Huaihai yingling ji bingji 淮海英靈集丙集, in: *Xuxiu siku quanshu* 續修四庫全書 j. 1682, Shanghai.
Shi Yunli 石云里. Forthcoming. The Astronomical Images in the First Chinese Treatise on the Telescope by Johann Adam Schall von Bell Revisited, *Journal for the History of Science, Technology and Medicine.*
Sivin, N. 1973. Copernicus in China, *Studia Copernicana* 6, 63–122.
Stafford, B.M. and Terpak, F. 2001. *Devices of Wonder: From the World in a Box to Images on a Screen*, Los Angeles.
Sun Chengsheng 孫成晟. 2007. Ming Qing shiqi xifang guangxue zhishi zai Zhongguo de chuanbo jiqi yingxiang: Sun Yunqiu *Jingshi* yanjiu 明清時期西方光學知識在中國的傳播機器影響：孫雲球《鏡史》研究, *Ziran kexueshi yanjiu* 自然科學史研究 26.3, 363–376.
Sun Yunqiu 孫云球. 1680. *Jingshi* 鏡史 [printed text preserved in the Shanghai Library].
Tang Ruowang 汤若望. 1626 (1995). Yuanjing shuo 遠鏡說, Ren Jiyu 任繼愈 (ed.), *Zhongguo kexue jishu dianji tonghui tianwenjuan* 中國科學技術典籍通彙天文卷 vol. 8, Zhengzhou.
Tao Xuan 陶煊. 1662–1722. *Shixi shichao* 石溪詩鈔, *Siku jinhui shu congkan* 四庫禁毀書叢刊, Beijing.
Vermeir, K. 2005. The Magic of the Magic Lantern (1660–1700): On Analogical Demonstration and the Visualization of the Invisible, *The British Journal for the History of Science* 38.2, 127–159.
Wade, N.J. 2004. Philosophical Instruments and Toys: Optical Devices Extending the Art of Seeing, *Journal of the History of the Neurosciences*, 13.4, 383–384.
Wang Guangchao 王廣超 *et al.* 2008. Mingqing zhiji wangyuanjin de chuanru dui zhongguo tian-wenxue de yingxing 明清之际望远镜的传入对中国天文学的影响, *Ziran kexueshi yanjiu* 自然科學史研究 27.3, 309–327.
Wang Shizhen 王士禎. 1982 (1691). *Chibei outan* 池北偶談, Jin Siren 靳斯人 (ed.), Beijing.

Wheatstone, C. 1827. Description of the Kaleidophone, or Phonic Kaleidoscope: A New Philosophical Toy for the Illustration of Several Interesting and Amusing Acoustical and Optical Phenomena, *The Quarterly Journal of Science, Literature, and Art* 23, 344–351.
Xi Zezong 席澤宗 *et al.* 1973. Rixindidongshuo zai Zhongguo: jinian Gebaini dansheng wubai zhounian 日心地動說在中國 – 紀念哥白尼誕生五百週年, *Zhongguo kexue* 中國科學 3.3, 270–279.
Xu Qianxue 徐乾學. 1631–1694. *Zhanyuan wenji* 憺園文集, *Xuxiu siku quanshu* 續修四庫全書 j. 1412, Shanghai.
Yin Guangren 印光任 *et al.* 1988 (ca. 1751). *Aomen jilue* 澳門紀略, Zhao Chunchen 赵春晨 (ed.), Guangzhou.
Yuan You 袁佑. 1717. Jixuan shichao 霽軒詩鈔, *Siku weishoushu jikan qiji* 四庫未收書輯刊柒輯 j. 27, Beijing.
Yun Lu 允祿. 1759. Huangchao liqi tushi 皇朝禮器圖式, *Yingyin qinding wenyuange siku quanshu* 影印文淵閣四庫全 j. 657, Taipei.
Zha Weiren 查為仁. 1741. Lianpo shihua 蓮坡詩話, *Xuxiu siku quanshu* 續修四庫全書 vol. 1071, Shanghai.
Zhang Chao. 1985 (1683). *Yu Chu xinzhi* 虞初新志, Shijiazhuang.
Zheng Fuguan 鄭復光. 1842 (reprint 1985). *Feiyin yuzhi lu* 費隱與知錄, Shanghai.
Zheng Fuguang 鄭復光. 1846 (reprinted 1937). Jingjing lingchi 鏡鏡詅癡, *Congshu jicheng chubian* 叢書集成初編, Shanghai.

Chapter 4

The Karakorum Highway: Gateway of Empires, Religions, and Commerce[1]

Saba Samee

The Karakorum Highway (KKH), a segment of the Silk Roads that runs through Pakistan, is known locally as *Shah-rah-e-Karakorum* (Fig. 4.1). The KKH branches southward from Kashgar to the region known in antiquity as Gandhara, which today incorporates the valleys of Hunza, Gilgit, Swat, Peshawar, Buner, Hund, and Taxila.[2] This road formed an important crossing point for merchants plying their wares on the Silk Roads. From Gandhara they could travel to Iran in the west, Central Asia and China in the north, the Indian subcontinent in the east, or the Arabian Sea in the south where they could link up with the maritime routes of the Indian Ocean.[3]

I took the opportunity to explore this segment of the Silk Roads in 2007. This paper constitutes part of my ongoing research undertaken at the Institute for Art and Culture, Lahore, in which I investigate those 'values' that form the relationship between people and places in one's heritage by employing a variety of methodologies. These involve preserving material remains, and retaining traditional crafts and techniques used in constructing historical monuments.

'Cultural Mapping' is the process by which places representing the cultural heritage of a region are identified along with its 'values', which the inhabitants of a particular region have given to them, or associated with them, through oral tradition: stories, myths, or legends. These embody socio-political and cultural interests, such as ancestral associations, ethnic identity, religious affiliation, and even the inclination to glorify the past for economic gain. It is for this reason that the historical assets along the KKH prove to be useful links in bridging the past with the present. The travelers of this highway left behind a rich and diverse repository of petroglyphs and other cultural expressions.

Every province in Pakistan was linked historically to trade and enjoyed a level of prosperity brought to it by merchants.[4] Words like *sodagar*, merchant, a character who appears in many myths and legends, and *serai*, a caravanserai where weary travelers

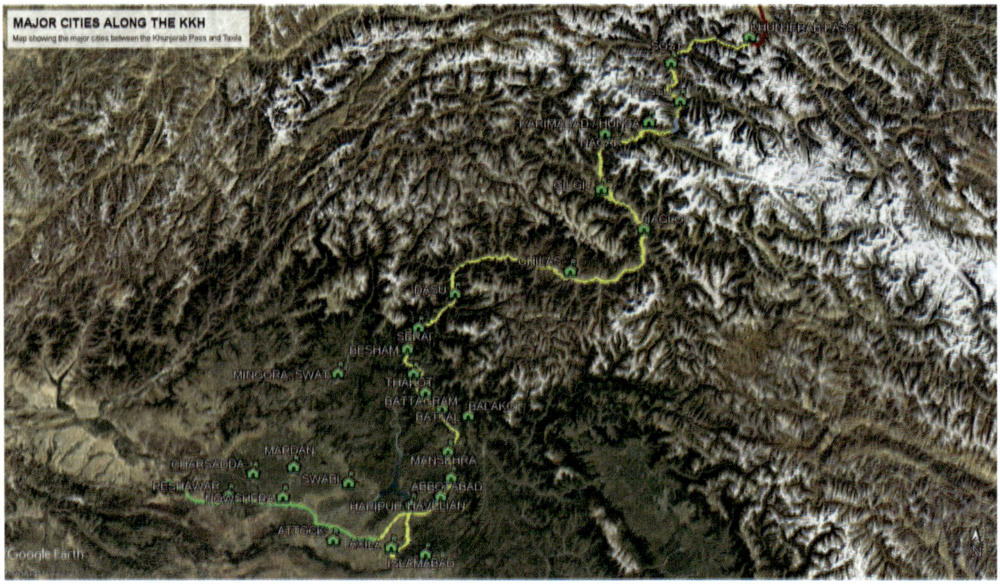

Fig. 4.1. Sites located along the KKH (Google Earth, *earth.google.com/web/*).

could rest, feed livestock, restock supplies, and even buy and sell merchandise, form part of the toponyms of many settlements along the KKH. The precursor of the KKH as a branch of the Silk Roads is described as a road filled with hardships that crossed numerous territories and terrains.

Routing the KKH

One segment of the KKH originates from the southern periphery of the historic city of Kashgar, which is located at the junction of the northern and southern Silk Roads. A 'quintessential caravan city',[5] Kashgar was known for its thriving bazaars, some of which still exist as Grand Bazaars. Upon passing through Kashgar, the road proceeds southward toward the Karakorum mountain range. The word *Karakorum* is Tajik, meaning 'Black Mountain',[6] referring to the long dark shadows the high mountains cast upon themselves and the valleys below. This segment of the Silk Roads then passes through Tashkurgan[7] (known as the 'Stone Tower' during the Pamir Kingdom of Sarikol (ca. 1300 BCE)), running northeast between the Pamir and the Karakorum, and enters Pakistan through the Khunjerab Pass. The word *Khunjerab* means 'Blood Valley', referring to the mountainous pass once controlled by the bandit tribes of Hunza and Wakhi. From the Khunjerab Pass onwards, the road becomes the Karakorum Highway proper, meandering through the Karakorum mountain range. It runs along the mighty River Indus and is the highest public road in the world.[8]

The KKH forks near the historic city of Attock, one section of which leads to the ancient city of Taxila mentioned in the ancient religious texts of *Ramayana* and

Mahabharata as the city of *Takshashila*.[9] Takshashila is the city that received Alexander and his army,[10] and was a Buddhist center during the Kushan Empire (early 1st century CE). This route provided access to two main plains: the Indus with its fertile fields of fruits, vegetables, cotton, wheat, barley, and other grains, and the Ganges[11] leading to areas rich in wheat, spices, and gems.

The second segment, known as the Torkham Highway (hereafter TKH), heads towards the valleys of Peshawar (ancient *Purushapura*), Swat, Buner, and Charsadda (*Pushkalavati* of the *Mahabharata*), comprising the seventh satrapy of Gandhara during the Achaemenid Empire (550–330 BCE), according to Herodotus (3.89–95), and later ruled by the Kushans.

The TKH meanders along the Kabul River into Afghanistan (ancient *Ariana*) through the Khyber Pass, situated within the Hindu Kush mountain range. Eventually, it makes its way to the cities of Bamiyan and Balkh in ancient Bactria, from where the road enters Iran in the west and Central Asia in the north.

The Merchants' Road

The KKH segment of the Silk Roads also referred to as the '*Shah-rah-e-Reesham*', a Persian name that was later infused in the Urdu language. '*Shah-rah*' literally means the 'King's Road'; its generic meaning may be taken as a stately or majestic road that led to prosperity. '*Reesham*', on the other hand, means 'Silk'. The Majestic Silk Road, or the Kingly Silk Road, acts as the proper translation of '*Shah-rah-e-Reesham*', a name reflective of the major trading commodity – silk.

In antiquity, silk was not regarded as a luxury but a necessity in the northern steppe regions, Central Asia and South Asia. In winter, woolen cloths were prone to moths and lice. Silk lining provided a natural barrier against body lice and skin sores, and was worn under woolen as well as animal skins or leather clothing. In summer, silk was more breathable and durable compared to linen, lawn, or muslin. Its other uses were as strings in bows, due to its strong, flexible, and elongated fiber, and in the production of flexible and long-lasting paper. Early Buddhist petroglyphs along the KKH depict stupa *chatras*, or umbrellas, flanked with silken banners flowing in the wind, their luster guiding the pilgrims and attracting new believers. In South Asia silk was associated with beauty, such as 'silky hair' or 'skin as soft as silk'. It also became connected with nature, so that the flow of a gentle stream was compared to the feel of silk against one's skin, or rice paddies which were likened to silk cloths waving in a breeze as they danced in the wind. For Central Asians, silk was also a form of currency, which they received from China as payment for their sturdy horses. They in turn sold their silk in the European market where 'China Silk' was a luxurious commodity and brought good money.[12]

The petroglyphs at the historic sites of Chilas point to another major trading commodity of this region – horses. These rock carvings show stylized horses with one knee bent, drawn in the Achaemenid style.[13] In the stupa of Butkara I, Swat valley, is a stone relief of a saddled horse in the same posture. Used as a commodity primarily to restock cavalries, the horse trade brought huge profits to the nomadic breeders of the steppe from Persians, Gandharans, and Sogdians, who purchased them with spices, medicine, and jewelry.[14]

There are numerous proverbs in South Asian culture associated with horses, such as '*ghoray baich key sow raha hai*', which connotes a person who sleeps as if he does not have a care in the world. The meaning is intended to reference a merchant, who, after having worked hard to raise his horses, has sold them for a neat profit. Another typical saying is '*zuban ko lagam lagao*', urging the speaker to control his/her tongue as one would restrain a horse with a bridle.

According to Tucker,[15] commerce along the Silk Roads revolved around three basic purposes: a source of profit, fascination with the exotic, or to enhance one's political power. Many such mercantile families once existed in this region along the KKH. For instance, the original inhabitants of the community of Bhabra settled in the regions surrounding the KKH for better trading prospects, economic prosperity, and growth. According to Tahir,[16] the Bhabra clan was made up of goldsmiths and gold dealers, belonging to an ancient Jain merchant community. Members of this community may still be found in Pakistan, as Bhabra is a common surname. The historic mercantile city of Sialkot contains two neighborhoods called Bhabrian Wala and Serai Bhabrian ('serai' is a resting place for traders and their horses). Lahore likewise has a neighborhood called Thari Bhabrian, and a road crossing known as the Bhabra stop on the main Ferozepur Road. In Rawalpindi, an ancient city near Taxila, there is a traditional market known as Bhabhra Bazaar.

Another such family is mentioned in an 18th-century story of Classical Urdu literature called *Sohni Mahiwal*. The family context of the male protagonist, Prince Mirza Izzat Baig (Mahiwal) of Balkh-Bukhara, is set in Tsarist Russia, nowadays Uzbekistan.[17] The story, apart from narrating various socio-cultural aspects of these regions, also indicates the cultural familiarity between the two regions acquired through continuous interaction through trading.

In Peshawar, the Sethi clan is linked with the mercantile activities that took place along the KKH. 'Sethi' derives from the Sanskrit word 'Sresthin' or 'Setthis',[18] used to describe a mercantile clan, the head of which is called the 'Seth', a word associated with wealth.[19] The Sethi Muhallah within the old walled city of Peshawar, established by the Sethi clan, is in close proximity to the oldest area within the walled city, Gor Khatri, famous for its religious structures and bustling bazaars.[20] One of the Sethi *havelis* has been conserved and adapted as a museum, showcasing the various objects the clan had acquired from its long history in trade and commerce.

The importance of trade along the KKH is further apparent once the number of *serai* is taken into account. The caravanserai built within the city of Peshawar alone amounts to 13, catering to the traders from a variety of regions.[21] These *serai* were located beside the gates of the city along with a fresh water well, or *hammam*, and stables for horses. The northern and western gates were reserved for visitors from Kabul, Central Asia, Iran, and elsewhere, whereas the eastern and southern gates received travelers and traders heading to Sialkot, Taxila, Lahore, Delhi, and South India.

The Pilgrims' Road

The KKH was also known as the 'Pilgrims' Road'[22] connecting Sindhu (the ancient name for Indus/India) and China. It was also the route taken by Fa Xian (399–412 CE) and Sung Yun (approx. 520–518 CE), two celebrated Buddhist monks who undertook pilgrimages from China to various Buddhist sites. Their records mention a number of regional kingdoms that had converted to Buddhism when these were ruled by the Indian Emperor Ashoka of the Mauriyan Empire (322–187 BCE), and later by King Kanishka of the Kushans (1st century CE). Another Buddhist monk Xuanzang (ca. 629–645 CE) journeyed through the region of Gandhara, passing the valleys of Peshawar, Buner, and Swat, returning to the road at the city of Hund. Both these routes pass through the ancient town of Takshashila, known by the Greeks as Taxila. The city later served as the capital of the Kushan Empire and became a thriving center of Buddhism from where Buddhists ventured to China.[23]

As a religious road, the KKH, according to archaeological evidence, dates as far back as the composition of *Rig Veda* in ca. 1500–1200 BCE.[24] This religious text mentions 16 ancient kingdoms – *Mahajanapadas*,[25] 'the strong tribal footholds'.[26] Chief among these were Gandhara, Purushapura, Pushkaravati, and Takshashila. More than 700 petroglyphs and inscriptions in various languages have been mapped along these valleys, recording generations of travelers from Sogdiana, Persia, China, and Tibet,[27] who, according to Tucker, created 'nothing less than a visitors' book of the centuries'.[28]

The ruins of Buddhist religious centers consisting of rock edicts, stupas, and monasteries still exist along the KKH. The valley of Hunza is home to more than 50 historic settlements, including Karimabad, Baltit, Altit, Ganesh, and Attabad. The village of Ganesh has the earliest recorded petroglyphs. The adjacent village of Chel Ganesh contains the 'sacred rocks of Hunza', also known as the 'Haldeikish' – 'the place of male ibexes'.[29] The village is scattered with watchtowers and ancient ruins. The meaning of 'Ganesh' in the local language of Brushaski is 'gold'. Both sites have Kharoshthi, Sogdian and Tibetan scripts along with carved images.[30] The city of Altit has the oldest fort in the valley of Hunza, known as the Altit Fort, whose settlement is associated with the first village of the White Huns in ca. 47 CE. In the 11th century, a new fort was constructed over the first one, and experienced intensive restoration lasting until 2007. The city of Karimabad has the second oldest fort along the KKH, known as the Baltit Fort. Constructed in the 8th century, this fort was situated at the highest point in the valley and underwent massive alterations in the 16th century.

The Gilgit Valley begins at the city of Gilgit and ends just before Chilas. As the Indus makes its way through the valley, both its banks contain numerous Buddhist monuments attesting to the region as a place of pilgrimage. There are 13 known sites containing such remains, namely Thak, Ba Das, Gukona, (ke)Ges, Shing Nala, Bargin, the rock boulders at the Partab and Alam bridges, the cities of Danyor, Kargah, Kuno Das, Naupura, and Henzal. Dominating the valley is the eighth highest mountain peak of the Karakorum, Nanga Parbat. At Kargah, a massive figure of a standing Buddha is carved on a living rock cliff surface. A natural spring waterfall is located beneath this cliff. The posture of the Buddha is one of bestowing blessings and warding off evil. The context is magical: the gentle stream of water, the blessing Buddha, the sun's warmth and the green valley beyond. The 'Gilgit Manuscripts' were also found in the remains of the nearby stupa and monastery.[31] Jaglot is another historic city in the valley. It is a meeting point of all three mountain ranges in this region: the Karakorum, the Hindu Kush, and the Himalayas. It is also the point where the rivers Gilgit and Indus meet. The rock boulders with petroglyphs are located near the Partab and Alam bridges, which span both rivers.

The favorable location of Chilas in the valley of the same name is situated where the crossing of the River Indus is at its narrowest and the river at its calmest. For this reason, the city received numerous visitors. Between the towns of Sazin and Chilas along the Indus are petroglyphs carved on rock boulders depicting Buddha and Buddhist stupas.[32] In addition, there are 16 important sites: Shatiyal, Darel, Harban, Khanbari, Bazeri Das, Minargah, Manro Das, Kino Kor, Thor, Helor, Oshibat, Domu Das, Hodur, Dadam Das, Gichi, and Thalpan. These sites also contain the earliest rock inscriptions in Kharoshthi[33] along with petroglyphs of Buddha and stupas. At Shatiyal/Shatial there is a boulder with Buddhist petroglyphs and inscriptions.[34] The site of Hodur has a boulder representing two smaller Buddha figures flanking two smaller stupas on either side of a larger stupa, which contains a lotus flower and a silk streamer atop the *chatri*, while another at Thalpan with its sacred rocks has a number of inscriptions in various languages.[35] The most celebrated petroglyph appears on a rock boulder with a Buddha under the Bodhi tree and above stands a grand stupa with banners and streamers flying in the wind.

The valley of Mansehra is famous for its rock boulders with inscriptions carved in Kharoshthi script, ordered by Emperor Ashoka of the Mauriyan Empire.[36] The message given on these rock edicts is of humanitarian nature, where King Ashoka is asking people to follow the teachings of Buddha.

According to Chinese Buddhist pilgrims, the area around Purushapura (Peshawar) had about 1,000 *sangharamas* which they found in ruins.[37] The Peshawar valley constitutes three historic cities mentioned as kingdoms in the *Mahajanapadas*. The city of Purushapura contains the ruins of Shah ji ki Dheri, which legend has it housed the *patra* (begging bowl) of Buddha, a large stupa of 'exquisite beauty' with a monastery believed to have been built by king Kanishka.[38] The stupa is said to have been decorated with 'every sort of precious substance'. Today only a small mound survives along with an ancient Pipal tree and the foundations of the monastery within the confines of the city's principal graveyard. The city of Pushkaravati/Pushkalavati (Charsadda) contains Buddhist ruins at 12 locations: Khanizaka Dheri/Kula Dheri,

Shahr-i-Napursan/Rajar, Palatu Dheri, Ghaz Dheri, Sarai Makhe Dheri (Hariti Stupa), Mamane Dheri/Sher Pao, and Periano Dheri/Palo Dheri.

Among the existing ruins around the historic city of Mardan are Sahri Bahlol, Shahbaz Garhi, Takht-i-Bahi, Jamal Garhi, Sawal Dheri, Hund, Asota, Surkh Dheri, Rani Ghat, and the Sanghao/Kashmir Ghar caves.[39] The Buddhist site of Shahbaz Garhi is mentioned as Polusha by Chinese pilgrims.[40] An Ashokan Rock Edict was also found along with a stupa and monastery built upon older structures by Kanishka. The site of Jamal Garhi possesses extensive remains of a large stupa surrounded by monastic cells and dwellings. The area between Shahbaz Garhi and Jamal Garhi is scattered with numerous stupas and cave dwellings.[41]

The valley of Swat (ancient Oddiyana) is scattered with the remains of archaeological sites. Locations of these sites are documented by a series of reports published by the Italian Archaeological Mission in Pakistan. The report titled *Buddhist Architecture in the Swat Valley*[42] documents 19 sites, of which four are *Viharas*. The best known are the stupas of Gharasa at Dangram, Jurjurai and Loebanr, Arap-khan-china at Shararai, Butkara I, Saidu Sharif I, Shnaisha, Shankardar, Amluk Dara, Abba-saheb-china and Tokar Dara at Najigram, Barikot, Top Dara at Haibatgram, Thana, Gumbatuna, Nemogram, and the Ghalegai Buddha statue carved on the roadside cliff face, and the rock carved Buddha of Jahanabad.

The valley of Taxila is located at the crossing point of three major trade routes of the Silk Roads network, the KKH from the north, the TKH from Kabul and the valley of Gandhara, and in the southeast the Grand Trunk Road or GT Road from the ancient city of Pataliputra in northeastern India. The most visited Buddhist religious center was the valley of Takshashila (Taxila), especially the cities of Bhir, Sirkap, and Sirsukh, and the stupa sites of Bhamala, Jinnan Wali Dheri, Julian, Mohra Muradu, the great Dharmarajika, Bhallar, Ratta Pind, Akhor, and Giri. The lone stupa of Mankiala, mentioned by the pilgrims as Singhapura, is located in the village of Mankiala, marking the southern extremes of this valley.

The Road of Temples and Observatories

Another element that has survived in some of the cities along the KKH is that of Hindu temples. Many have been destroyed, others deteriorated, while others still were incorporated into newer monuments for a variety of reasons. A number have survived in Peshawar. The majority of these temples are associated with *Jogis* and *Pirs*. Some date back thousands of years. Their presence is indicative of the persistence of Hinduism, as seen in the area known as the Gor Khatri,[43] and of the area's propensity for religious tolerance that allows for different belief systems to coexist.

The oldest temples along the KKH, however, belong to the Hindu Shahi Dynasty (843–1026 CE). The region south of Gandhara and Taxila, known as the Potohar/Potwar Plateau, was the center of this dynasty. The plateau rises abruptly from the plains of the Punjab between the rivers Jhelum and Indus. This region comprises of numerous riverine valleys, spring fed streams, jagged peaks, and a massive escarpment of salt mines. The area is also rich in fruit trees, flowering shrubs with medicinal properties, and green prairies. With peace established, trade between Sindhu and Khurasan once again flourished under this dynasty. A number of structures survive along this stretch of the KKH.

Kallar Kahar contains the mystical peacock valley and a central lake. This shallow valley is home to numerous peacocks protected by the Sufi saints Ahoo Bahoo, Syed Ishaq, and Pir Yaqoob.

Katas Raj is the mythical pond surrounded by temples: the ancient religious site of Katas has its roots in Brahmanical history, where it appears as Kataksha. This is where Lord Shiva was inconsolable after the death of his wife Sati. Tears fell from his divine eyes as rain, creating two ponds on earth, from each eye. One was created at Pushkara (Ajmir) and the other at Kataksha (Katas). The word Kataksha literally means 'spring formed from raining eyes'.[44] The site contains a central pond fed by spring water surrounded by a group of seven temples containing *Shiva lingam*, called the *Satgarah*[45] (lit. 'Seven temple houses') established by the Pandavas of the *Mahabharata.* The palace or *haveli* structure of Hari Singh Nalwa is a site consisting of the Buddhist ruins of a stupa and is the spot where a number of minor rock edicts[46] of Ashoka have been found. The city of Nandna, where Al-Bīrūnī's Observatory is situated, served as the capital of the Hindu Shahi Dynasty. Nandna means 'Indra's Garden', indicating the divine garden of the *devas.*[47] It is located along the southern ridge of the Potohar Plateau, between two peaks that form a natural gorge, called the Nandna Pass. The remains of a massive bastion along with the ruins of a fort and its defensive walls, and an elaborate temple structure are visible along the western edge of this gorge. The settlement of Baghan Wala marks the access road to Nandna. The site is associated with Al-Bīrūnī, the master astrologer from the courts of Mahmud Ghazni. His travels in India took him along what would later become the KKH. Legend has it that in Nandna the great scholar compiled his literary masterpiece, *Al Hind*, and where he calculated the circumference of the earth.[48]

Passageway for Invaders

The northern road between the Hindu Kush and Karakorum mountain ranges has always served as an important route taken by invaders to reach the fertile lowlands in the south. Thapar[49] provides the most comprehensive analysis of the volatile political situation of this region. In her book, she describes various local dynasties and foreign invaders who came to conquer this region whenever there was a dynastic or power void in the region. Recorded history indicates that these invasions began with the coming of the Indo-Aryans, first mentioned in the *Rig Veda.* These invaders, referred to as the 'conquerors of the land of the Dasas',[50] may have come from Iran, Sogdiana, or Ariana in Afghanistan. They brought with them such practices as horse riding,[51] unique religious beliefs, and their own language, which they used to compose the

Vedas. The various rivers recorded in this text indicate the migratory routes that they took to reach India. The Kabul River is mentioned in the beginning verses of the *Rig Veda*, followed by other northern rivers like the Gumal, Panjkora, Swat, Kunar, and Indus. The text then mentions a second river network flowing from the northeastern Punjab plains and joining Indus near Attock: Sohan, Jhelum, Chenab, Ravi, Beas, and Hakra-Saraswati. A third river network mentioned is the southeastern rivers of Yamuna and Ganga.[52] The regions which these invaders initially inhabited constituted the three Gandhara valleys of Peshawar, Swat, and Buner, before they reached the plains of Punjab. These Aryan conquerors used the same northern route, which later formed part of the Silk Roads network.[53]

Almost a millennium later, Darius I (ca. 550–487 BCE) conquered Gandhara and made it his seventh province. According to Herodotus, Darius I conquered the Indians to 'find out where the Indus joins the sea'.[54] The earliest mention of Gandhara as a 'country' (*dahyus*) is mentioned by Darius I (ca. 520 BCE) in his Bīsotūn (*Behistun*) inscription.[55] Both provinces were prosperous, providing a wealth of gold dust and wheat. The Achaemenid *satrapies* were overthrown by the armies of Alexander of Macedon, including the regions from Gandhara to Taxila in 326 BCE. According to the historians of his expeditions, the route he took to enter the lands of the Indus ran from the city of Kunar to Gandhara, into the Swat valley and subsequently to Taxila. The remains of a temple in Jandial, having Greek architectural features, such as the plan and Ionic columns, testify to the influence brought by the Greek colonists.[56]

Following Alexander's campaign, Gandhara enjoyed a few peaceful and prosperous centuries under his successors and the Indian Mauryan dynasty, founded by Chandragupta Maurya, known to the Greeks as Sandracottus.[57] He is credited for having expelled the Macedonians from the Punjab shortly after Alexander left the area. The region continued to be ruled by the Mauryas until the death of Chandragupta's grandson Ashoka in the mid-2nd century BCE, along with the Indo-Greeks, Parthians (Pahlavas), and Scythians (Shakas) among others. Eventually, the Kushans in the early 1st century CE, who had established their dynasty north of the Hindu Kush, pushed southward to Gandhara[58] and eventually into southeast India. Kanishka I in the 2nd century CE made Takshashila the new Kushan capital of the dynasty.

The 3rd century CE witnessed the decline of the Kushans, resulting in both banks of the Indus River to be governed by different rulers: the west bank was ruled by the Persian Sasanids (254–651 CE) and east bank by the Indian Guptas (350–550 CE).[59] The Sasanids dominated the trade that passed along the KKH, since they controlled the mountain passes of the Hindu Kush and Karakorum along with Sogdiana, Persia, and Bactria.

The Arab invasions into this region changed how trade was carried out. Their priority was to capture and secure the trading routes to 'Al-Hind', which in turn caused the coastal kingdoms of Debal and Maka/Mahi Khoran (Makran) to become important ports along the maritime Silk Routes.[60] It is from these ports that Islam spread into South Asia. The land route of the KKH had no further encounters with invaders, until the conquests of the Mongols.[61] Genghis Khan (1162–1227 CE) initiated his invasions from Beijing, capturing all of southern China and reaching Central Asia through Kashgar.[62] Retracing the Mongol footsteps of his ancestors, Timur (1336–1405 CE) invaded the regions of Gandhara, Taxila, and Punjab, crossed the Indus River from Attock and reached Delhi.[63] Keeping with family traditions, the Mughal warrior Babar (1483–1530 CE) left his homeland of Ferghana in Central Asia, traversed through Kabul and entered the Indus region from the TKH, passing through the Hindu Kush at the Khyber Pass, and founded the Mughal dynasty in this region.[64]

The Silk Roads, the 'Spine of Asia', acted as a strategic borderline for the British colonizers of Hindustan when the region gained independence, resulting in the formation of two nations, India and Pakistan. The KKH formed the 'Northern Tier' during the Cold War between the US and the Soviet Union, however, this time not for trade or invasion, but for control and espionage.[65]

At present, the KKH forms part of the 'Silk Road Economic Corridor' stemming from the Chinese 'One Belt One Road' policy. In Pakistan, this route proper is termed 'CPEC': the China-Pakistan Economic Corridor. This corridor will follow the route of KKH, extending it to the Makran/Gawadar coastal regions.

Physical Manifestation of Cultural Hybridity in Architecture

According to Maeda,[66] the Silk Roads were 'cultural networks' consisting of a network of 'highways, connecting the vast cultural worlds of the East and the West' and functioned as an 'intricate web of byways connecting many regions, towns and villages'. The architectural vocabulary that developed along the KKH is one of many components of the cultural heritage hybridity of this region.

The theory of a 'Central Eurasian Culture Complex' proposed by Beckwith,[67] posits that cultural exchange within 'Euro-Asia' occurred four millennia ago. One of the physical manifestations of such a cultural exchange along the KKH can be observed through the architectural diversity and adaptation prevalent in this region. The fusions of local architectural style combined with foreign elements are a manifestation of the cultural receptivity of the region that led to its peculiar cultural hybridity.

What follows are some of the many examples of cultural hybridity present in the architectural vocabulary of this region, which I observed during my travels along the KKH.

The wooden lantern ceilings of Baltit Fort, Hunza, and those of the Bamiyan Cave tombs[68] represent a common practice of corbelling the roof using wooden beams. The same roofing style is also prevalent in the Kalasha valley. The ancient feature of projected wooden balconies constructed at Baltit Fort, is similar to the projected wooden balconies of the Walled City of Lahore and that of ancient Persian cities.

The wooden decorative pillars at Baltit Fort and at the shrines of Sufi saints in South Punjab have similar structural proportions. Likewise, the architectural feature of a wooden geometric ceiling found in the shrines of South Punjab is also found at the Idgah Mosque in Kashgar.

The art of *Kashikari,* or glazed tile work, used as an architectural element for building ornamentation is one of the widespread architectural influences. Examples from South Punjab, Lahore, Kashgar, and Kashan, present a common craft behind the intricately decorative glazed tile work. Fresco wall panels in various historic structures of Pakistan even contain images of Chinese porcelain vases, indicating their cultural familiarity. The same treatment of a wooden pillar at Kashgar is detected in the marble pillar at Lahore whose capital is decorated with muqarnas inlaid with floral motifs.

When I see the stupas along the KKH as well as the many shrines and tombs of Sufi saints and other revered individuals, I cannot help but notice that their design, architectural vocabulary, and purpose share many similarities: a square terrace whose basement is ornamented on which is erected a square or octagonal stupa base, similar to a shrine hall containing a sarcophagus, on top of which is a semi-hemispherical dome, similar to the *patra* of a stupa. The decorative pinnacle resembles the *chatri* on the top of a stupa, the purpose of which is to honor the Buddha in the same way that a Sufi Saint is honored.

The inherent cultural hybridity observed through the architectural vocabulary of this region indicates the development of a consistent socio-cultural value system, which is reflective of diverse cultural influences. For Pakistanis, the architecture found along the length of the KKH is a record of this region's rich and vibrant past, of infinite exchange and innovative hybridity.

Notes

1 Images for this article except the map have been uploaded as an online PDF accessible at http://stromeon.com/archive/publications/silkroad-ss.pdf.
2 Thapar 2002, 222.
3 Beckwith 2009, 85.
4 Rehman 1997, 39–84, 108–123.
5 Tucker 2015, 153.
6 Tucker 2015, 193.
7 Tucker 2015, 190–191. The Alexandrian geographer Ptolemy described the location of Tashkurgan as a mid-point between China and the West.
8 Tucker 2015, 192–193.
9 Marshall 2013, 10.
10 Marshall 1951, 12.
11 Rehman 1997, 55.
12 Nebenzahl 2004, 8–9.
13 Hauptmann 2005, 28–29.
14 Tucker 2015, 64.
15 Tucker 2015, 3.

16 Tahir 2017. 9.
17 Sekhon and Duggal 1992, 85. The classical folktale of Sohni Mahiwal was written in the 18th century by Syed Hashim Shah. Later variations of the same story were written by Hafiz Barkhuddar and Mian Muhammad Bakhsh. This was explained to me by Professor Mushtaq Soofi, a Punjabi writer and expert on Folktales of the Punjab, in an interview that I conducted at Lahore in 2018.
18 Thapar 2002, 161.
19 Lari 2015, 1.
20 Jaffar and Salim 2008, 57–64.
21 Imran 2012, 56.
22 Beal 2003, x–xv.
23 Tucker 2015, 192.
24 Pal 2012, 15; Radice 2014, 18.
25 Dar 2006, 161–162.
26 Chakrabarti 2006, 396–397.
27 Beckwith 2009, 85.
28 Tucker 2015, 193–194.
29 Tucker 2015, 195.
30 Hauptmann 2005, 135.
31 Tucker 2015, 197.
32 Hauptmann 2005, 22.
33 Hauptmann 2005, 136.
34 Dar 2006, 283.
35 Hauptmann 2015, 135.
36 Tucker 2015, 203; Dani 1995, 23.
37 Beal 2003, 98.
38 Jaffar and Salim 2008, 19–22.
39 Jaffar and Salim 2008, 31–37.
40 Beal 2003, 111–112.
41 Beal 2003, 113.
42 Faccenna *et al.* 2014.
43 Jaffar and Salim 2008, 62–163. Note: Gor Khatri, 'house of Hindu' – *Gor*, 'place' or 'house'; *Khatri*, 'Hindu'.
44 Masih 2015, 188–189.
45 Singh 2016, 288.
46 Masih 2015, 204.
47 Masih 2015, 203.
48 Sachau 2002, 235.
49 Thapar 2002.
50 Nasr 1974, 51, 54.
51 Radice 2000, 85–95.
52 Pal 2012, 23–24.
53 Nasr 1974, 51.
54 Herodotus 4.44.
55 Magee *et al.* 2005, 711.
56 Thapar 2002, 160.
57 Hamilton 1971, 263.
58 Thapar 2002, 221.
59 Thapar 2002, 235–245, 282.
60 Thapar 2002, 407–425.

61 Thapar 2002, 425.
62 Robinson 2007, 8–9.
63 Robinson 2007, 44–102.
64 Robinson 2007, 115–116.
65 Frankopan 2015, 421.
66 Maeda 2007, 20.
67 Beckwith 2009, 28.
68 Lena 2007, 46.

Bibliography

Baumer, C. 2014. *The History of Central Asia - The Age of the Silk Roads*, London.

Beal, S. (transl.) 2003. *Si-Yu-Ki Buddhist Records of the Western World - Translation from the Chinese of Hiuen Tsiang (AD 629-645), vols. 1-2*, New Delhi.

Beckwith, C. 2009. *Empires of the Silk Road*, Oxford.

Burn A.R. (transl.) 1972. *Herodotus: The Histories*, London.

Chakrabarti, D. 2006. *The Oxford Companion to Indian Archaeology - The Archaeological Foundations of Ancient India*, New Delhi.

Dani, A.H. 1995. *Human Records on Karakorum Highway*, Lahore.

Dar, S.R. 2006. *Historic Routes Through Gandhara (Pakistan)*, Lahore.

Dwivedi, S. 2009. *The Quest of the Budhha - A Journey on the Silk Road*, New Delhi.

Faccenna, D., Spagnesi, P. and Olivieri, L.M. 2014. *Buddhist Architecture in the Swat Valley, Pakistan: Stupas, Viharas, a Dwelling Unit*, Lahore.

Fisher, R.E. 1993. *Buddhist Art and Architecture*, Singapore.

Hamilton, J.R. 1971. Introduction, in: de Sélincourt, A. (transl.), *Arrian: The Campaigns of Alexander*, London, 13–40.

Hansen, V. 2012. *Silk Road - A New History*, New Delhi.

Harle, J.C. 1994. *The Art and Architecture of the Indian Subcontinent*, London.

Hauptmann, H. 2005. Pre-Islamic Heritage in the Northern Areas of Pakistan, in: Bianca, S. (ed.), *Karakoram - Hidden Treasures in the Northern Areas of Pakistan*, Turin, 21–40.

Hauptmann, H. 2015. *Pakistan: The Construction of Hydropower Projects in the Upper Indus Valley of Northern Pakistan and their Threat to the Rock Art Galleries.* Heidelberg. https://journals.ub.uni-heidelberg.de/index.php/heritage/article/view/19968/13761.

Imran, R. 2012. *Peshawar - Faseel-i-shehr awr Darwazey*, Peshawar.

Jaffar, S.M. and Salim, A. 2008. *Peshawar - City on the Frontier*, Karachi.

Lari, Y. 2015. *Sethi Haveli: A Gem in the Heart of Historic Peshawar*, Karachi.

Lena, K. 2007. Chinese and Central Asian connections in the Koguryo Mural Paintings, in: Yamauchi, K., Taniguchi, Y. and Uno, T. (eds), *Mural Paintings of the Silk Road: Cultural Exchanges Between East and West: Proceedings of the 29th Annual International Symposium on the Conservation and Restoration of Cultural Property, National Research Institute for Cultural Properties, Tokyo, January 2006*, London, 45–53.

Magee, P., Petrie, C., Knox, R., Khan, F. and Thomas, K. 2005. The Achaemenid Empire in South Asia and Recent Excavations in Akra in Northwest Pakistan, *American Journal of Archaeology* 109, 711–741.

Marshall, J. 1951. *A Guide to Taxila, vol 1, 2, and 3*, Karachi.

Marshall, J. 1960. *A Guide to Taxila*, New York.

Masih, F. 2015. *Temples of the Salt Range - Detailed Analysis of their Architecture and Decorative Designs*, Lahore.

Nadeem, I. 2008. *Buddhist Gandhara - History, Art and Architecture*, Lahore

Nasr, T. 1974. *The Eternity of Iran*, Tehran.

Nebenzahl, K. 2004. *Mapping the Silk Road and Beyond – 2000 Years of Exploring the East*, London.
Pal, T. 2012. Physical Geography during Vedic Civilization: A Literature Survey, *International Journal of Academic Research in Progressive Education and Development, vol 1, no. 1*, Greece. http://hrmars.com/admin/pics/576.pdf.
Radice, B. (ed.) 2014. *The Rig Veda: An Anthology: One Hundred and Eight Hymns.* Selected, translated and annotated by W.D. O'Flaherty, New Delhi.
Rashid, S. 2005. *The Salt Range and the Potohar Plateau*, Lahore.
Rehman, A. 1997. *Historic Towns of Punjab – Ancient and Medieval Period*, Karachi.
Robinson, F. 2007. *The Mughal Emperors and the Islamic Dynasties of India, Iran and Central Asia*, London.
Sachau, E.C. (transl.) 2002. *Alberuni's India – An Account of the Religion, Philosophy, Literature, Geography, Chronology, Astronomy, Customs, Laws and Astrology of India about 1030 AD*, New Delhi.
Sekhon, S.S. and Duggal, K.S. 1992. *A History of Punjabi Literature*, Delhi. http://apnaorg.com/books/sekhon-1/sekhon.php?fldr=book&page=84.
Singh, A. 2016. *Lost Heritage – The Sikh Legacy in Pakistan*, New Delhi.
Tahir, S. 2017. Walking through Rawalpindi's Bhabra Bazaar was a Journey into a Majestic Past, *DAWN*, May 6, Karachi.
Thapar, R. 2002. *Early India: From Origins to AD 1300*, New Delhi.
Thubron, C. 2006. *Shadow of the Silk Road*, New York.
Tucker, J. 2015. *The Silk Road – China and the Karakorum Highway*, London.
UNESCO 1979. Rock Carvings at the Sacred Rock of Hunza, and Near Gilgit and Chilas, *World Heritage List Nomination Submitted by Pakistan, Identification Number 142*, Paris.
Wood, F. 2003. *The Silk Road – Two Thousand Years in the Heart of Asia*, London.
Yamauchi, K., Taniguchi, Y. and Uno, T. (eds) 2007. *Mural Paintings of the Silk Road: Cultural Exchanges Between East and West: Proceedings of the 29th Annual International Symposium on the Conservation and Restoration of Cultural Property, National Research Institute for Cultural Properties, Tokyo, January 2006*, London.

Section Two

Understanding Spice Through Interdisciplinarity

Chapter 5

A TRP Along the Silk Roads: How and Why We Detect and Use Spices

Wayne L. Silver and Cecil J. Saunders

For millennia humans have used spices as flavorings, preservatives, and medicines. Some of the oldest evidence of cooking with spices comes from 6,100 years ago.[1] Remains of clay cooking pots in northern Europe suggest that the garlic mustard plant was used to flavor meat dishes. The importance of the spice trade in history is well documented.[2] The pursuit of spices led to an age of exploration both from West to East and East to West, presumably initiating global trade. Here we endeavor to both describe the unique neurophysiological properties of spices in a way that is accessible to interested researchers from fields outside of the biological sciences and to summarize biological and psychological theories that attempt to address what it is that has made these particular chemicals so sought after throughout human history.

Humans have been using spices for medicinal purposes and for flavoring food for thousands of years. As a result the term spice – defined commonly and for the purposes of this paper as a plant substance used to season or preserve food – is a culinary term with some inherent peculiarities. For example, despite being characterized by their use in seasoning and preserving food this definition precludes the single most common substance humans have used for those purposes because it does not have a botanical origin: salt. Similarly, this definition also excludes plants, which were similarly sought after and traded along the Silk Roads primarily for their psychoactive properties, such as *Papaver somniferum* or the opium poppy. Here we argue that the unifying property of all spices is that they are botanicals and potent activators of the arcane chemical sense of chemesthesis.

The Chemical Senses

Spices are primarily used to enhance the flavor of food. Flavor is a complex multi-module sensation produced when chemicals interact with the neural sub-systems responsible for smell, taste and even touch (see below). When chemicals fire up our chemical senses, our brains integrate these signals into a single sensation that untrained individuals often confuse as flavor. The first step in understanding how the flavor sensations created by spices are unique is to have a precise understanding of the differences between each of the chemical senses.

Olfaction

Olfaction, usually called the sense of smell, is considered a distance chemical sense. It involves the detection of volatile (dissolved in air) chemicals in the environment. Chemicals that we breathe in through the nose must stimulate olfactory sensory neurons (OSNs) located in the olfactory epithelia in the roof of the nasal cavity to be perceived as odor (Fig. 5.1). This is called orthonasal olfaction. Although sniffing can be used to increase the number of odorant molecules reaching the OSNs, only about 5–15% of inspired air normally reaches these cells. In humans, the right and left olfactory epithelia each contain about 6 million OSNs packed into a 4 cm^2 area. Each OSN has a cell body with tiny hair-like cilia reaching into the mucus layer lining the nasal cavity. Expressed on the cilia are olfactory receptor proteins, which recognize odorant molecules.[3] In the simplest terms, the sensation of an odor is caused by a chemical binding to one of these olfactory receptors, triggering a signal in the OSN, which is then passed to the brain. Humans have about 400 different olfactory receptor proteins; each receptor protein recognizes a different odorant, that is, each recognizes a chemical whose atoms occur in a particular shape.

Each OSN only expresses one of the 400 olfactory receptor proteins. Once the right odorant binds with the right receptor protein (in a lock and key type arrangement) a biochemical transduction cascade is established, which causes the OSN to fire and send information in the form of electrical signals (action potentials) via fine nerve fibers (axons), which terminate in a part of the brain called the olfactory bulb. Here the information is processed using a combinatorial code and sent to higher brain centers. This mechanism allows us to recognize the odorant and determine whether we like it. The coding of information in the olfactory system is referred to as combinatorial because a single chemical can have multiple molecular patterns, which bind to different olfactory receptor proteins and thus activates a different combination of OSNs. Combinatorial coding allows many more odors to be detected than there are receptors. One recent report suggests this anatomical arrangement allows the 400 olfactory receptor proteins in humans to detect 1 trillion different odors.[4] This is analogous to spelling any word in the English language with a combination of only 26 letters.

While people commonly think of odors entering the nose during inhalation (*e.g.* orthonasal olfaction), during eating and drinking, volatile compounds may also enter the nose retronasally (from the mouth through the nasopharynx) and stimulate OSNs (Fig. 5.1). In this case, our perception is erroneously attributed to the taste or gustatory system (see below). One of the purposes of chewing food is to break it down and cause the release of more volatile chemicals to stimulate OSNs. The importance of olfactory stimulation in the pleasurable and unpleasurable appreciation of food and drink is shown clearly when there is a loss of olfactory function. This loss is a common, and usually temporary, result of a simple head cold and stuffy nose. When this occurs, we usually say that our food does not taste good, but this is actually due to our inability to smell it.

Humans have been said to have a poor sense of smell compared to other animals. Dogs, by comparison, have about 30 percent more OSNs than humans and are reported to be 10,000–100,000 times better than humans in detecting odorants.[5] However, this idea has been challenged. J.P. McGann recently wrote that 'it arose not from empirical studies of human olfaction but from a famous 19th-century anatomist's hypothesis that the evolution of human free will required a reduction in the proportional size of the brain's olfactory bulb'.[6] In addition, Gordon Shepherd, a noted olfactory scientist and the founder of the field of Neurogastronomy, has suggested that dogs may outperform humans in orthonasal smelling, but humans are the animal kingdom champions of retronasal olfaction.[7] The diet of humans, consisting of cooked food, spices, and fermented foods and liquids, offers a much greater and refined assortment of odorants than the diets of other animals. Considered in this context, it is probably unsurprising that the incredible potpourri of sensations detected by the olfactory system is often confused by untrained individuals as 'taste'.

Gustation

Gustation, usually called the sense of taste, is considered a contact chemical sense. In order for a substance to be 'tasted' it must come in direct contact with gustatory sensory cells (GSCs) located in the oral cavity. GSCs are found on the tongue, back of the mouth, pharynx, epiglottis, and upper esophagus and are characteristically grouped together in taste buds. Humans have around 4,000 taste buds. About 75% of all taste buds are found on the top surface of the tongue (Fig. 5.1), arranged on papillae, which appear as blunt protrusions on the surface of the tongue. There are several types of papillae, and these have different distributions on the tongue surface.

As in olfaction, chemicals that activate GSCs do so by binding to gustatory receptor proteins. Unlike olfaction, activation of GSCs leads to only a small number of sensations. One reason for this is there are fewer gustatory receptor proteins than olfactory receptors proteins by at least an order of magnitude. These gustatory receptor proteins allow the detection of five gustatory qualities: salty, sweet, sour, bitter, and the newest quality, umami. Umami is the sensation elicited by glutamate and is particularly recognizable in Parmesan cheese and mushrooms. Each cell in a taste bud is activated by only one of the five taste qualities. For years, these modalities were reported to be mapped on the tongue. Sweet was supposed to be localized on the tip, salt to the front sides, sour to the back sides and bitter on the back of the tongue. However, we now know that there is no taste map and all five taste qualities can be distinguished from all parts of the tongue.[8]

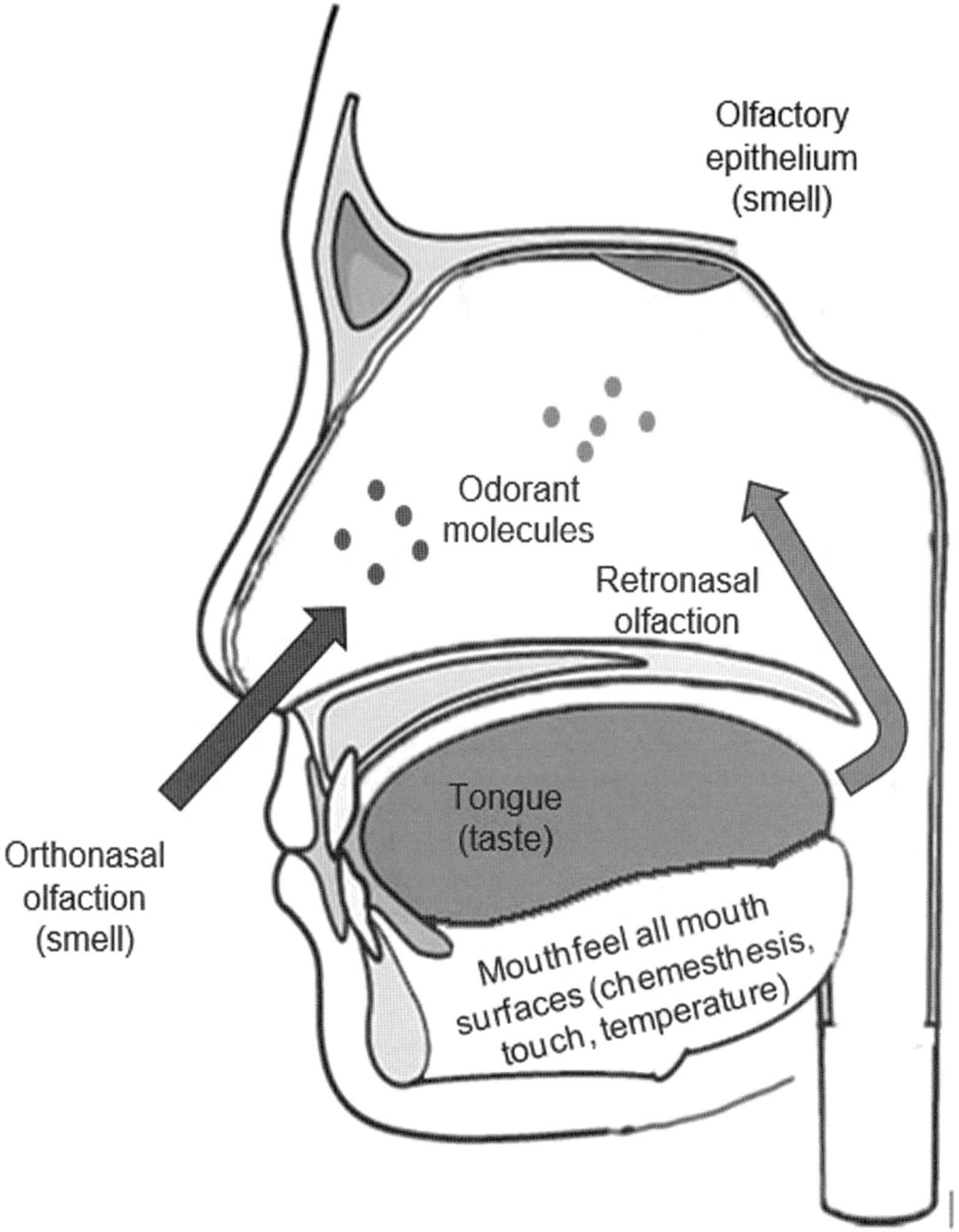

Fig. 5.1. Schematic diagram of the nose and mouth, showing the sensory systems that spices may activate to contribute to the flavor of food.

Chemesthesis

In addition to the commonly known senses of olfaction and gustation, there is a third chemical sense which spices primarily modulate called chemesthesis. Originally, the ability to detect irritating chemicals was called the common chemical sense.[9] However, the common chemical sense implied a separate chemical sensory system. We now know that irritating and pungent chemicals work by activating nerve fibers belonging to the somatosensory system (the nerves responsible for detecting touch, vibration, and temperature along the skin).[10] After these discoveries were made, the neologism, chemesthesis, was coined to refer to the chemical stimulation of the somatosensory system and is now the preferred term.[11]

The somatosensory system for the head and face (including mouth and nose) is mediated by the trigeminal nerve (the Vth of the 12 cranial nerves). In addition to sensing irritating chemicals, trigeminal nerve fibers also detect mechanical stimuli (touch) and thermal stimuli (temperature). In the mouth, these sensations comprise 'mouthfeel', which plays a role in determining flavor (see Flavor section). The trigeminal nerve has three branches – the mandibular, maxillary, and ophthalmic – that innervate the oral cavity, the nasal cavity, the nasopharynx, and the cornea respectively. When we eat chili peppers, sniff vinegar, or cut onions, chemicals in those substance activate trigeminal nerve endings in our mouth, nose, and eyes respectively producing irritation. Similarly, an ice cream headache or 'brain freeze' is triggered by stimulating nociceptive (from the Latin *nocere* to hurt) trigeminal nerve fibers leading to a change in cerebral blood flow, which causes pain; although it is temperature, not chemicals, that stimulates the nerves in this case.[12]

Activation of nasal and oral trigeminal nerve fibers is often accompanied by sweating, runny nose, sneezing, and tearing.[13] These protective responses are among the most powerful reflexes in the body, which serve to reduce the effects of extreme temperatures and chemical irritants and protect the body from further exposure. As in the olfactory and gustatory system, a unique class or family of protein receptors is responsible for recognizing many of the chemicals, which produce the sensations detected by the trigeminal nerve.

Transient Receptor Potential (TRP) Channels

Although there are a number of receptor proteins found on nociceptive nerve endings, the receptor proteins most commonly associated with chemesthesis are called Transient Receptor Potential (TRP: pronounced trip) channels. These are typically the receptors activated by compounds found in spices. In humans, there are 28 different TRP channels that fall into six main subfamilies. The TRP channels that are most relevant to spice detection are TRPV1, TRPM8, and TRPA1, which respond to the chemicals found in chili peppers, mint, and horseradish, respectively. Once the receptor proteins are activated (the chemical binds with a site on the receptor protein in a lock and key arrangement), they change their shape and a channel or pore is opened allowing ions to cross the cell membrane. The influx of ions leads to the triggering action potentials, which are transmitted from neuron to neuron, providing the brain information as to the location and nature of the stimulation.

Many of these TRP channels which are modulated by compounds found in spices are referred to as thermo-TRP channels. This name refers to the phenomenon that temperature as well as chemicals can activate these channels.[14] Thus, in the above example of the ice cream headache, the pain is caused by low temperature triggering a TRP channel and a similar sensation might be created by inhaling a chemical which would activate the same TRP channel to create the sensation of painful cold. Most TRP channels will only respond to being above or below a particular temperature but might be activated by a number of different compounds. This is the case for the three best studied hermo-TRPs: TRPV1, TRPA1, and TRPM8.

TRPV1 was the first thermo-TRP channel discovered and characterized. It is activated by capsaicin, the active ingredient in chili peppers, as well as, noxious heat, above 43°C (109°F). When TRPV1 channels on nociceptive nerve fibers are stimulated, we detect a painful, burning sensation. The temperature of the mouth does not change; rather neurons containing activated TRPV1 send information to the brain, which is interpreted as noxious heat.

TRPM8 channels are considered cold detectors and normally respond to temperatures below 26°C (79°F).[15] These channels also respond to menthol – the active principle in mint. Again, menthol does not change the temperature of the mouth; the brain interprets signals from the activated trigeminal nerve endings as cold. Noxious cold temperatures below 10°C (50°F) may also be mediated TRPM8 channels. Interestingly, menthol products are often used to open up 'stuffy' nasal passages. There is, however, no evidence that menthol actually affects nasal mucus. Instead, when we take a breath with menthol in the nasal cavity, the air 'feels' colder, which the brain reads as greater air flow.[16]

TRPA1 is known as a promiscuous receptor as it is activated by over 100 different compounds of various structures.[17] The archetypal TRPA1 chemical activator is allyl isothiocyanate (AITC), which is found in mustard, horseradish, and wasabi. Interestingly, olive oil contains the compound oleocanthal, which activates TRPA1 channels on the back of the throat.[18] This is what elicits the pungent sensation olive oil often elicits from this area. TRPA1 channels have been reported to be activated by temperature, and may serve to contribute to the detecting of noxious heat.[19]

TRP channels are not only found on trigeminal nerve endings, which innervate the head and face region, but also are found throughout the skin on spinal nerve fibers originating in the dorsal root ganglia of the spinal cord. These spinal nerve fibers respond to the same stimuli that activate trigeminal nerve fibers. Capsaicin (TRPV1) and menthol (TRPM8), for example, heat and cool the skin respectively. Topical application of products containing capsaicin and menthol is often used to treat mild aches and pains in joints and muscles, although their efficacy is still in question.[20]

while these compounds are commonly used as analgesics, they are more often used to enhance the flavor of food.

Flavor

As we have seen, olfaction, gustation, and chemesthesis all detect chemicals in the mouth. The information provided to the brain by these three chemical senses plays a major role in determining the flavor of food. ASTM International, formerly known as American Society for Testing and Materials, defines flavor as the '(1) perception resulting from stimulating a combination of the taste buds, the olfactory organs, and chemesthetic receptors within the oral cavity; (2) the combined effect of taste sensations, aromatics, and chemical feeling factors evoked by a substance in the oral cavity.'[21]

Many people untrained in the chemical senses are shocked to learn of the essential role that 'touch'[22] or the somatosensory system plays in the perception of flavor. However, an important contributor to flavor is mouthfeel (Fig. 5.2). This term is used to describe the physical sensations elicited by food in the oral cavity. They include temperature and texture. The importance of mouthfeel on flavor can be observed by eating soggy, cold French fries. The chemicals in the French fries have not changed, although the flavor is now decidedly off-putting.

The sight of food and even the sound of chewing food can also contribute to flavor. For example, in one experiment, subjects rated the odor of colored liquids more intense than liquids with the same odorant but no color.[23] In another experiment, researchers manipulated the sound of crunching potato chips through headphones as subjects chewed.[24] When the crunching sound was increased, participants rated the potato chips as tasting both significantly crisper and significantly fresher. When the crunching sound was decreased, participants rated the potato chips as tasting both staler and softer. These findings underlay the argument that flavor is a complex multimodal sensation created from the interplay of our senses when we eat.

It should be apparent at this juncture that spices activate all of our chemical senses. Most, if not all, spices have an odor when eaten. For example, thyme and turmeric have a woodsy smell while saffron has a floral odor and sage smells like pine.[25] Thus one way spices contribute to our enjoyment of food is by stimulating OSNs in our nasal cavity by retronasal olfaction. In addition to being pungent and odoriferous, spices may also have a taste. For example, cloves, horseradish, and turmeric are bitter, tamarind and sumac are sour, and fennel, caraway, and nutmeg are sweet.[26] Thus another way spices contribute to our enjoyment of food is by stimulating GSCs in our oral cavity. However, the thing that makes spices unique as a food additive is that they are among the most efficient activators of the receptor proteins found on trigeminal nerve fibers and specifically are particularly good at stimulating the thermo-TRPs.

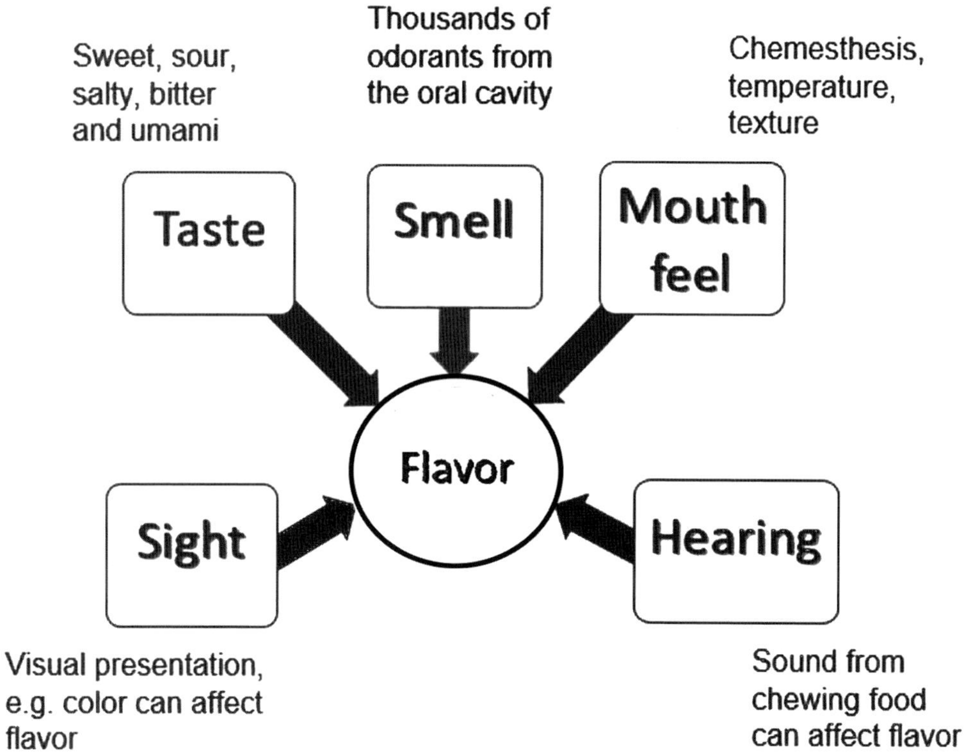

Fig. 5.2. The sensory systems that combine to produce the sensation of flavor.

Spices

Spices originated in both the Eastern and Western Hemispheres. Table 5.1 lists a number of spices, their area of origination, their principal chemesthetic ingredient, and the TRP channel they are believed to activate. Some spices were grown only in very specific places, which made them extremely valuable. For example, nutmeg and cloves only grew in the Moluccas (Spice Islands).[27] Cloves were traded from the Spice Islands to the Middle East as early as 1700 BCE.[28] The sultans of two of the Spice Islands, Ternate and Tidore became wealthy and powerful by controlling the trade of these spices. Europeans fought over the islands in the 16th and 17th centuries.[29] After one of these 'Spice Wars' the British exchanged New Amsterdam (now New York) with the Netherlands for the Moluccan island of Run.

Chili peppers are indigenous to Central and South America.[30] Columbus encountered chili peppers on his second voyage and brought them back to Europe. The Portuguese spread chili peppers to Asia and Africa. Many cuisines, including Thai, Mexican, Indian, Korean, and Chinese, now use chili peppers extensively.

Table 5.1. Information about spices.

Spice	*Area of Origination*	*Active Chemical*	*TRP Channel Activated*
Mint	Asia and Mediterranean	Menthol	TRPM8
Horseradish	Central Europe	Allyl Isothiscyanate	TRPA1
Mustard	Mediterranean	Allyl Isothiscyanate	TRPA1
Ginger	China	Gingerol	TRPV1, TRPA1
Nutmeg	Moluccas	A-pinene	TRPV3
Cloves	Moluccas	Eugenol	TRPV1, TRPA1, TRPM8
Cinnamon	Sri Lanka	Cinnamaldehyde	TRPA1
Black Pepper	India	Piperine	TRPV1
Cumin	Middle East	Cuminaldehyde	TRPA1
Chili Pepper	Mexico	Capsaicin	TRPV1
Turmeric	India	Curcumin	TRPA1
Cardamom	India	Cineole	TRPV1, TRPA1
Oregano	Mediterranean	Carvacrol	TRPA1

Black pepper has a fascinating history, dating back to 4,000 years ago. It originated in India and was so valuable it was often used as currency.[31] An often cited example is the thousands of pounds of pepper (in addition to gold and silver) demanded and received by Alaric I to end his siege of Rome in 409 CE.[32] One goal of the voyages of Vasco De Gama and Christopher Columbus was to find sailing routes to India to obtain black pepper, in addition to other riches.

Sichuan pepper is a spice that contains the chemical hydroxy-alpha-sanshool. This compound is actually one of the few chemesthetic agents, which stimulates a touch receptor not included in the TRP channel family. However, like the thermal-TRPs this receptor is found on trigeminal nerve endings in the mouth where it produces a unique tingling sensation. This property has made Sichuan peppers sought after as both a remedy for toothaches in traditional Chinese medicine and a culinary novelty.

Evidence suggests that Neanderthals used medicinal spices as long as 60,000 years ago.[33] Indeed, many animals and all of the higher primates use plants to self-medicate.[34] Many spices are used today to protect from acute and chronic diseases. In addition to pungent chemicals, spices contain a number of different bioactive molecules which are antioxidative (chili pepper, cinnamon, ginger, black pepper, turmeric, rosemary), anti-inflammatory (chili pepper, cinnamon, black pepper, turmeric, rosemary, garlic), antimicrobial (chili pepper, cinnamon), and anticarcinogenic (capsaicin, ginger, turmeric).[35] Spices (chili pepper, cinnamon, turmeric, fenugreek, rosemary, garlic) may also be used to lower glucose and cholesterol levels (garlic).[36]

A recently published study, completed in China, found that people eat spicy foods to improve their health.[37] Participants who ate spicy foods six to seven times a week had a 14% lower risk of death compared to those who consumed spicy foods less than once a week. Frequent consumption of spicy foods also decreased the chances of having ischemic heart disease, respiratory ailments, and even cancer. However, despite the widespread cross-cultural use of spices for their medicinal properties, spices are still primarily and most often used to enhance the flavor of food. This stands in contrast to the botanically sourced compounds that are primarily used as pharmaceuticals (*e.g.* aspirin/acetylsalicylic acid from willow tree bark, morphine from the opium poppy, tetrahydrocannabinol from *Cannabis*) but are not used as flavor enhancers.

Thus, when we consider all the botanical substances that met our original culinary definition of what constitutes a spice, the most essential biological or chemical characteristic all these items share is that they are potent modulators of chemesthesis. Furthermore, while these agents are capable of triggering sensations anywhere in the body – there is a Hungarian aphorism about good paprika burning twice (*e.g.* once when eaten and a second time when excreted) – they are particularly well served to stimulating trigeminal nerve fibers in the oral and nasal cavity. Finally, while spices may trigger chemesthetic sensations in these nerves through any mechanism, most of them do so by modulating TRP channels.

Why Do We Consume Pungent Spices?

Now having described the essential neurophysiological characteristics of spices, we turn to the puzzling questions of why humans would willingly choose to hurt themselves using spices.

Most spices contain pungent chemesthetic compounds, which activate the TRP channels in nociceptive (pain) nerve fibers. Botanists designate these pungent chemicals as secondary plant compounds, meaning they are chemicals that have no known direct role in the basic biological functions of plants.[38] Particular secondary plant compounds may only be found in some plants but not others. As we discussed above, capsaicin is found in chili peppers but not mint. It is 'expensive' for the plant to expend energy to make compounds, which do not contribute to their everyday functioning. One hypothesis is that plants expend energy to produce these pungent, pain-producing chemicals to deter herbivores from damaging or destroying them. Humans have taken advantage of these defense chemicals and developed numerous consumer products that contain chemesthetic compounds that activate TRP channels and prevent animals such as squirrels and deer from devouring gardens and crops.

Very few mammals are known to seek out and eat spices like chili pepper, black pepper, horseradish, and cinnamon. A recent study reported that the preferred food for tree shrews in China was chili peppers, but these animals have a mutation in their TRPV1 channels that decreases their sensitivity to capsaicin.[39] Birds also have TRPV1 channels that are insensitive to capsaicin. Capsaicin is often included in bird feed to repel mammals like squirrels.[40] However, many humans seek out chili peppers and other pungent spices, even though their TRP channels function normally. Why is this so? Why do humans go to extreme trouble to get spices, which contain chemicals that are supposed to prevent us from wanting to eat them? We do not know the answer to this question, but there are numerous conjectures.

The Learned Taste Preference Theory

It is well known that after ingesting food that makes a human (or other animals) sick, that individual is unlikely to eat that food again, because they associate the food's flavor with feeling bad. Individuals will not like any food that has this flavor. This is called a learned taste aversion.[41] The opposite is also true. If a human or other animal eats something that makes them feel good, they are likely to eat the food again because they associate its flavor with feeling good. These individuals will like any food that has this flavor and are said to have developed a learned taste preference for that flavor.

Spices are known to have many beneficial health effects in humans, some acute and some chronic (see above). This suggests that eating spices could make humans feel better, which in turn could develop into a learned taste preference driving a liking for spices in the future.

The Thermoregulation Theory

The use of spices is greatest in hot climates, especially in the tropics.[42] One of the reflexes associated with eating spices is sweating. When we sweat, the water on the skin dissolves into the air (evaporation). The water takes 'heat' with it, making an individual feel cooler. This may go along with the learned taste preference theory above in that eating spices in hot weather cools off an individual, making them feel better.

The Endogenous Opioid Theory

When we encounter a painful stimulus, which activates TRP channels, our brains subsequently release endogenous opiates (endorphins), which help alleviate the pain. The chemical in pungent spices like chili peppers, black pepper, and horseradish activates TRP channels producing pain and presumably causes the release of endorphins. Increasing levels of endorphins would decrease the sensation of pain elicited by the spices, producing a positive feeling and converting the pain to pleasure.[43] The pungent spices in the future would be associated with a pleasant experience.

The Antimicrobial Theory

The antimicrobial theory is based on the idea that the irritating chemicals in spices inhibit the growth of foodborne bacteria and fungi.[44] It has been called an example of Darwinian gastronomy. Over the course of human history, spices came to be added to food, especially meat, which destroyed potentially harmful pathogens. This meant the people who used the spices would be healthier and store foods for longer periods of time than non-spice users. As Paul Sherman, the main author of this theory, wrote, 'Families that used appropriate spices would presumably rear more healthy offspring, who would then learn spice-use traditions from their parents.'[45] Eventually, 'food likings would be associated with spicy dishes, especially in climates where foods spoil rapidly.'[46]

The Benign Masochism Theory

In order for a theory to explain why humans like to eat chemicals found in spices that activate TRP channels and produce pain sensations, it must take into account that very few mammals willingly eat spicy food. Paul Rozin, who has been studying this question since the 1970s, has called the liking of spices a uniquely human phenomenon.[47] In a series of experiments he was unable to get other mammals to prefer spicy food. Therefore, he suggested that the preference for pungent food must be due to a specifically human quality. This quality is the human enjoyment of a controlled risk. Humans enjoy participating in activities that produce sensations that indicate danger, but are not actually dangerous. These include riding a roller coaster, watching frightening movies, and taking hot baths. Eating pain-inducing spices also falls into this category. He called this benign masochism, 'It is as if the mind realizes that these activities are actually safe, but the body does not.'[48] A recent study thoroughly examined the relationship between sensation seeking and liking the burn of chili peppers.[49] It clearly demonstrated that risk-related personality traits were related to the predilection for pungent spices.

Summary

Humans have ingested spices for millennia, as flavorings for food as well as for medicinal purposes. The desire for spices was so great it led to an age of exploration from West to East and East to West, presumably helping initiate global trade. Derived from plants, spices contain chemicals used primarily for repelling herbivorous predators. The active ingredients in most spices are repellant because they stimulate pain receptors. Many of these chemicals activate receptor proteins found on nerve cells in the mouth, nose, and eyes called Transient Receptor Potential (TRP) channels. TRP channels provide the sensory input for chemesthesis, which we perceive, primarily as painful. Chemesthesis, a component of mouthfeel, along with olfaction, gustation, and to some extent vision and audition are responsible for determining the flavor of food. When eaten, the chemicals found in spices provide significant health benefits, in part, acting as anti-inflammatories and antioxidants. Several theories have been suggested to explain why humans have a preference for pungent spices. Any successful theory explaining why humans seek out and cultivate spices must take into account that the primary function of these chemicals is to prevent mammals from consuming these same plants. Liking of spices, therefore, appears to be a principally human phenomenon and appears to be a form sensation seeking behavior.

Notes

1 Saul 2013, e70583.
2 Lawton, 2004, 127; Turner 2005, 384; Keay 2006, xv, 288.
3 Richards *et al.* 2010, 45–64.
4 Bushdid 2014, 1370–1372.
5 Jenkins *et al.* 2018, 29.
6 McGann 2017, 6338.
7 Shepherd 2012, 288.
8 Bartoshuk 1993, 251–262.
9 Green 2016, 1–7.
10 Saunders and Silver 2016, 77–91.
11 Green 2016, 1–7.
12 Selekler and Budak 2004, 6–9.
13 Saunders and Silver 2016, 77–91.
14 Castillo *et al.* 2018, 021001.
15 Bautista *et al.* 2007, 204–208.
16 Pereira *et al.* 2013, E1–E4.
17 Zygmunt and Högestätt 2014, 583–630.
18 Cicerale *et al.* 2009, 333–339.
19 Hill and Bautista 2018, 591–592.
20 Lisi 2019, 15–21.
21 ASTM International 2009, E253–09a.
22 We place 'touch' in quotes here to acknowledge that the term touch is not entirely synonymous with somatosensation, just as taste and smell are colloquial approximations for the precise terms of gustation and olfaction.
23 Koza *et al.* 2005, 643–649.
24 Dacremont *et al.* 1992, 443–456.
25 Christensen n.d.
26 Spices Inc. n.d.
27 Van Gils and Cox 1994, 117–124.
28 Turner 2005, 384.
29 Nilius and Appendino 2013, 1–76.
30 Halikowski Smith 2008, 389–418.
31 Nilius and Appendino 2013, 1–76.
32 Loewe 1971, 166–179.
33 Hardy *et al.* 2012, 617–626.
34 Shurkin 2014, 17339–17341.
35 Jiang 2019, 395–411.
36 Jiang 2019, 395–411.
37 Lv *et al.* 2015, h3942.
38 Fraenkel 1959, 1466–1470.
39 Han *et al.* 2018, e2004921.
40 Tewksbury *et al.* 2008, 11808–11811.
41 Myers and Sclafani 2006, 380–388.
42 Sherman and Billing 1999, 453–463.
43 Rozin 1990, 231–269.
44 Sherman and Billing 1999, 453–463.
45 Sherman and Billing 1999, 453–463.
46 Sherman and Billing 1999, 453–463.

47 Rozin 1990, 231–269.
48 Rozin 1990, 231–269.
49 Byrnes and Hayes 2016, 411–422.

Bibliography

ASTM International. 2009. *Standard Terminology Relating to Sensory Evaluations of Materials and Products*, E253-09a, Pennsylvania.

Bartoshuk, L.M. 1993. Genetic and Pathological Taste Variation: What Can We Learn From Animal Models and Human Disease?, *Ciba Foundation Symposium* 179, 251–262.

Bautista, D.M., Siemens, J., Glazer, J.M., Tsuruda, P.R., Basbaum, A.I., Stucky, C.L., Jordt, S.E. and Julius, D. 2007. The Menthol Receptor TRPM8 is the Principal Detector of Environmental Cold, *Nature* 448, 204–208.

Bushdid, C., Magnasco, M.O., Vosshall, L.B. and Keller, A. 2014. Humans Can Discriminate More Than 1 Trillion Olfactory Stimuli, *Science* 343.6177, 1370–1372.

Byrnes, N. and Hayes, J.E. 2016. Behavioral Measures of Risk Tasking, Sensation Seeking and Sensitivity to Reward May Reflect Different Motivations for Spicy Food Liking and Consumption, *Appetite* 103, 411–422.

Castillo, K., Diaz-Franulic, I., Canan, J., Gonzalez-Nilo, F. and Latorre, R. 2018. Thermally Activated TRP Channels: Molecular Sensors for Temperature Detection, *Physical Biology* 15.2, 021001. doi: 10.1088/1478-3975/aa9a6f.

Christensen, E. n.d. *Quick Guide to Every Herb and Spice in the Cupboard*. https://www.thekitchn.com/quick-guide-to-every-herb-and-spice-in-the-cupboard-108770 (accessed July 2019).

Cicerale, S., Breslin, P.A.S, Beauchamp, G.K. and Keast, R.S.J. 2009. Sensory Characterization of the Irritant Properties of Oleocanthal, a Natural Anti-inflammatory Agent in Extra Virgin Olive Oils, *Chemical Senses* 34.4, 333–339. doi:10.1093/chemse/bjp006.

Dacremont, C., Colas, B. and Sauvageot, F. 1992. Contribution of Air- and Bone-conduction to the Creation of Sounds Perceived During Sensory Evaluation of Foods, *Journal of Texture Studies* 22, 443–456.

Fraenkel, G.S. 1959. The Raison D'etre of Secondary Plant Substances, *Science* 129, 1466–1470.

Green, B.G. 2016. Introduction: What is Chemesthesis?, in: McDonald, S.T., Bolliet, D.A. and Hayes, J.E. (eds), *Chemesthesis: Chemical Touch in Food and Eating*, Chichester, 1–7.

Halikowski Smith, S. 2008. 'Profits Sprout Like Tropical Plants': A Fresh Look at What Went Wrong with the Eurasian Spice Trade Between c. 1550–1800, *Journal of Global History* 3, 389–418.

Han, Y., Li, B., Yin, T.-T., Xu, C., Ombati, R., Luo, L., Xia, Y., Xu, L., Zheng, J., Zhang, Y., Yang, F., Wang, G.-D., Yang, S. and Lai, R. 2018. Molecular Mechanism of the Tree Shrew's Insensitivity to Spiciness, *Public Library of Science Biology* 16, e2004921.

Hardy, K., Buckley, S., Collins, M.J., Estalrrich, A., Brothwell, D., Copeland, L., García-Vargas, S., de la Rasilla, M., Lalueza-Fox, C., Huguet, R., Bastir, M., Santamaría, D., Madella, M., Wilson, J., Cortés, A.F. and Rosas, A. 2012. Neanderthal Medics? Evidence for Food, Cooking, and Medicinal Plants Entrapped in Dental Calculus, *Naturwissenchaften* 99, 617–626.

Hill, R.Z. and Bautista, D.M. 2018. A Trio of Ion Channels Takes the Heat, *Nature* 555, 591–592.

Jenkins, E.J., DeChant, M.T. and Perry, E.B. 2018. When the Nose Doesn't Know: Canine Olfactory Function Associated with Health, Management, and Potential Links to Microbiota, *Frontiers in Veterinary Science* 29. doi: 10.3389/fvets.2018.00056.

Jiang, T.A. 2019. Health Benefits of Culinary Herbs and Spices, *Journal of Association of Official Analytical Chemists International* 102, 395–411.

Keay, J. 2006. *The Spice Route: A History*, Berkeley.

Koza, B.J., Cilmi, A., Dolese, M. and Zellner, D.A. 2005. Color Enhances Orthonasal Olfactory Intensity and Reduces Retronasal Olfactory Intensity, *Chemical Senses* 30, 643–649.

Lawton, J. 2004. *Silk, Scents & Spice: Tracing the World's Great Trade Routes*, Paris.

Lisi, D.M. 2019. OTC Transdermal Analgesic Patches in Pain Management, *US Pharmacist* 44, 15–21.

Loewe, M. 1971. Spices and Silk: Aspects of World Trade in the First Seven Centuries of the Christian Era, *The Journal of the Royal Asiatic Society of Great Britain and Ireland* 1971.2, 166–179.

Lv, J., Qi, L., Canqing, Y., Yang, L., Guo, Y., Chen, Y., Bian, Z., Sun, D., Du, J., Ge, P., Tang, Z., Li, Y., Chen, J., Chen, Z. and Li, L.2015. Consumption of Spicy Foods and Total and Cause Specific Mortality: Population Based Cohort Study, *British Medical Journal* 351, h3942. doi: 10.1136/bmj.h3942.

McGann, J.P. 2017. Poor Human Olfaction is a 19th-century Myth, *Science* 356, 6338.

Myers, K.P. and Sclafani, A. 2006. Development of Learned Flavor References, *Developmental Psychobiology* 48, 380–388.

Nilius, B. and Appendino, G. 2013. Spices: The Savory and Beneficial Science of Pungency, *Reviews of Physiology, Biochemistry and Pharmacology* 164, 1–76. doi:10.1007/112_2013_11.

Pereira, E.J., Sim, L., Driver, H.S., Parker, C.M. and Fitzpatrick, M.F. 2013. The Effect of Inhaled Menthol on Upper Airway Resistance in Humans: A Randomized Controlled Crossover Study, *Canadian Respiratory Journal* 20, E1–E4.

Richards, P.M., Saunders, C.J. and Silver, W.L. 2010. Functional Neuroanatomy of the Upper Airway in Experimental Animals, in: Morris, J.B. and Shusterman, D.J. (eds), *Toxicology of the Nose and Upper Airways*, New York.

Rozin, P. 1990. Getting to Like the Burn of Chili Pepper: Biological, Psychological and Cultural Perspectives, *Chemical Senses, Irritation* 2, 231–269.

Saul, H., Madella, M., Fischer, A., Glykou, A., Hartz, S. and Craig, O.E. 2013. Phytoliths in Pottery Reveal the Use of Spice in European Prehistoric Cuisine, *PLoS One* 8.8, e70583.

Saunders, C.J. and Silver, W.L. 2016. Anatomy and Physiology of Chemesthesis, in: McDonald, S.T., Bolliet, D.A. and Hayes, J.E. (eds), *Chemesthesis: Chemical Touch in Food and Eating*, Chichester.

Selekler, H.M and Budak, F. 2004. Idiopathic Stabbing Headache and Experimental Ice Cream Headache (Short-lived Headaches), *European Neurology* 51.1, 6–9.

Shepherd, G.M. 2012. *Neurogastronomy: How the Brain Creates Flavor and Why it Matters,* New York.

Sherman, P.W. and Billing, J. 1999. Darwinian GAstronomy: Why We Use Spices. Spices Taste Good Because They Are Good for Us, *BioScience* 49, 453–463.

Shurkin, J. 2014. News Feature: Animals That Self-medicate, *Proceedings of the National Academy of Sciences of the United States of America* 111.49, 17339–17341. doi:10.1073/pnas.1419966111.

Spices Inc. n.d. Flavor Characteristics of Spices. *SpicesInc.com*. https://www.spicesinc.com/p-3743-flavor-characteristics-of-spices.aspx (accessed July 2019).

Tewksbury, J.J., Reagan, K.M., Machnicki, N.J., Carlo, T.A., Haak, D.C., Calderón Peñaloza, A.L. and Levey, D.J. 2008. Evolutionary Ecology of Pungency in Wild Chilies, *Proceedings of the National Academy of Sciences of the United States of America* 105, 11808–11811.

Turner, J. 2005. *Spice: The History of a Temptation*, New York.

Van Gils, C. and Cox, P.A. 1994. Ethnobotany of Nutmeg in the Spice Islands, *Journal of Ethnopharmacology* 42, 117–124.

Zygmunt, P.M. and Högestätt, E.D. 2014. TRPA1. *Handbook of Experimental Pharmacology* 222, 583–630. doi: 10.1007/978-3-642-54215-2_23.

Chapter 6

Silk Road Pharmacy: Debating Theriac and Defining the Natural World

Monique O'Connell

In the second half of the 16th century, there was fierce rivalry in the Italian Renaissance field of botany. Controversies over plant identifications often turned personal, and in letters and printed works the insults flew back and forth. One of the quickest to cast aspersions on the character of his colleagues was Pietro Andrea Mattioli (1501–1578).[1] Originally from Siena, Mattioli took a degree in medicine at Padua and went on to a career as a town physician and eventually as personal physician to the Holy Roman Emperor. His contribution to botany came through his commentaries on Dioscorides, the ancient Greek pharmacologist whose encyclopedic work *De Materia Medica* was the leading classical source for plant names and descriptions of their medical uses.[2] Mattioli used his published works as well as his correspondence to accuse other Renaissance botanists of jealousy and ignorance, labeling his enemies as 'eel-skinners', 'beasts', 'great pigs', 'a man who could not tell basil from lettuce' and 'so ignorant of botany that it was impossible to revile him more than he deserved'.[3] In 1561, the famous Paduan anatomist Gabriele Fallopia (1523–1562) warned a young Bolognese naturalist away from botany altogether, writing that 'You will enter a playing field in which there are nothing but angry people who contradict one another, who continuously upset each other and write against one another.'[4]

Two questions arise: why was botany the site of such contention in 16th-century Italy, and what do these quarrels have to do with the Silk Roads? Renaissance botanical controversies offer a window into the large-scale changes in European understandings of the natural world in the early modern era. There were a number of elements that drove these quarreling botanists: professional rivalries, public acknowledgement of botanical expertise, the creation of a scientific community, and the role of printing in the dissemination of new knowledge.[5] But the most pressing reason that motivated botanical investigation was medical: plants were used as the basis for most medicines, and more accurate plant identifications could lead to more effective treatments. One of the best ways to trace changing ideas about medical botany is in the 16th-century debates over theriac, a universal cure-all with roots in the classical world. One could take it preventatively, to maintain health, or as a cure for a specific disease, including bubonic plague. If one could master the recipe to theriac, the queen of compound medicines, other medicinal preparations would also benefit. Debates over theriac reveal shifts in medical authority, the role of state regulation in medicine, and the clash between new and old conceptions of the human body and the natural world.

The specific case of Renaissance botany takes on new dimensions when placed in the global context of Eurasian commercial and cultural exchange enabled by Silk Road networks of trade and travel. Medicinals – plant, animal, and mineral products that were sold as simple remedies or compound medicines – traveled across the Silk Roads alongside luxury textiles, ivory, jade, tea, and horses.[6] But the trade in medicinals was different than the exchange of a physical object like a ceramic vase or jade sculpture, because medicinal products traveled with ideas about health and healing attached.[7] The trade in medicinals thus involved intellectual as well as economic exchange, recipes and promises of healing as well as exotic plants and animal parts. Theriac, a universal panacea with global reach, can also serve as an example of these kinds of medicinal exchanges along Silk Roads. Like alchemists' search for the philosophers' stone, arguments about the efficacy of theriac rest on the quality and accuracy of the ingredients and their combination. In multiple Silk Road societies, the overall idea of theriac as a universal cure-all persisted while the actual recipe adapted to particular local cultures of natural philosophy and institutional medicine. Theriac recipes along the Silk Road were entwined with state medical regulation, commerce and empire, and natural philosophy, and they point to the interplay of the global and local in intercultural exchange.[8]

The controversies over theriac in 16th-century Italy can be compared to other cases where theriac recipes sat at the intersection of long-distance trade and local medical culture. In the first centuries of the Roman Empire, recipes for theriac and the related drug mithridatium multiplied, indicating the medicine's popularity in Imperial Rome. From its classical roots, theriac entered the Islamic medical tradition, and scholars in Abbasid Baghdad and Ghazni recorded recipes for the drug. From Central Asia, theriac spread both east, to Tang China, and west, to Islamic Spain and western Europe. In these cases, just as in Renaissance Italy, an increase in Eurasian commerce brought a flood of new substances, texts, and ideas. Physicians and pharmacists struggled to incorporate these novelties into pre-existing medical practices. Valerie Hansen has called the Silk Road 'one of the most transformative superhighways in human history – one that transmitted ideas, technologies, and artistic motifs'.[9] The winding roads of theriac demonstrates that intercultural contact and the transmission of new ideas was not a one way journey. Theriac – the drug itself and recipes for creating it – transformed multiple times as it traveled the Silk Road, while still retaining a basic essence as a universal cure-all.

Poisons and Antidotes from the Roman Empire to the Islamic World

The origins of theriac lay in the classical world, with Mithradates, king of Pontus (120–63 BCE), nicknamed 'the Poison King'.[10] Mithradates, justifiably concerned about assassination, wanted an antidote that could fight any poison – he tested remedies on condemned criminals and drank antidotes himself daily, using the ancient belief that every natural poison had a natural antidote.[11] Legend has it that his experiments were so successful that when he tried to commit suicide by drinking poison, it was ineffective and he had to die by the sword instead.[12] While the exact recipe for his drug was a closely guarded secret during his life and was lost after his death, his remedy combined natural antitoxins like sulfur, charcoal, St John's wort, and garlic with small amounts of the poisons themselves.[13] Roman imperial physicians claimed to possess the secret recipe, and formulas for the ultimate antidote proliferated in a Rome obsessed with the very real danger of political poisoning. Totelin has compared the multiple recipes that claimed to reproduce or improve on Mithridates' original, arguing that as the number of recipes grew, 'Mithridatium became a word to designate a kind of medicinal preparation and not a fixed recipe'.[14] Andromachus, the imperial physician to Nero, added viper flesh and opium poppy seeds to his recipe for theriac, a combination which he claimed could be used not only as an antidote but as a cure for all kinds of illnesses. Playing on the popularity of compound medicines in antiquity, all of the recipes had a high number of ingredients: up to 50 for Mithridatium and as many as 70 for theriac, with large numbers of expensive ingredients from distant locations: cassia, cinnamon, cardamom, nard, pepper, and storax.[15] The ingredients required for imperial Roman theriacs thus relied on Roman commercial connections in East Africa, Arabia, and the Indian Ocean, and the recipes highlighted the geographical reach, calling for Syrian balsam, Indian cassia, Istrian beaver, and Cretan dittany.[16]

With such exotic ingredients, theriac in Rome was a drug for emperors and the elite. It is not the case that everyone in Rome was enthusiastic about its potential: Pliny the Elder included a description of the drug in his *Natural History*, focusing on the proper method for preparing vipers. Pliny also offered a blistering critique of the irresponsibility of Roman physicians and their irresponsibility in relying on luxurious foreign medicines, using theriac as an example. He wrote:

> There is an elaborate mixture called *theriace*, which is compounded of countless ingredients, although Nature has given as many remedies, any one of which would be enough by itself. The Mithridatic antidote is composed of fifty-four ingredients, no two of them having the same weight, while of some is prescribed one sixtieth part of one denarius. Which of the gods, in the name of Truth, fixed these absurd proportions? No human brain could have been sharp enough. It is plainly a showy parade of the art, and a colossal boast of science.[17]

Pliny continued on to question the accuracy of the ingredients included in the medicine. He claimed that physicians put their patients in peril through their ignorance, writing 'I have discovered that instead of Indian cinnabar there is commonly added to medicines, through a confusion of names, red lead, which, as I shall point out when I discuss pigments, is a poison.'[18]

By the 3rd century, the popularity of theriac and mithridatium was waning among the Roman elite, but the writings of the Greco-Roman physician Galen of Pergamon (129–216 CE) ensured that the drug outlived the Roman empire.[19] Galen was the personal physician to emperor Marcus Aurelius, and he offered a detailed description of his own recipe in his treatises on antidotes as well as in *On Theriac to Piso*.[20] His formula included 64 ingredients, with viper flesh, opium, honey, wine, and cinnamon featured prominently.[21] While Galen made no efforts to keep his recipe secret, he was at pains to point out that one needed an expert to prepare the medicine properly, highlighting the difficulty of identifying and using the correct ingredients. He wrote, 'There is a great deal of inexperience about the drugs which go into theriac. For example, there is much disagreement about the finest of ingredients, I mean cassia and true cinnamon, and this deceives many blenders. And what is called false cinnamon is similar to the real thing, but in taste and smell is found to be much inferior.'[22] Galen publicized his own formula while at the same time asserting the importance of expertise in its preparation, a marketing combination Renaissance apothecaries also employed.

Galenic medicine, which was based on balancing the hot, cold, dry, and moist humors of the human body, gained a dominant position in the medical school of Alexandria in Late Antiquity and from there became an organizing principle for Byzantine Christian and Arab Islamic medicine.[23] Hunayn ibn Ishaq (d. 873 CE), the influential Arab Christian leader of the translation school in Abbasid Baghdad, searched out as many Galenic texts as he could find in the Byzantine world and translated the Greek works into Syriac and into Arabic.[24] These Arabic texts became the basis for subsequent medical understandings of Galen in both Christian and Islamic societies, and were the source for Byzantine and Islamic theriac recipes. One of the foremost physicians and pharmacologists in 9th-century Baghdad, Sābūr ibn Sahl, illustrates the cross-cultural influences that made up early Islamic medical traditions. Sābūr ibn Sahl (d. 869 CE) came from Gondēšāpūr, a town in southwestern Iran settled with Graeco-Syrian prisoners of war in the Sassanian era that by the 6th century CE had become a major center of Nestorian Christian medical learning.[25] After taking charge of the hospital of Gondēšāpūr, he traveled to Baghdad, eventually becoming court physician to the Abbasid caliphs and writing pharmaceutical treatises, including the *Small Dispensatory* that contains a recipe for a 'four ingredients theriac' said to be useful against snakebite.[26] In the Nestorian Christian transformation of the Greek medical tradition into Arabic, theriac remained a prominent antidote, but the contents of the drug changed and recipes multiplied.

As physicians across the medieval Islamic world absorbed and commented on the Galenic tradition of medicine, two issues emerged: the proper identification of plants named in classical works and the incorporation of new, previously unknown products like camphor and musk.[27] This debate created space for experimentation and innovation in pharmacology in general and with theriac recipes in particular.

Galen's treatise *On Theriac to Piso* was translated into Arabic and became a part of the Arabic tradition of polypharmacy, or compound medicines.[28] The Central Asian polymath and physician Al-Biruni (d. 1048), included a number of medicines called *tiryaq* in his pharmacopoeia, with methods for determining which version of theriac one had on hand. He wrote 'It is said that a pure theriac is identified by means of garlic. If a person eats *tiryaq-i-faruq* having already eaten garlic, the smell of the garlic disappears.'[29] Ibn Sina (d. 1037), the Persian Muslim physician known in the Latin West as Avicenna, described the preparation and use of *tiryaq* in his influential *Canon of Medicine*, and in the medieval period *tiryaq* appeared in multiple pharmaceutical treatises and was regularly prepared in Islamic hospitals.[30] Al-Zahrawi (d. 1013) and Ibn Djuldjul of Cordoba (d. 984), both physician-botanists, listed seven prescriptions for theriac of varying degrees of complexity, and Ibn Djuldjul claims that he had made the original Galenic concoction twice and found it to be the best.[31]

Theriac across Silk Roads: from Islamic Central Asia to China

Theriac underwent a parallel process of translation and transformation in Tang China. The drug was introduced into Tang medical culture by Byzantine travelers who crossed Central Asia via the Silk Roads. Sources report that in 667 Byzantine ambassadors presented the Tang emperor with the 'true universal antidote', theriac, a drug that was useful against 'the hundred ailments'.[32] Theriac took its place in a Chinese medical culture that was increasingly open to foreign cures and medical materials.[33] It seems that the theriac compound was sometimes transported to China in a pre-made formula, which would make sense as the galenic recipe calls for theriac to be aged for 12 years before consumption, a process that could happen during travel.[34] Recipes for making theriac also appear in Tang pharmaceutical catalogs, under the title *Di-ye-jia* 底野(也)迦, although a new ingredient has appeared: gall-bladder. One such manual lists the compound as

> Di-ye-jia, taste: acrid, bitter [thermo-influence:] neutral. No marked medicinal effectiveness. Masters a hundred illnesses, if one is hit by the malevolent, and evil influences, as well as abdominal obstructions. Origin: Western countries. It is said this [drug] is made from gall-bladders [of pigs]. Externally, it resembles pills that have been rotten for a long time. The color is red-black. Occasionally, [this drug] is brought here by foreigners.[35]

This recipe was part of the flourishing of Tang pharmaceutical texts, many now lost, that attempted to catalog and make sense of the new exotic materials introduced by the Silk Road trade.[36] While the story of theriac's arrival points to Byzantine ambassadors, the most intense Tang intellectual and commercial exchanges were with the Islamic world, through interaction with Persia.

The Tang Dynasty's imperial expansion into Central Asia, the cosmopolitan culture at the capital of Chang'an, and the active trading networks that linked China to the rest of Eurasia led to an influx of about 100 new plant, animal, and mineral products that found their way to Chinese pharmacies and into pharmaceutical literature.[37] Theriac, and these other medical products, had to be incorporated into a well-established system of Chinese beliefs and practices about health and healing. The synthesis of healing practices that emerged during the Han dynasty rested on the framing principles of *qi* (the essential energy of the universe), the balance of yin and yang energies, and the Five Agents, used for systemic correspondences.[38] Recently, scholars of Chinese medicine have emphasized the innovation and change over time brought by trade as well as internal political and cultural developments.[39] One of the more substantial changes came with the incorporation of Indian Buddhist healing traditions, which together with Daoism exercised an influence over both drug therapy and Chinese pharmaceutical literature.[40] Chen has demonstrated that the Tang elite read *materia medica* literature for literary as well as medical purposes, and that there was considerable interaction between Nestorian Christians and Daoists in the culture that produced the classics of Tang pharmacy.[41]

Theriac might have come with Byzantine ambassadors, but its continued presence in Chinese pharmacy came through the influence of Islamic medicine, which enjoyed a prominent status among Mongols and which was institutionalized in China by the Yuan dynasty.[42] The Yuan government founded an Imperial Academy of Medicine in 1260, which oversaw medical education and regulated the production of pharmaceuticals; this institution included the Office of Western Medicines, which focused on Islamic medicine and drugs.[43] The impact of Islamic medicine on Chinese recipes for theriac is clear in a late Yuan/early Ming pharmaceutical text, the *Huihui yaofang*, which gives drug names in both Persian and Chinese scripts and which categorizes theriac as a type of antidote, labeling it as an Islamic cure for poison.[44] There was a transformation in the ingredients from 'various kinds of gall bladders' to 'pig gall bladders' in a 15th-century compilation of formulas entitled *Puji fang* 普濟方 (Formulas for Universal Benefaction), compiled by Zhu Su 朱橚 (d. 1425) and others.[45] The late Ming scholar Li Shizhen (1518–1593) recorded theriac among his remedies in his encyclopedia of natural history *Bencao gangmu*, again with the key ingredient listed as pig gall.[46] Nappi has used this transformation of a complicated compound substance of exotic ingredients into a simple animal byproduct to argue that the change communicated that Western 'medicine was no more powerful than indigenous Chinese remedies'.[47] Theriac in China continued to be written about as a foreign or exotic remedy, but the content of the medicine changed, assimilating into local traditions of healing.

From Medieval to Early Modern Theriac in the Venetian Republic

An increasingly globalized commerce also affected the fierce debates in 16th-century Italy over theriac in particular and medical botany in general. As in the earlier cases, pharmacists and physicians debated how to incorporate new ideas and strange substances into pre-existing medical traditions. The controversies reveal struggles between antiquity and innovation as well as clashes between global and local understandings of medicine. Theriac's revival as a drug of choice in medieval western Europe is connected to the recovery of the Galenic medical tradition through the translation of Arabic medical texts into Latin in the 11th century at the medical school at Salerno.[48] Theriac had not disappeared in the Latin West, as a recipe in Bald's Leechbook, a 10th-century compendium of Anglo-Saxon healing demonstrates.[49] But the translation of the Greek medical tradition from Arabic texts into Latin, particularly Avicenna's *Canon of Medicine*, solidified the authority of Galenic medicine and cures like theriac in university medical curriculum in the Latin West.[50] So when Venetian apothecaries began manufacturing theriac by the 12th century, they were drawing on an Arabic tradition of polypharmacy developed in 10th-century Central Asia as much or more than classical Mediterranean roots.[51] Medieval medical treatises mention theriac as a drug that could 'cure fevers, prevent internal swellings and blockages, alleviate heart problems, treat epilepsy and palsy, induce sleep, improve digestion, strengthen limbs, heal wounds' as well as its original function as an antidote to poisons.[52] In the 14th century, with the massive mortality brought by the bubonic plague, writers about theriac elaborated its potential as plague prevention or even cure, advocating taking a dose daily, washing one's face with a diluted version of the drug, or applying it near the heart to draw out the poison.[53]

Medieval Venetian pharmacies were well-known for the excellence of their theriac preparation, a reputation that rested on apothecaries' access to high quality exotic materials as well as their expertise. At the high point of production, Venetian apothecarics produced 60,000 pounds of theriac annually.[54] The manufacture of theriac in Venice was an annual public spectacle: each spring, large numbers of poisonous vipers were chopped up by apothecaries' apprentices in streets and squares in front of the apothecary shops.[55] The shops also displayed the numerous ingredients required to make the drug in the street, and apothecaries and their assistants compounded the mixture in brass mortars while poets and musicians praised the drug's many powers. This ceremony went on over the course of the 40 days it took to produce the drug, which then had to be aged before it could be consumed. Mozzato's work on the inventory of 15th-century apothecary Agostino Altucci demonstrates the reach of the trade that enabled this drug's production.[56] The ritual of production emphasized the balance and harmony of the Venetian republic itself in addition to the city's commercial connections, so it was subject to significant state regulation.[57] When Venetian apothecaries created their own guild (1565), regulations on the production of theriac dominated their statutes.[58]

Venice had more than 50 apothecary shops in the early modern era; these shops used spices in medicines and at the same time acted as centers of knowledge, museums of exotic and strange material from around the world, offering medical advice, theological speculation, and healthy doses of gossip and news.[59] When it came to theriac, Venetian pharmacists used their city's reputation as a long-standing center of production to market their own remedies for an international export trade.[60] The theriac produced at the Struzzo d'Oro (Golden Ostrich), was particularly famous, in part because of the publications and marketing efforts of its founder Georg Melich.[61] The vibrant Venetian printing industry also contributed to the commercialization of pharmaceutical remedies and of theriac cures.[62] Both university trained physicians like Prospero Borgarucci and guild trained apothecaries included lengthy discourses on theriac as part of their vernacular publications.[63] There were some clear disputes here based on professional identity, as in the physician Maranta's *On Theriac* where he claimed modern theriac was ineffective because apothecaries were producing it wrong.[64] But overall, while physicians, apothecaries, and natural historians often had different educational backgrounds, their frequent correspondence, travel experiences, and common aims linked them into a scientific community that used debate and disagreement as well as discovery to create shared knowledge about the natural world.

Public manuals and printed recipes for theriac, as well as other drugs, grew out of a half-century of debate over the value of ancient remedies. The rise of medical humanism was an effort to return to the original classical sources; by returning to the original Greek texts, medical humanists aimed to purge their field of Arab influence and return to first principles.[65] This movement was not without its own debates: one of the fiercest raged over the value of Pliny's *Natural History*.[66] Niccolò Leoniceno, professor of medicine at Ferrara, in 1492 published an attack on Pliny entitled *On the Many Errors of Pliny and others on Medicine,* using his own library of Greek texts and knowledge of the language to critique Pliny's Latin translations of Greek botanical terms as well as his medical knowledge.[67] Unlike earlier commentators, who had attributed errors to careless copyists, Leoniceno insisted that the errors were Pliny's own fault. This assault on the authority of an ancient author caused others to leap to Pliny's defense, notably Pandolfo Collenuccio, who struck back with his 1493 *Defense of Pliny against the Accusations of Nicolo Leoniceno.*[68] As the controversy developed, Leoniceno continued to write on the proper identification of medical substances, including the flesh of the vipers used in the preparation of theriac, arguing for the efficacy of local vipers and stating that Avicenna had mistranslated the names of snakes from Galen.[69]

The larger problem for medical humanists was not vipers but plants, and the difficulties over plant identifications were not limited to Pliny and his Natural History. As classical authors like Dioscorides were re-translated from the Greek and began to circulate more widely in print, errors and textual uncertainties became more apparent. The first Renaissance translators of Dioscorides were literary scholars and humanists with little knowledge of the natural world, but botanists struggled to match the plants in their gardens to what was described in classical texts.[70] As early as the 1540s, there were long lists of missing or misidentified drugs like balsam, cinnamon, and Petroselinum.[71] The controversy was not only an academic exercise but went to the healing power of medicines: how could one trust that the brown powder the pharmacist was adding to one's theriac was actually the true cinnamon called for by the recipe? Some producers of Renaissance theriac, in line with the medical humanist tradition, went on a quest to reproduce the classical medicine exactly. The Veronese apothecary and collector Francesco Calzolari (1522–1609), for instance, sought a return to Galenic purity in his mixture, using only six substitute ingredients in 1561 and dropping to three substitutes by 1566.[72] The fame of this preparation attracted critics. In 1566 Calzolari published *A Letter on Certain Lies and Calumnies Made Against his Theriac*, defending himself against another Veronese apothecary Ercolano Scalcina, who had attacked his style of preparation and decision to include re-discovered simples.[73] The medical and botanical community closed ranks around Calzolari and his theriac, and Scalcina not only lost his position but was frozen out of the profession.

Others looked outside the classical tradition, looking to new – or newly identified – botanical specimens and emphasizing the innovative nature of their cures. Some held that those new ingredients were to be found in New World plants but others looked closer to home.[74] Ambassadors, travelers, doctors, and botanists pioneered a more rigorous, systematic approach to plant identification, going on herborizing expeditions in the Italian countryside as well as the Mediterranean.[75] Botanical gardens were founded in Pisa, Padua, and Florence where samples were grown, collected, and documented.[76] The overall goal of these enterprises was twofold – the more abstract, natural philosophical goal of observing and classifying the natural world and the more practical goal of using these plants to create medicines to cure the sick- or at least not to poison them by using the wrong ingredients. While the race to create an ideal theriac was not the only engine driving these developments, it was part of the puzzle. At least one author attributed the founding of the botanical garden at Padua to the search for a better understanding of theriac in particular. Marco Guazzo explained that Francesco Buonafede, professor of medicine at Padua, 'after reading the exposition of Galen on the Theriac of Andramaco, physician to Nero, knew the great difficulty in acquiring the cognition of such a drug' and thus petitioned university and state authorities to create a medicinal garden.[77]

While personalities had some effect on the so-called Theriac Wars of the 1560s, the debate over the 'correct' recipe for theriac raised the question of classical authority versus contemporary innovation. Mattioli and his *Commentaries* became one of the primary intellectual spaces for balancing claims of ancient and modern botany. These books were among the greatest publishing successes of the 16th century: both the Latin and Italian versions went into numerous revised and illustrated editions and German, French, Spanish, and Czech translations.[78] Findlen has characterized Mattioli's commentaries as a 'paper garden' that deliberately sought to create a scientific community of botanical knowledge.[79] Mattioli's successive editions included botanical insights and plant identifications gathered not only from his own observations of nature, but from other prominent botanists like Luca Ghini (1496–1556) and Ulisse Aldrovandini (1522–1605).[80] His informants were rewarded with a prominent place in the book's letter of dedication; those who disagreed with him were exiled from his botanical republic through critical marginal notes as well as more directly in private correspondence. The Portuguese Amatus Lusitanus (1511–1568) and the Italian Luigi Anguillara (ca. 1512–1570) both disagreed with Mattioli and were the targets of Mattioli's wrath. He accused Lusitanus, a Jewish *converso*, of heresy as well as bad botany, writing, 'Just as there is no faith and no religion within you, so in truth you are completely blind as to the medical art which you unworthily profess.'[81]

Mattioli also entered into debate with received knowledge. By the 1568 edition of his work, Mattioli had added a two-column arrangement highlighting the differences between his own remedies and plant identifications and those of Dioscorides.[82] He describes New World plants such as American chili peppers, arguing that they were much hotter and thus not the equivalents of Asian peppers. New understanding of the natural world allowed for challenges to ancient authority. Based on the scope of his plant identifications, Mattioli declared that theriac of his own day was the equal to that made by Galen for the emperors.[83] He also invented new remedies such as Scorpion Oil, a combination of both theriac and mithridatium with 123 ingredients.[84] While Mattioli was free to propose new theriac remedies in his paper garden, the apothecaries and physicians who worked in Italian cities and towns had to face the fact that their medicines were increasingly regulated by professional organizations and the 16th-century state.[85]

In the 1590s, one physician traveler turned back to the Islamic world for insight into a more effective theriac. Prospero Alpini (1553–1617), who lived in Cairo as physician to the Venetian consul, and used his time abroad to observe Egyptian medical practices and plants. On his return, he published two books based on his experiences: *On the Medicine of Egypt* and *On the Plants of Egypt*.[86] Alpini's body of botanical work included several important advances for European botany: he was the first European to observe and document sexual difference in plants, using the datepalm as his example, and he was also among the first to write about the coffee plant.[87] According to Alpini, Egyptian theriac was prepared publically but the recipe was a closely kept secret; nontheless, he managed to obtain it and published a detailed list of ingredients.[88] Like many authors before him, Alpini's assessment of this theriac's effectiveness rested on the quality of the ingredients. He explained that the Egyptians distinguished between three types of cinnamon: cinnamon, canella, and cassia. There were subtle differences in smell and taste between each variety, but Italians were not able to distinguish between cinnamon and canella because some of the differences in smell faded during long sea voyages, and merchants imported cannella under the name of cinnamon because it transported better.[89] The medical humanists of the Italian Renaissance had tried, unsuccessfully, to eliminate the influence of medieval Arabic polypharmacy; Alpini turned to contemporary practice in the Islamic world in search of more effective medicines.

Alpini's text was written as a dialogue with his mentor Melchior Guilandino (1520–1589). Guilandino (Wieland) was originally from Prussia; his Italian and Mediterranean career was dramatic enough to be ripped from the pages of a chivalric romance. After studying medicine and Botany in Bologna and Rome, he formed a connection with the prominent Venetian diplomat Marino Cavalli (1500–1573). Cavalli was interested in agricultural reform and the problems of provisioning the cities of Venice's empire, and from 1550–1558 served as one of the overseers of the University of Padua.[90] Cavalli introduced Guilandino to Gabriele Falloppia, professor of anatomy at Padua, and the two formed a very close friendship – a relationship that gave rise to scurrilous gossip spread by none other than Pietro Andrea Mattioli.[91] By the mid-1550s, Mattioli's commentaries on Dioscorides had been published in both Italian (1544) and Latin (1554) editions, but Guilandino did not agree with some of Mattioli's plant identifications and Greek translations and did not hesitate to say so in print. In a 1557 letter to the Swiss botanist Conrad Gessner, Guilandino accused Mattioli of textual mistakes and of being insufficiently attentive to Galen, and in 1558 he published a treatise entitled *Defense against Pier Andrea Mattioli*.[92] Mattioli published an open letter to Fallopia in his own defense, which Guilandino replied to, describing Mattioli's work as 'a dung-heap' and a 'lurid rag bag, constantly being retouched but never complete'.[93]

In the middle of this controversy, Guilandino left Italy and traveled the Mediterranean in search of plant specimens. But this herborizing trip to the Middle East from 1558–1561 went from bad to worse: after collecting specimens in Constantinople, Arabia and Egypt, he was captured by Algerian pirates, lost all his specimens, and enslaved on a galley.[94] Eventually he was ransomed by Fallopia, and on his return to Italy in 1561 he was appointed Prefect of the Botanical Garden in Padua. In 1572, he published a commentary on Pliny's chapters on papyrus that drew on his own first-hand observations of the plant growing in Egypt as well as ancient and modern authors to demonstrate that Pliny was correct in stating the origins of papyrus manufacture pre-dated the founding of Alexandria.[95] This publication led to an exchange of invectives with the French scholar Joseph Scaliger (1540–1609), who declared he would 'expose the errors of this man who raves against all worthwhile authors'.[96] In 1574, Guilandino became embroiled in a Venetian state plot to poison a suspected Ottoman spy, thanks to his patronage connection to the Cavalli family.[97] The Ottoman government sent Mustafà dai Cordoani to Venice in 1574, ostensibly on a diplomatic mission but in the eyes of the Council of Ten for the purpose of acting against Venice and the Council of Ten. The Ten voted to poison Mustafà, and Marino Cavallo's son Sigismond was dispatched to bring Guilandino to Venice under the somewhat flimsy pretext that a fire in the Ducal Palace had destroyed their poison recipes and that they needed to be replaced, together with some poisons to be kept in storage.[98] Guilandino did as requested, but when the poison found its way from storage to Mustafà it did not work as anticipated; Mustafà was strangled on a subsequent trip to Venice and Guilandino's name disappears from the Council of Ten records.

While no other botanists' careers involve pirates and poison plots, Guilandino's controversial and colorful career brings us back to the quarrels and insults of Renaissance botanists. Scholars have interpreted many of these disputes from an Italian or a European perspective, but when seen in a more global and comparative context, Renaissance botanical controversies take on a new aspect. While the insults they hurled were particular to their own time and place, Italian botanists of the 16th century share circumstances with pharmacists and physicians in other times and other places. Long distance trade, intercultural exchange, and geographical discoveries impacted philosophers' understandings of the natural world, and pharmacists' beliefs about the effectiveness of cures. Chinese, Arabic, and Latin writers of pharmaceutical treatises all took Greco-Roman recipes for the universal cure-all theriac as their source material, but each culture transformed the medicine in its own way, to fit local ingredients and cultural beliefs about health, healing, and the natural world. Theriac thus offers an example of the type of complex and multi-directional transformation that occurred as a result of cross cultural commerce along Silk Roads.

Notes

1 Preti 2008; see also the essays in Ferri 2001.
2 Beck 2005.
3 All the quotations are from Palmer 1985, 150; see also Findlen 1999, 385–392.
4 Cited and translated in Findlen 1999, 385.
5 Palmer 1985a, 149–157; Findlen 2017, 155–195.
6 Lo and Cullen 2005; Hansen 2017, 5–6.
7 Whitfield 2018, 3.
8 Wear 1999, 149–151; Totelin 2015.
9 Hansen 2017, 5.
10 Watson 1966, 1–91; Fabbri 2007, 253–254; Mayor 2010.
11 Mayor 2010, 239–247.
12 Mayor 2010, 347–352.
13 Watson 1966, 33–44; Totelin 2004, 1–19; Mayor 2010, 240–241.
14 Totelin 2004, 10.
15 Totelin 2004, 11.
16 Totelin 2015, 152.
17 Pliny *Natural History* 29.8.21–23 in Jones 1963, 199.
18 Pliny *Natural History* 29.8.26 in Jones 1963, 199.
19 Gourevitch 1998, 123–138; Totelin 2004, 14.
20 Leigh 2016; for a comparison of this recipe with Galen's other formulations, see Watson 1966, 75–93.
21 Ciraldi 2000, 90–98, offers archeological evidence for the preparation of this medicine in the Italian countryside.
22 Leigh 2016, 125.
23 Strohhmaier 1998.
24 Savage-Smith 2013, 144–145.
25 Dols 1984, 4–6; Kahl 2003, 11–12.
26 Kahl 2003, 33.
27 Johnstone 1981, 197–211.
28 Johnstone 1981, 208.
29 Nappi 2009, 749.
30 Dols 1984, 140; Chipman 2007, 272; Nappi 2009, 749; Savage-Smith, 2013, 146–151.
31 Johnstone 1981, 208; Ricordel 2000a and 2000b.
32 Schafer 1963, 184.
33 Chen 2007.
34 Leigh 2016, 137.
35 Unschuld 1986, 47. The transliteration of 'Theriac' here has been altered to Pinyin, thanks to my colleague Qiong Zhang.
36 Nappi 2009, 741–742.

37 Chen 2007, 253.
38 Unschuld 1985, 51–99; Lo 2018, 578.
39 Hsu 2001.
40 Unschuld 1985, 111–116; Salguero 2014; Lo 2018, 593–594.
41 Chen 2007.
42 Buell 2007, 282; Shinno 2016, 48–53.
43 Chipman 2007, 269; Nappi 2009, 743–744.
44 Nappi 2009, 744, 755.
45 Shih-Hsun 劉世珣 2014, 15. I owe this reference to Qiong Zhang, who located the citation and kindly consulted the Chinese language source for me.
46 Nappi 2009, 753.
47 Nappi 2009, 753–754.
48 Jacquart 1998, 206.
49 Watson 1966, 97; Griffin 2004, 318.
50 Siriasi 1987, 43–63.
51 De Vos 2013.
52 Fabbri 2007, 254; Parrish 2015, 77–78.
53 Fabbri 2007, 263–267.
54 Laughran 2003, 102.
55 Stössl 1983; Parrish 2015, 1; Findlen 1994, 242 points out the ritual was not unique to Venice but common in other Italian cities as well.
56 Mozzato 2014, 165–171.
57 Laughran 2003, 103
58 Laughran 2003, 102.
59 De Vivo 2007.
60 Griffin 2004.
61 Palmer 1985b, 105–108; Parrish 2015, 123–131; Prosperi 2017, 239–240.
62 Eamon 2003; Parrish 2015, 109–141.
63 Prosperi 2017, 241.
64 Maranta 1572.
65 Nutton 1997.
66 Nauert 1979, 81–83; Ogilvie 2006, 121–132.
67 Leoniceno 1492.
68 Collenucio 1493.
69 Watson 1966, 101.
70 Palmer 1985a, 150–152.
71 Palmer 1985b, 109.
72 Palmer 1985b, 109.
73 Findlen 1994, 273–275.
74 Worth-Estes 1995, 3–23; Pereira *et al.* 1996.
75 Findlen 1994, 155–192.
76 Visentini 1984.
77 Cited and translated in Parrish 2015, 61.
78 Findlen 1999, 373–374.
79 Findlen 1999, 382.
80 Findlen 1999; 2007.
81 Findlen 1999, 387.
82 Mattioli 1568, available at https://biodiversitylibrary.org/page/51721025.
83 Mattioli 1568, 3; Palmer 1985b, 108.
84 Findlen 1994, 269.

85 Findlen 1994, 261–287.
86 Alpini 1591; 1592.
87 Ongaro 1963.
88 Alpini 1591, 133–136.
89 Alpini 1591, 138.
90 Oliveri 1979.
91 Findlen 1999, 387.
92 Ferrari 1959, 385–417; Palmer 1985a, 154.
93 Palmer 1985a, 154.
94 Trevisan 1995, 61.
95 Ferrari 1959, 424–436.
96 Grafton 1979, 167.
97 Preto 1989–1990, 233–236.
98 Preto 1989–1990, 234.

Bibliography

Alpini, P. 1591. *De medicina Aegyptiorum libri quatuor*, Venice.

Alpini, P. 1592. *De plantis Aegypti liber*, Venice.

Azzi Visentini, M. 1984. *L'Orto botanico di Padova e il giardino del Rinascimento*, Milan.

Borgarucci, P. 1566. *La fabrica de gli spetiali: partita in XII distintioni*, Venice.

Buell, P.D. 2007. How Did Persian and Other Western Medical Knowledge Move East, and Chinese West? A Look at the Role of Rashīd al-Dīn and Others, *Asian Medicine* 3.2, 279–295.

Chen, M. 2007. The Transmission of Foreign Medicine via the Silk Roads in Medieval China: A Case Study of Haiyao Bencao, *Asian Medicine* 3.2, 241–264.

Ciaraldi, M. 2000. Drug Preparation in Evidence? An Unusual Plant and Bone Assemblage from the Pompeian Countryside, Italy, *Vegetation History and Archaeobotany* 9.2, 91–98.

Collenucio, P. 1493. *Defensio Pliniana adversus Nicolai Leonecenu accusationem*, Ferrara.

De Vivo, F. 2007. Pharmacies as Centres of Communication in Early Modern Venice, *Renaissance Studies* 21.4, 505–521.

De Vos, P. 2013. The 'Prince of Medicine': Yūḥannā Ibn Māsawayh and the Foundations of the Western Pharmaceutical Tradition, *Isis* 104.4, 667–712.

Dioscorides, P. 2005. *Pedanius Dioscorides of Anazarbus. De Materia Medica*, translated by L.Y. Beck, Altertumswissenschaftliche Texte Und Studien 38, Hildesheim.

Dols, M.W. 1984. *Medieval Islamic Medicine: Ibn Ridwān's Treatise,'On the Prevention of Bodily Ills in Egypt'*, Berkeley.

Eamon, W. 2003. Pharmaceutical Self-Fashioning, or How to Get Rich and Famous in the Renaissance Medical Marketplace, *Pharmacy in History* 45.3, 123–129.

Fabbri, C.N. 2007. Treating Medieval Plague: The Wonderful Virtues of Theriac, *Early Science and Medicine* 12.3, 247–283.

Ferrari, G. 1959. Le opere a stampa del Guilandino. Per un paragrafo dell'editoria scientifica padovana del pieno cinquecento, in: Bellini, G. (ed.), *Libri e Stampatori in Padova*, Padova, 377–463.

Ferri, S. (ed.) 2001. *Pietro Andrea Mattioli: Siena 1501 - Trento 1578: la vita, le opere: con l'identificazione delle piante*, Perugia.

Findlen, P. 1994. *Possessing Nature: Museums, Collecting, and Scientific Culture in Early Modern Italy*, Berkeley.

Findlen, P. 1999. The Formation of a Scientific Community: Natural History in Sixteenth Century Italy, in: Grafton, A and Sirasi, N.G. (eds), *Natural Particulars: Nature and the Disciplines in Renaissance Europe*, Cambridge, MA, 369–400.

Findlen, P. 2017. The Death of a Naturalist: Knowledge and Community in Late Renaissance Italy, in: Manning, G and Klestinec, C. (eds) *Professors, Physicians and Practices in the History of Medicine: Essays in Honor of Nancy Siraisi*, New York, 155–195.

Firpo, L. 1971. 'Boragarucci.' *Dizionario Biografico Degli Italiani* 12, http://www.treccani.it/enciclopedia/borgarucci_(Dizionario-Biografico).

Gourevitch, D. 1998. The Paths of Knowledge: Medicine in the Roman World, in: Grmek, M.D., Fantini, B. and Shugaar, A. (eds), *Western Medical Thought from Antiquity to the Middle Ages*, Cambridge, MA, 104–138.

Grafton, A. 1979. Rhetoric, Philology and Egyptomania in the 1570s: J.J. Scaliger's Invective against M. Guilandinus's Papyrus, *Journal of the Warburg and Courtauld Institutes* 42, 167–194.

Griffin, J.P. 2004. Venetian Treacle and the Foundation of Medicines Regulation, *British Journal of Clinical Pharmacology* 58.3, 317–325.

Hansen, V. 2017. *The Silk Road. A New History with Documents*, Oxford.

Hsu, E. (ed.) 2001. *Innovation in Chinese Medicine*, Needham Research Institute Studies 3, Cambridge.

Jacquart, D. 1998. Medical Scholasticism, in: Grmek, M.D., Fantini, B. and Shugaar, A. (eds), *Western Medical Thought from Antiquity to the Middle Ages*, Cambridge, MA, 197–241.

Johnstone, P. 1981. Galen in Arabic: The Transformation of Galenic Phmaracology, in: Nutton, V. (ed.), *Galen: Problems and Prospects*, London, 197–211.

Kahl, O. 2003. *Sābūr ibn Sahl, The Small Dispensatory*, Islamic Philosophy, Theology, and Science 53, Leiden.

Laughran, M.A. 2003. Medicating with or without 'Scruples': The 'Professionalization' of the Apothecary in Sixteenth-Century Venice, *Pharmacy in History* 45.3, 95–107.

Leigh, R. (ed.) 2016. *On Theriac to Piso, Attributed to Galen: A Critical Edition with Translation and Commentary*, Studies in Ancient Medicine, Leiden.

Leoniceno, N. 1492. *De Plinii et aliorum in medicina erroribus*, Ferrara.

Lo, V. 2018. Medicine and Healing in Han China, in: Jones, A. and Taub, L. (eds), *The Cambridge History of Science*, Cambridge, vol. 1, 574–594.

Lo, V. and Cullen, C. (eds) 2005. *Medieval Chinese Medicine: The Dunhuang Medical Manuscripts*, London.

Maranta, B. 1572. *Della theriaca et del mithridato libri due*, Venice.

Matthioli, P.A. 1568. *I discorsi di M. Pietro Andrea Matthioli sanese.... Venice*. Accessed at https://www.biodiversitylibrary.org/item/211436#page/5/mode/1up (accessed 26 May 2020).

Mayor, A. 2010. *The Poison King: The Life and Legend of Mithradates, Rome's Deadliest Enemy*, Princeton.

Mozzato, A. 2014. Oppio, triaca e altre spezie officinali a Venezia nella seconda metà del Quattrocento, in: Knapton, M., Law, J.E., Andrews Smith, A. and Kohl, B.G. (eds), *Venice and the Veneto during the Renaissance: The Legacy of Benjamin Kohl*, Florence, 155–184.

Nappi, C. 2009. Bolatu's Pharmacy Theriac in Early Modern China, *Early Science and Medicine* 14.6, 737–764.

Nauert, C.G. 1979. Humanists, Scientists, and Pliny: Changing Approaches to a Classical Author, *The American Historical Review* 84.1, 72–85.

Nutton, V. 1997. The Rise of Medical Humanism: Ferrara, 1464–1555, *Renaissance Studies* 11.1, 2–19.

Ogilvie, B.W. 2006. *The Science of Describing: Natural History in Renaissance Europe*, Chicago.

Oliveri, A. 1979. Cavalli, Marino, *Dizionario Biografico Degli Italiani* 22. http://www.treccani.it/enciclopedia/marino-cavalli_(Dizionario-Biografico)/.

Ongaro, G. 1963. *Contributi alla biografia di Prospero Alpini*, Padua.

Palmer, R. 1985a. Medical Botany in Northern Italy in the Renaissance, *Journal of the Royal Society of Medicine* 78.2, 149–157.

Palmer, R. 1985b. Pharmacy in the Republic of Venice in the Sixteenth Century, in Wear, A., French, R.K. and Lonie, I.M. (eds), *The Medical Renaissance of the Sixteenth Century*, Cambridge, 100–117.

Parrish, S.D. 2015. *Marketing Nature: Apothecaries, Medicinal Retailing, and Scientific Culture in Early Modern Venice, 1565-1730*. PhD Thesis, Duke University.

Pereira, N.A., Jaccoud, R.J. and Mors, W.B. 1996. Triaga Brasilica: Renewed Interest in a Seventeenth-Century Panacea, *Toxicon* 34.5, 511–516.
Pliny the Elder. 2014. *Natural History. Eight Books 28-32*, translated by W.H.S. Jones, Loeb Classical Library 418, Cambridge, MA.
Preti, C. 2008. Mattioli, Pietro Andrea, *Dizionario Biografico degli Italiani* 72. http://www.treccani.it//enciclopedia/pietro-andrea-mattioli_(Dizionario-Biografico).
Preto, P. 1989–1990. Un infortunio professionale di Melchiorre Guilandino, direttore dell'Orto Botanico di Padova, *Quaderni per La Storia Dell'Università Di Padova* 22–23, 233–236.
Pugliano, V. 2017. Pharmacy, Testing, and the Language of Truth in Renaissance Italy, *Bulletin of the History of Medicine* 91.2, 233–273.
Ricordel, J. 2000a. Ibn Djuldjul: 'propos sur la thériaque', *Revue d'Histoire de La Pharmacie* 325, 73–80.
Ricordel, J. 2000b. Le Traité sur la thériaque d'Ibn Rushd (Averroes), *Revue d'Histoire de La Pharmacie* 325, 81–90.
Savage-Smith, E. 2013. Medicine in Medieval Islam, in: Lindberg, D. and Shank, M.H. (eds), *Cambridge History of Science* 2, 139–167.
Shih-Hsun, L. 劉世珣, 2014. Di-ya-jia de zaixian: Kangxi nianjian gongting xiyao de-li-ya-ge chutan 底野迦的再現: 康熙年間宮廷 西藥德里鴉噶初探 (Reappearance of Theriac: The Use of Deriyaka in the Imperial Court in the Kangxi Period) Qingshi yanjiu 清史研究, *The Qing History Journal* 2014.3, 12–22.
Shinno, R. 2016. *The Politics of Chinese Medicine under Mongol Rule*, London.
Siraisi, N.G. 1987. *Avicenna in Renaissance Italy: The Canon and Medical Teaching in Italian Universities after 1500*, Princeton.
Stössl, M. 1983. *Lo Spettacolo della Triaca: produzione e promozione della 'droga divina' a Venezia dal dinque al settecento*, Venice.
Strohmaier, G. 1998. Reception and Tradition: Medicine in the Byzantine and Arab World, in: Grmek, M.D., Fantini, B. and Shugaar, A. (eds), *Western Medical Thought from Antiquity to the Middle Ages*, Cambridge, MA, 139–169.
Totelin, L.M.V. 2004. Mithradates' Antidote: A Pharmcological Ghost, *Early Science and Medicine* 9.1, 1–19.
Totelin, L.M.V. 2015. The World in a Pill. Local Specialities and Global Remedies in the Graeco-Roman World, in Kennedy, R.F. and Jones-Lewis, M. (eds), *The Routledge Handbook of Identity and the Environment in the Classical and Medieval Worlds*, London, 151–170.
Trevisan, R. 1995. Melchiorre Guilandino, in: Minelli, A. (ed.), *The Botanical Garden of Padua: 1545–1995*, Venice, 59–62.
Unschuld, P.U. 1985. *Medicine in China: A History of Ideas*, Comparative Studies of Health Systems and Medical Care, Berkeley.
Unschuld, P.U. 1986. *Medicine in China: A History of Pharmaceutics*, Comparative Studies of Health Systems and Medical Care, Berkeley.
Watson, G. 1966. *Theriac and Mithridatium; a Study in Therapeutics*, Wellcome Historical Medical Library, n.s. vol. 9, London.
Wear, A. 1999. The Early Modern Debate about Foreign Drugs: Localism versus Universalism in Medicine, *The Lancet* 354.9173, 149–151.
Whitfield, S. 2018. *Silk, Slaves, and Stupas: Material Culture of the Silk Road*, Berkeley.
Worth-Estes, J. 1995. The European Reception of the First Drugs from the New World, *Pharmacy in History* 37.1, 3–23.

Chapter 7

Spice and Taste in the Culinary World of the Early Modern Mediterranean

Eric Dursteler

The importance of spices in the medieval diet, and society more broadly, is well known and has been amply documented by a rich body of scholarly work. The place of spices in the early modern diet, in contrast, has attracted much less attention. According to the still dominant narrative, beginning in France in the mid-17th century, and spreading to the rest of Europe over the course of the next decades, before ultimately conquering the world, the exotic eastern flavors and alimentary combinations that had characterized medieval cuisine were supplanted by more subtle, natural flavorings produced from indigenous herbs and aromatic plants. There is a certain teleological quality to this narrative, which both presumes and accelerates the eventual conquest of medieval extravagance by early modern culinary finesse. Among many problematic qualities, this paradigm is highly localized to and privileges the foodways of western Europe, and in particular the culinary cultures of 17th- and 18th-century France and England. Culinary traditions from other eras and other areas are often viewed as occupying the prehistory or periphery of the inevitable triumph of French *haute cuisine*.[1] Taking a cue from Fernand Braudel, what if we look at this familiar account 'upside down', that is what if we consider it from the vantage point of the Mediterranean, in what ways does this inverted perspective alter the narrative, and what insights does such a vantage point bring to our understanding of early modern culinary culture, in Europe and beyond?[2] This perspective, I will argue, reveals that while foodways are in constant flux, and there were undoubtedly significant changes that began to stir in 17th-century Paris, when viewed with a wider angle lens, the abandonment of spices was both gradual and definitely not universal. The persistence of spice usage, even as foodways in this period experienced important changes as a result of the introduction of new foods and flavors from both the New and the Old Worlds, is particularly evident when viewed from the vantage point of the Mediterranean world.

Medieval Spice

The role of spice as food, medicine, and status symbol in medieval Europe and the Mediterranean has been both richly chronicled and hotly debated, and I will make no attempt to survey that literature here.[3] Braudel aptly describes the monumental increase in the commerce and consumption of spices that characterized this era as 'la folie des épices', and there is ample evidence to support his view.[4] In his famous commercial manual from the first half of the 13th century, *La practica della mercatura*, Francesco Pegolotti lists 288 spices used for a wide range of culinary, pharmaceutical, and industrial purposes.[5] A century later, in a single year the Talbot family of Shropshire, England consumed an astonishing 17 lbs of pepper, 14 lbs of ginger, and 17 lbs of other assorted spices, including 3 lbs of saffron.[6] A single 14th-century Italian recipe that served 12 called for 'a half pound of spice',[7] while a 15th-century recipe for roasting a lamprey included an unbelievable 2 lbs of cinnamon, 1 lb of ginger, and at least another half dozen spices in smaller quantities.[8] Finally, a recipe for a medieval sauce included pepper, ginger, cinnamon, saffron, three nutmegs, and 26 g of cloves. These are staggering quantities, given that 28 g of cloves is enough to make an effective anesthetic, and excessive amounts of nutmeg can actually be poisonous.[9]

Beyond mind-boggling quantities, medieval cuisine was also noteworthy for its 'enormous diversity of spices' and the broad 'spectrum of flavor' created through unique combinations, as well as their 'ubiquity' during all stages of the meal. Spices appear in 75–90% of all recipes contained in the era's cookbooks, and according to a popular 14th-century collection, Taillevent's *Le Viandier*, 20 spices comprised the 'standard repertoire'. Saffron, ginger, sugar, and cinnamon were the most common, but pepper, cloves, nutmeg, mace, galingale, grains of paradise (melegueta pepper), spikenard, cassia, and cumin were also popular. Bruno Laurioux has argued that 'never before and never since were so many spices used in European cooking'.[10]

While we may generally speak of a pan-European, international gothic 'taste for highly spiced food', we must be cautious in asserting some generalized medieval culinary model or universal taste that ignores the great range and variety of regional alimentary inflections that characterized this age.[11] These differences were a product of both natural resources and culinary techniques, but also culturally formed 'tastes and distastes'. For instance, the 'vogue' for grains of paradise in France 'was only feebly imitated elsewhere', while the popularity of mace in England contrasted with the preference for nutmeg in other regions. If 'the characteristic cuisines of Italy, Germany, France, Spain, and Britain' began very gradually to take form, no manifestation of some proto-nationalist culinary identity or uniformity can be projected back onto this era.[12] National culinary cultures, to the degree that they do exist, are largely products of the 19th and even 20th centuries. This medieval regional diversity is evident in Italy: contemporary texts make clear that there were 'distinctive Venetian, Tuscan, and even Roman attitudes toward the use of spices'. It is also evident in the great diversity of types and quantities of spices called for in the burgeoning literature of cookery in this period.[13]

Early Modern Taste for Spice

According to the traditional narrative, the spicy Middle Ages (which had deep roots in the culinary and medical culture of classical antiquity) were subverted by a more understated flavor palette during the early modern period. What one contemporary described as the 'spices from hell', went from being 'adored' to becoming 'an object of revulsion'.[14] According to Paul Freedman, this sea change represents 'one of the most distinctive' transformations in the history of taste.[15] It grew out of a number of different factors: one was the slowly evolving shift away from the traditional Hippocratic and Galenic medical theories that had deeply informed alimentary practices since antiquity.[16] Another component was a significant increase in the supply of spices after 1600, which led to a notable drop in prices, which in turn produced a decline in the social currency of spice.[17] A key moment in this process occurred during the middle decade of the 17th century, which saw the publication of François Pierre de La Varenne's deeply influential 1651 work, *Le Cuisinier françois*,[18] followed by Nicholas de Bonnefons' *Les Délices de la campagne* three years later. These cookbooks are considered the first to make a 'decisive break' with the medieval culinary past.[19] This marked the opening bell of the '*grand siécle*' of French cuisine, and these works were followed in short order by a flurry of other French books of cookery published over the subsequent decades.[20] Unlike Renaissance food writing, which offered extensive medical and dietetic advice, these texts focused more on food as food, how to cook it, and most importantly, how to make it taste delicious.[21]

During the Middle Ages and Renaissance, the copious use of spice combined with the discordant (to some sensibilities) mixture of sweet, sour, and bitter in a single dish, was intended to fuse a variety of flavors into a 'unitary whole' that masked and transformed the natural flavoring of individual ingredients, rendering them 'unidentifiable'. French culinary revolutionaries usurped and ultimately overturned this flavor regime by insisting on the primacy of the natural flavors of unadorned ingredients, what one contemporary protagonist described as 'the art of cooking foods delicately'.[22] The objective of this 'new culinary aesthetic' was subtlety and simplicity: this was achieved by preparing foods in ways that 'accentuated and intensified' flavors through the use of 'fresh herbs and aromatics', such as onions, shallots, chives, truffles, and mushrooms. Another key component of this *nouvelle cuisine* was the extensive use of butter or other fats.[23] The objective, as Bonnefons wrote, was that 'food should taste like what it is'.[24]

As this shifting tide of taste took hold in France, it also began to be manifested in burgeoning notions of French culinary superiority. Bonnefons recommended leaving 'depraved ragouts' to foreigners, who 'never enjoy good fare except when they have cooks from France'.[25] The chef François Massialot similarly declared in 1691, 'Man is not everywhere capable of such discernment, which is an emanation of his reason and intellect ... Only in Europe do cleanliness, good taste, and skillful seasoning of meat prevail ... in France we can pride ourselves on our superiority in this regard to all other nations.'[26]

This sense of superiority was evident in the culinary critiques by French travelers of the spicy foods they encountered abroad. They considered the enduring use of spices by non-French chefs as variously 'baroque' or 'gothic', which is to say culturally retrograde.[27] One of the classic examples is the 1691 narrative composed by the French noblewoman and noted writer, Marie-Catherine Le Jumel de Barneville, Baroness d'Aulnoy, which recounts in engaging detail her travels in Spain, and represents one of the most important early modern accounts of that land. D'Aulnoy was deeply interested in cuisine, and made numerous, detailed references to meals she ate during the course of her travels. A recurring critical theme is the overly spiced character of Spanish cuisine: She noted 'they eat every thing so very high season'd with Pepper and Spice, that one would think it impossible they are not burnt up.'[28] Fish pasties 'would be good, were they not stuff'd with Garlick, Saffron, and Pepper'; a pheasant ragout she was served 'burn[ed] the Mouth, it [was] so excessively pepper'd'.[29] In Toledo she enjoyed 'a very plentiful Meal' at the archbishop's house, though 'every thing was so full of Amber, that I never tasted any Sawces so rich, and less pleasant. I fancy'd my self at this Table, like Tantalus, in the midst of Victuals, and yet starv'd with hunger, because I could not eat; for either all the Meat was perfum'd, or else full of Saffron, Garlick, Onion, Pepper and Spices.'[30] D'Aulnoy was not alone in her critical assessment of Spanish cuisine: two decades previous, in his multi-volume *Le voyageur d'Europe* (1672–76), the French cartographer Albert Jouvin de Rochefort struck a more objective tone, but still was critical of Spanish culinary abuse of spice. He commented on a 'horrible stew, full of garlic and pepper', and in another instance was starving but unwilling to eat the available meat since it was 'all flavored and stuffed with saffron, garlic, onion, pepper and spices'.[31]

There is no question, as d'Aulnoy's supercilious asides suggest, that spice consumption in Spain (and elsewhere) differed from that in at least some parts of France. Turning to Italy, for example, we find that the French revolution in cooking made its mark only slowly and very gradually.[32] La Varenne's cookbook was published in translation in Bologna in 1682, and was reprinted 13 times over the next 150 years, but there are significant differences between the Italian and French editions, including different recipes and very different uses of flavorings and spices.[33] Despite this early translation, however, in general 17th-century Italian cuisine preserved its medieval and Renaissance character, with 'ingredients, methods of cooking and favorite dishes' remaining largely unchanged.[34] The outlines of this traditional cuisine are evident in the most famous Italian cookbook of the age, the monumental 1570 work, *Opera dell'arte del cucinare*, by the papal chef, Bartolomeo Scappi. Spices occupy a dominant position in the *Opera*: out of a total of about 1000 recipes, cinnamon appears in over 600, pepper over 550, saffron and cloves more than 200, and sugar 'almost collapses' Scappi's spice shelf, appearing in over 900 recipes.[35] In Book I there is a section entitled 'To identify every variety suitable for cooking', which includes Scappi's recipe for his common spice mixture that could be 'made up ahead of time for use at will in a variety of dishes'. He calls for '4½ ounces of cinnamon, 2 ounces of cloves, 1 ounce of ginger, 1 ounce of nutmeg, ½ ounce of grains of paradise [meleguetta pepper], ½ ounce saffron, 1 ounce of sugar'.[36] The result was a 'sweet, sour and spicy flavor ... not ... unlike a good barbecue sauce'.[37]

Moving forward a century, *L'arte di ben cucinare* (1662), by Bartolomeo Stefani, chef to the dukes of Mantua, while clearly rooted in the same tradition as Scappi, nonetheless evidences a culinary aesthetic that seems to be informed to a degree by what was happening among his French contemporaries. For instance, Stefani proposes a beef stew flavored simply with rosemary and garlic: he writes, 'Do not add spices, for when it is cooked it will be good.' Overall, however, the complex contrasts of flavor characteristic of medieval cuisine, remained prominent. Stefani advocates for the 'moderate use of spices', which he believes will make a dish more 'dignified'.[38] This is evident in several preparations he proposes for turkey: turkey 'alla Svizera' includes flakes of cinnamon and mace ... and pepper,[39] while a dish that molds ground turkey into the shapes of eagles or lions, includes 'odoriferous herbs', but also 'sugar' and a dusting of 'cinnamon'.[40] Stefani's straddling of the old and the new is also evident in his recipe for butter sauce, in which 'a pound of vinegar, sugar and spices' is added to melted butter.[41]

The enduring popularity of spices in Italian cookery is also evidenced in another influential culinary text of the era, Antonio Latini's weighty two-volume work, *Lo scalco moderno* (1692–94).[42] Based on his long career as a cook and steward in Naples and other parts of Italy, *Lo scalco* is considered 'the last of the great Italian culinary books published since the Renaissance', and is justifiably famous because it includes the first printed recipes to utilize tomatoes and chili peppers, two New World plants which transformed Italian, European, and indeed world cuisines.[43] Latini is also interesting for the ways in which he straddles the traditions of medieval and Renaissance cuisine and the emerging culinary ideas in France. In this important transitional work, Latini seems to go beyond Stefani when he proposes the radical idea that it is 'possible to cook and flavor food without using spices', instead using various herbs to achieve more subtle flavorings. Like Stefani, however, Latini proves less than committed to the culinary 'new wave'. He includes many traditional recipes that contain combinations of 'suitable' spices, such as 'cinnamon, coriander, nutmeg, cloves, and pepper',[44] and like his predecessors, he also includes his own version of a basic spice mixture to be used with 'all the above foods'.[45] He also uses sugar in most of his sauces, though not his famous recipe for tomato sauce, 'salsa di Pomadoro, alla Spagnola'.[46] Latini's use of spice has tapered significantly, however: in a collection of recipes for a banquet for six, he uses spice, but in somewhat reduced quantities: 1½ ounces of cinnamon, 2 ounces of pepper, 1 ounce of cloves, and 2 ounces of nutmeg are prescribed for an entire meal. Latini also tends to segregate sweet and savory flavors, and he uses herbs 'to accentuate rather than contrast with the main ingredient'. Overall, his tastes and some of his recipes are simpler, and 'fewer are doused with spices and sugar'. This allows the natural flavors of ingredients to occupy center stage.[47]

Shifting the focus away from western Europe, we find further evidences of spice's continued culinary role. In Wallachia, for example, spice 'remained in vogue' and a recurring presence in cookbooks, letters, and account records.[48] Importation of spice into Moldavia 'increased massively' starting at the end of the 16th century as a result of growing political connections with the Ottoman Porte.[49] In Transylvania, spices were heavily used into the 17th century, and the region was a nexus for the spice trade coming from both east and west, which increased significantly over the course of the century.[50] Gaspard d'Hauteville, who lived in Poland for 20 years, observed in 1686 that Polish sauces were 'very different from the French', in part because the Poles 'add lots of sugar, pepper, cinnamon, cloves, nutmeg, olives, capers and raisins' to their food.[51] Braudel also contended that 'Germany, Scandinavia, and Eastern Europe remained prodigious consumers of spice' well into the 17th century.[52]

Looking into the eastern Mediterranean, we see a similar persistence of spice in Persian and Ottoman cuisine. One of the few extant Safavid cookbooks is *Maddat ol-hayat: resaleh dar elm-e tabbakhi* (*The Substance of Life: A Treatise on the Science of Cooking*), compiled in 1594 by Nurollah, chef to Shah Abbas I. Nurollah uses a wide range of spices in his recipes for *buraq* (meat pastries), *palav* (rice dishes), and *qelyeh* (stews). These include copious amounts of what he terms 'hot spices' – ginger, cinnamon, pepper, cloves – as well as nutmeg, cardamom, sugar, and saffron. His recipe for *Behbahan Palav* (mutton rice) is representative: 'The meat of a fat yearling is stewed, and a large amount of whole chickpeas, cinnamon sticks, and peppercorns, as much as needed, is added to it; after the rice is prepared, ginger, nutmeg, cardamom, and cloves are all pounded and sprinkled on top of the rice, and a small amount of saffron is added. This type is cooked by the one pot method.'[53] Elsewhere he indicates what he means by 'copious amounts': 20 *mesqual*s (100 g) of cinnamon, and 10 *mesqual*s each of whole ginger, pepper, cardamom, and cloves.[54]

The immense 17th-century Ottoman travel narrative of Evliya Çelebi is filled with observations on food from all the places he visited during his many travels, and spices of all sorts appear with regularity. In Trebizond, anchovies were prepared in a 'stoneware pan called pileki'. The cleaned fish are arranged in rows of 10, then covered with a mixture of chopped 'parsley, celery, onion and leeks', which are then mixed together with 'cinnamon and black pepper'. Layers of fish and vegetables alternate, and are covered with the local olive oil, which Evliya describes as 'like the Water of Life'. The *pileki* is then placed over a hot fire. The resulting dish, Evliya writes, is 'like congealed light, and one who eats it is full of light ... This fish is indeed a table from heaven, fit to be a favourite; glory be to the eternal Creator!'[55] He similarly reports on a rice dish made in Diyarbekir out of the local melons, which 'are huge, very juicy and sweet, and with a delicious aroma as of musk and ambergris'. These are 'flavoured with cinnamon and cloves, according to the recipe of Caliph Mu'awiya,' Evliya writes, and 'if you eat one of them the fragrance of melon will not leave your brain for an entire week.[56] He also notes that well-off Ottomans ate bread that was enriched with aniseed and sesame, or even 'a liquid seasoned with exotic spices, such as cinnamon and cloves'.[57]

Regulations for the many eating houses in Istanbul's commercial districts provide passing glimpses at the use of spice in dishes popular among the common folk of the capital, though neither types nor quantities are specified. Popular milk-based hot beverages sold by street vendors for breakfast were served sprinkled with cinnamon or ginger.[58] Stews containing different varieties of fish inevitably included cinnamon, and occasionally saffron, and kebabs were flavored with cinnamon and cumin.[59] Sheep's head and trotters were roasted and served with 'vinegar, melted butter and spices', and various types of bread were baked with cumin, poppyseeds, and other spices, that was either mixed into the dough or sprinkled on top of the loaves. Tripe soup, 'served with garlic, spices and vinegar', was believed to cure hangovers, and was sold by Ottoman Greeks from hundreds of establishments that opened early in the morning to serve the needs of inebriated denizens of the city's many taverns.[60]

Spice continued to occupy a prominent position in elite cuisine as well. A wide range of spices for culinary and medical uses were regularly delivered to the imperial palace in the 18th century. Pepper and cinnamon predominated, but cloves, saffron, cumin and other spices were also popular, particularly among the women of the harem.[61] During a 15-day festival in Istanbul in 1720, banquets held for dignitaries utilized 118 kilos of pepper, over 52 kilos of saffron, and several kilos each of cinnamon and cardamom, plus more than a kilo of mastic.[62] The palace confectionery produced syrups and pastes that were consumed both as health supplements and delicacies: some of these included between 40 and 60 different spices.[63] Paralleling its growing popularity in early modern European cuisines, pepper was the 'indispensable spice' in the Porte: by the 18th century, one scholar of Ottoman food has claimed that 'black pepper was used in the preparation of nearly every dish'.[64] Compared with medieval Europe, Ottoman consumption of spices was generally more moderate, and this moderation continued into the early modern period.[65] A 17th-century report states that except for pepper, in Istanbul at least, the Ottomans used 'little spice'.[66] Scholars of Ottoman cuisine have argued that 'spices and aromatics were used with restraint', and food was not 'very spicy'. Indeed, after the 18th century it became 'quite plain', perhaps mirroring trends in France and other parts of Europe.[67]

Still spicy, as is true of all tastes, is a highly subjective notion that is heavily dependent on the palate of the taster. Writing of her culinary experiences in the Ottoman Empire in the early 18th century, Lady Mary Wortley-Montagu noted to one correspondent that 'they use a great deal of rich spice'. While quite open to diverse cultural experiences, Wortley-Montagu found that over time the intense flavors of Ottoman cuisine wore on her to the point that she was unable to eat all the food that was offered to her. During the three weeks she spent in the house of an important official in Belgrade, she was fed 'very magnificent dinners ... which the first week pleased me extremely but, I own I then began to grow weary of it and desired our own cook might add a dish or two after our manner. But I attribute this to custom.'[68]

Conclusion

To bring this overview to a conclusion, we should be cautious in accepting the traditional narrative of the early modern decline of spice for a number of reasons. First, it is likely that medieval spice consumption has been overstated. Since gastronomes, cooks, and writers began denouncing medieval foodways in the 17th and 18th centuries, scholars have accepted at face value the claim that spices were used 'in enormous quantities' during the medieval centuries.[69] Culinary historians have in recent years convincingly shown that while medieval tastes differed from those of subsequent eras, 'the Middle Ages represent a substantial culinary historical period' that was neither impoverished nor 'absurdly elaborate'. Medieval cuisine had 'complexity, subtlety, variety', and its own 'aesthetic sensibility'.[70] Chefs in this period, in fact, used spices 'quite conscientiously'. Recipe instructions caution cooks to add spices 'with care', or as one important 15th-century Neapolitan recipe collection states, '*specie a discretione*'. The 14th-century *Libro di cucina* has numerous recipes which call for various spices, but quantities range from 'two ounces of mixed spices' for a dish to serve 12 people, to recipes that call for only 'a pinch or a touch' of spice.[71] The point is that medieval dishes had a range of spicing which should not be exaggerated as part of a teleological narrative that anticipates shifts in early modern tastes.

Second, changing attitudes toward spice consumption 'permeated society at different speeds'. When spices began to be procured directly from their eastern sources in the 16th and 17th centuries, Europe was inundated by 'aromas and flavours', which eventually caused fatigue. In addition, as spices became available to almost anyone, the rich and powerful sought 'signs of distinction' elsewhere.[72] If ultimately some French aristocrats and *haute bourgeoisie* in Paris largely abandoned spices, this was not the case among all of French (or European) society.[73] So for example, where La Varenne included pepper in only about 20% of his recipes, it figured in over 60% of the recipes in *La nouvelle maison rustique*, 'a housewife's handbook' published a century later.[74] And indeed, not even all French aristocrats rejected spice: to the end of his long life and reign, Louis XIV sprinkled spice liberally on all his food, 'at least twice as much as was usual for any ordinary dish', according to Saint-Simon. There was a regional quality to this as well: a contemporary reported that in Provence food was 'prepared in the Italian style, with abundant spices and extravagant, strongly flavored sauces'.[75]

Third, the way these changes played out differed depending on spatial context. In the case of saffron, the alleged French 'abhorrence' for spices was unique: according to the 18th-century Jesuit traveler, botanist, ethnographer, and gourmand, Père Jean-Baptiste Labat, 'Spanish, Italians, Portuguese, Germans, all the western nations like the Swedes, Danes, Muscovites and even the Dutch, English and Flemish consume [spice] in great amounts. They put it in their sauces, their ragoûts, their pastries.'[76] In the case of Italy, medieval-like spice use persevered among the common people in certain regions until well into the 19th century.[77]

Fourth, it is critical that we not look at the process of change expressed in the recipes of La Varenne as some sort of 'sudden culinary revolution' that rejected or entirely abandoned 'the old ways'. *Le Cuisinier françois* provided a home for just as many 'medieval descendants lingering around' as it did to the new ideas and techniques for which the text is rightly celebrated.[78] In 17th-century French cookbooks, over half of all recipes still include 'oriental spices', and indeed in the famed *Art de bien traiter* (1674) and *La nouvelle maison rustique* (1755), fully 71% of the recipes still include one or more of these.[79] Spices did not disappear from French cuisine, rather there was a shift in both the types and quantities of spices used: for example, cloves went from figuring in only between 4 and 16% of late medieval French recipes, to 40% in *Le Cuisinier françois*. Pepper and nutmeg were also used more frequently, though in lesser quantities. The most popular spices – saffron, ginger, cinnamon, but not sugar – largely disappeared.[80] Indeed, the ongoing persistence of spice led the author of the *Art de bien traiter* (1674) to attack his contemporaries 'for their "antique" and "rustic" ways of cooking', including their continued use of 'confused mixtures of diverse spices'. He criticized La Varenne for his 'ridiculous recipes': they were 'wretched things with which one would put up more readily among Arabs and Levantines than in a pure climate such as ours, where cleanliness, refinement, and good taste are matters of utmost concern'.[81]

Examining early modern culinary history from a Mediterranean perspective casts doubt on the utility of the traditional narrative of a sea change in culinary taste in the mid-17th century. Freedman has described the decline of spice in the early modern diet as 'a seismic shift in taste', but it must be said that the earthquake developed very gradually.[82] What we are talking about was not a 'revolution at the table,' so much as an evolutionary process that played out at different paces in different places.[83] We have to avoid the impulse to trace European culinary history teleologically and to project backward ideas and practices that began only haltingly, locally, and then spread slowly and intermittently. Nor should we assume that these changes represent inevitable modern progress against some sort of benighted medieval culinary backdrop. Indeed one could argue that '*le goût natural*', which was at the core of French cuisine, is in fact significantly 'less universal than the taste for spicy complexity' that characterized ancient and medieval cuisine, and which effectively 'continues to dominate the culinary aesthetics of most of the planet' to this day.[84] The presence in cities all over the world of many more Indian, Mexican, Szechuan, and Indonesian restaurants than ones serving up French cuisine, may provide ample evidence for the taste profile which ultimately won the day.

Notes

1 Neumann, 2003, 129.
2 Braudel 1972, 450.
3 Laurioux, 2005, 157–158 ; Woolgar 2007, 177; Freedman 2008, 49; 2012, 32.
4 Braudel 1967, 1, 163.

5 Heyd 1923; Evans 1936, 411–435; Lopez and Raymond 2001, 108–114.
6 Freedman 2008, 4.
7 Wright 1999, 510.
8 Woolgar 2007, 176.
9 Montanari 1996, 61.
10 Flandrin 1984, 76; Laurioux 1985, 45; 2005, 159–160; Freedman 2008, 20, 24.
11 On this generally, see Adamson 2002.
12 Flandrin 1981, 99; Woolgar 2007, 178–181; Freedman 2008, 27; 2012, 33.
13 Laurioux 1985, 47; 2005.
14 Flandrin 1981, 101; Halikowski Smith 2007, 250–251.
15 Capatti and Montanari 1999, 111.
16 Albala 2002, 241–294.
17 Pinkard 2009, 64–72; Gentilcore 2016, 47.
18 See La Varenne 2006.
19 Albala 2003, 155–156.
20 Revel 1982, 153–154; Wheaton 1983, 115; Mennell 1996, 72–74.
21 Cowan 2007, 224–225.
22 Pinkard 2009, xii; Astarita 2014, 4.
23 Pitte 2002, 93–94; Albala 2003, 155–156.
24 Pinkard 2009, 62.
25 Ferguson 2004, 36.
26 Flandrin 1989, 297.
27 Flandrin 1981, 101; 1989, 272.
28 d'Aulnoy 1697, 151–152.
29 d'Aulnoy 1697, 47, 234–235.
30 d'Aulnoy 1697, 254–255.
31 Martínez Llopis 1995, 309.
32 Capatti and Montanari 1999, 112.
33 Capatti and Montanari 1999, 197; von Hoffmann 2016, n. 16.
34 Albala 2003, 133.
35 Scappi 2008, 60.
36 Scappi 2008, 114.
37 Albala 2003, 130.
38 Albala 2003, 133; Freedman 2008, 218–219.
39 Stefani 1662/2000, 38.
40 Stefani 1662/2000, 119–126.
41 Montanari 2012, 105.
42 See Astarita 2019 for a complete translation into English.
43 Astarita 2014, 2–3, 5.
44 Freedman 2008, 218.
45 Latini 1694, 20.
46 Latini 1694, 444; Freedman 2008, 218–219.
47 Albala 2003, 137–138.
48 Lazăr 2018, 329–330.
49 Aslan 2018, 105.
50 Pakucs-Willcocks 2018, 307–308.
51 Gentilcore 2016, 72.
52 Pinkard 2009, ns. 50 and 66.
53 Nurollah 2017, 26.
54 Nurollah 2017, 23.

55 Dankoff and Kim 2010, 46–47.
56 Dankoff and Kim 2010, 115.
57 Faroqhi 2016, 157.
58 Işın 2018, 40.
59 Yerasimos 2007, 80, 160–167.
60 Dursteler 2014, 209–211; Işın 2018, 127, 131–132.
61 Neumann 2003, 135–140; Işın 2018, 29.
62 Faroqhi 2000, 167.
63 Yerasimos 2007, 274; Bilgin 2008, 88–89.
64 Yerasimos 2007, 51.
65 Reindl-Kiel 2003, 83.
66 Bon 1996, 99.
67 Yerasimos 2007, 52; Işın 2018, 29.
68 Montagu 1994, 87–88.
69 Laurioux 1985, 60.
70 Freedman 2012, 25.
71 Scully 1995, 84; Jotischky 2011, 170.
72 Montanari 1996, 119.
73 Laurioux 1985, 66.
74 Halikowski Smith 2007, 255.
75 Pinkard 2009, 79–80, 125–126; Watts 2011, 111.
76 Flandrin 1981, 100.
77 Capatti and Montanari 1999, 114.
78 Albala 2003, 155–156.
79 Flandrin 1981, 101.
80 Flandrin 1981, 101; 1989, 272; Laurioux 1985, 45; Halikowski Smith 2007, 252.
81 Flandrin 1989, 297; Cowan 2007, 127.
82 Freedman 2008, 222–223.
83 Levenstein 2003.
84 Pinkard 2009, 67.

Bibliography

Adamson, M. 2002. *Regional Cuisines of Medieval Europe: A Book of Essays*, Abingdon.

Albala, K. 2002. *Eating Right in the Renaissance*, Berkeley.

Albala, K. 2003. *Food in Early Modern Europe*, Westport.

Aslan, M. 2018. Turkish Flavours in the Transylvanian Cuisine, in: Jianu, A. and Barbu, V. (eds), *Earthly Delights: Economies and Cultures of Food in Ottoman and Danubian Europe, c. 1500–1900*, Leiden, 99–126.

Astarita, T. 2019. *Antonio Latini's the Modern Steward, Or the Art of Preparing Banquets Well: A Complete English Translation*, Kalamazoo.

Astarita, T. 2014. *The Italian Baroque Table: Cooking and Entertaining from the Golden Age of Naples*, Tempe.

d'Aulnoy, M. 1697. *The Ingenious and Diverting Letters of the Lady – Travels into Spain*, London.

Bilgin, A. 2008. *Turkish Cuisine*, Ankara.

Bon, O. 1996. *The Sultan's Seraglio: An Intimate Portrait of Life at the Ottoman Court*, London.

Braudel, F. 1967. *Civilisation Matérielle et Capitalisme: (XVe–XVIIIe siècle)*, vol. 1, Paris.

Braudel, F. 1972. Personal Testimony, *Journal of Modern History* 44, 448–467.

Capatti, A. and Montanari M. 1999. *Italian Cuisine: A Cultural History*, New York.

Cowan, B. 2007. New Worlds, New Tastes: Food Fashions after the Renaissance, in: Freedman, P. (ed.), *Food: The History of Taste*, Berkeley, 197–232.
Dankoff, R. and Kim, S. (trans. and eds) 2010. *An Ottoman Traveller: Selections from the Book of Travels of Evliya Çelebi*, London.
Dursteler, E. 2014. Bad Bread and the 'Outrageous Drunkenness of the Turks': Food and Identity in the Accounts of Early Modern European Travelers to the Ottoman Empire, *Journal of World History*, 25, 203–228.
Evans, A. (trans. and ed.) 1936. *Francesco Balducci Pegolotti: La Pratica della mercatura. A Study of the Man and His Work*, Cambridge.
Faroqhi, S. 2000. *Subjects of the Sultan: Culture and Daily Life in the Ottoman Empire*, London.
Faroqhi, S. 2016. *A Cultural History of the Ottomans: The Imperial Elite and its Artefacts*, London.
Ferguson, P.P. 2004. *Accounting for Taste: The Triumph of French Cuisine*, Chicago.
Flandrin, J.-L. 1981. Différence et différentiation des goûts: réflexion sur quelques exemples européens entre le 14ème et le 18ème siècle, in: Davidson, A. (ed.), *National & Regional of Cookery: Proceedings of the Oxford Symposium on Food and Cookery*, London, 99–115.
Flandrin, J.-L. 1984. Internationalisme, nationalisme et régionalisme, dans la cuisine des XVIe et XVe siècles: le témoignage des livres de cuisine, in: *Manger et boire au Moyen Âge: Acts du colloque de Nice, 15–17 octobre 1982*, t.2, Paris, 75–91.
Flandrin, J.-L. 1989. Distinction through Taste, in: Chartier, R. (ed.), *A History of Private Life, vol 3, Passions of the Renaissance*, Cambridge, 265–307.
Freedman, P. 2012. Food Histories of the Middle Ages, in: Claflin, K. and Scholliers, W.P. (eds), *Writing Food History: A Global Perspective*, New York, 24–37.
Freedman, P. 2008. *Out of the East: Spices and the Medieval Imagination*, New Haven.
Gentilcore, D. 2016. *Food and Health in Early Modern Europe: Diet, Medicine and Society, 1450–1800*, London.
Halikowski Smith, S. 2007. Demystifying a Change in Taste: Spices, Space, and Social Hierarchy in Europe, 1380–1750, *The International History Review* 29, 237–257.
Heyd, W. 1923. *Histoire du Commerce du Levant au Moyen Âge*, Leipzig.
von Hoffmann, V. 2016. *From Gluttony to Enlightenment: The World of Taste in Early Modern Europe*, Champaign-Urbana.
Işın, P.M. 2018. *Bountiful Empire: A History of Ottoman Cuisine*, London.
Jotischky, A. 2011. *A Hermit's Cookbook: Monks, Food and Fasting in the Middle Ages*, London.
de La Varenne, F.P. 2006. *La Varenne's Cookery: The French Cook; the French Pastry Chef; the French Confectioner*, Scully, T. (trans. and ed.), Totnes.
Latini, A. 1694. *Lo scalco alla moderna*, Naples.
Laurioux, B. 1985. Spices in the Medieval Diet: A New Approach, *Food and Foodways* 1, 43–76.
Laurioux, B. 2005. *Une histoire culinaire du Moyen Âge*, Paris.
Lazăr, G. 2018. The Food Trade in 18th-Century Wallachia between Daily Subsistence and Luxury, in: Jianu, A. and Barbu, V. (eds), *Earthly Delights: Economies and Cultures of Food in Ottoman and Danubian Europe, c. 1500–1900*, Leiden.
Levenstein, H. 2003. *Revolution at the Table: The Transformation of the American Diet*, Berkeley.
Lopez, R.S. and Raymond, I.W. 2001, *Medieval Trade in the Mediterranean World: Illustrative Documents*, New York.
Martínez Llopis, M. 1995. *Historia de la gastronomía española*, Val de Onsera.
Mennell S. 1996. *All Manners of Food: Eating and Taste in England and France from the Middle Ages to the Present*, Urbana.
Montagu, M.W. 1994. *The Turkish Embassy Letters*, Jack, M. (ed.), London.
Montanari, M. 1996. *The Culture of Food*, Oxford.
Montanari, M. 2012. *Medieval Tastes: Food, Cooking, and the Table*, New York.

Neumann, C. 2003. Spices in the Ottoman Palace: Courtly Cookery in the Eighteenth Century, in: Faroqhi, S. and Neumann, C. (eds), *The Illuminated Table, the Prosperous House: Food and Shelter in Ottoman Material Culture*, Wurzburg, 127–160.
Nurollah. 2017. *Dining at the Safavid Court [Madatolhayat]*, Costa Mesa.
Pakucs-Willcocks, M. 2018. Spices and Exotic Foods in 17th-Century Transylvania: The Customs Accounts of Sibiu, in: Jianu, A. and Barbu, V. (eds), *Earthly Delights: Economies and Cultures of Food in Ottoman and Danubian Europe, c. 1500–1900*, Leiden, 295–310.
Pinkard, S. 2009. *A Revolution in Taste: The Rise of French Cuisine, 1650–1800*, Cambridge.
Pitte, J.-R. 2002. *French Gastronomy: The History and Geography of a Passion*, New York.
Reindl-Kiel, H. 2003. The Chickens of Paradise Official Meals in the Mid-Seventeenth century Ottoman Palace, in: Faroqhi, S. and Neumann, C. (eds), *The Illuminated Table, the Prosperous House: Food and Shelter in Ottoman Material Culture*, Wurzburg, 59–88.
Revel, J.-F. 1982. *Culture and Cuisine: A Journey Through the History of Food*, New York.
Scappi, B. 2008. *The Opera of Bartolomeo Scappi (1570): L'arte et prudenza d'un maestro cuoco (The Art and Craft of a Master Cook)*, Scully, T. (trans. and ed.), Toronto.
Scully, T. 1995. *The Art of Cookery in the Middle Ages*, Woodbridge.
Stefani, B. 1662/2000. *L'arte di ben cucinare*, Mantova/Bologna.
Watts, S. 2011. Enlightened Fasting: Religious Conviction, Scientific Inquiry, and Medical Knowledge in Early Modern France, in: Albala, K. and Eden, T. (eds), *Food & Faith in Christian Culture*, New York, 105–123.
Wheaton, B.K. 1983. *Savoring the Past: The French Kitchen and Table from 1300 to 1789*, New York.
Woolgar, C.M. 2007. Feasting and Fasting: Food and Taste in Europe in the Middle Ages, in: Freedman, P. (ed.), *Food: The History of Taste, Berkeley*, 163–195.
Wright, C.A. 1999. *A Mediterranean Feast: The Story of the Birth of the Celebrated Cuisines of the Mediterranean, from the Merchants of Venice to the Barbary Corsairs*, New York.
Yerasimos, M. 2007. *500 Years of Ottoman Cuisine*, Istanbul.

Section Three

Tradition as Continuity and Change

Chapter 8

Devotional Prints and Practice: Woodcuts from the Library Cave at Dunhuang[1]

Bernadine Barnes

In both Europe and China, most of the earliest surviving printed images are religious woodcuts. In the West, these date from the 15th century, but in China we have woodcuts that date to around 700 CE – about 700 years earlier. Our knowledge of these prints and how they were used is largely due to their almost miraculous survival in a small cave within the Mogao complex near Dunhuang in western China. The oasis at Dunhuang, located at the point where the northern and southern routes of the Silk Road converge, was a thriving and cosmopolitan city for traders and pilgrims for hundreds of years. It became famous for the hundreds of meditation caves which were carved into the soft rock between the 4th and 14th century; the site was largely abandoned in the 15th century. A mingling of cultures at Dunhuang brought with it a myriad of faiths and religious practices. However, Buddhism was the dominant religion there, and the woodcuts found at Dunhuang give rich evidence of the use of prints in Buddhist practice.

The Chinese woodcuts I will discuss were found in Cave 17, the so-called Library Cave. A stele from the cave states that it was dedicated in 862 as a memorial chapel to the monk Hongbian; a statue of the monk has now been returned to its original location in the cave.[2] However, the more interesting use of the cave for my purposes happened between the 9th and 11th centuries, when the statue was moved out and the cave began to be used to store manuscripts and other objects – eventually approximately 50,000 objects were put into the cave. The cave was sealed around the year 1000, and the doorway was later covered with plaster and paintings that date to the Song dynasty (960–1279). We do not know why the cave was sealed – a number of invasions around that time suggest that it was for protection of the documents, but they were not hastily stashed there; instead, they were neatly rolled and densely packed in the dry and nearly airless cave. Many of the items from the cave are very well-worn religious works, and some scholars believe that the space served as a repository for sacred texts awaiting repair or as a way to bury them away in anticipation of the end of the world.[3] Yet, not all the objects found in the cave were texts – a very significant collection of paintings on silk and many small bronze Buddhist statues were also found. There were also many secular works, including almanacs, star charts, and documents on various mundane matters like boundary disputes, advice for women, and textbooks on writing and math. Sometimes these secular items were used to back or repair later Buddhist scrolls.[4] Sketches and stencils used in making the wall paintings were also found in the cave, and these provide precious information about the artisans and their working methods.[5] Of the documents found in the cave, manuscript scrolls far outnumber the printed works, and even fewer items contain images – only a few dozen exist in various collections.

The cave remained sealed until 1900, when it was rediscovered by Wang Yuanlu, a Daoist monk who had appointed himself abbot of Mogao. The Hungarian-born archeologist, Aurel Stein was the first European to explore the contents of the cave on his second expedition to the area in 1907, funded by the British and Indian governments. Because of the way the objects were brought to Stein and the manner in which he catalogued them (by type of object), it is impossible to know whether the scrolls and packets were placed in the cave in any sort of order; indeed, by the time Stein arrived Wang Yuanlu had repacked the entire cave.[6] Stein convinced Wang to allow him to remove many scrolls (including the famous *Diamond Sutra*) and take them to England; these form the core of the collections at the British Museum and the British Library. The following year the French Sinologist Paul Pelliot visited the cave and secured many more items for French collections. Since then, many other items have been dispersed to collections in Asia, Europe, and the United States.[7]

The techniques and materials used in these prints were ancient even by the time the earliest examples from the cave were made. The process of woodcut printing is very simple. It involves spreading ink on the upper surface of a block that has been carved in relief and then transferring that ink to another surface like paper or fabric. The use of seals is a related form of relief printing; we know that signature seals, made from materials like bronze or jade, were used as early as the 2nd century BCE in China.[8] These or similar stamps could make impressions on many types of materials, including fabric, wax, or earth, and textual sources point to the use of impressions in ritual practice around 600 CE.[9] However, these practices produced ephemeral images, or in the case of printing on fabric only a limited number of works. The widespread use and distribution of woodcut prints depended upon the availability of paper. According to ancient Chinese annals, paper was invented in China in 103 CE so that it could be used for taking rubbings of the Confucian texts that had been carved on stone. However, paper-making certainly existed for two or three centuries before that.[10] By the 2nd or 3rd century very fine paper was being made in several centers in China. It would still be some time before paper was inexpensive enough to allow it to be used extensively for printing; this happened by the 7th or early 8th century.[11]

Images and texts printed from woodblocks would need to be carved into a block of wood in reverse before printing. However daunting this may seem to us, the process was simple and could be carried out by even an illiterate block cutter because the characters would be transferred in reverse to the block from paper; the carver only needed to work around the outlines. The woodblocks could print as many as 15,000 identical copies – but it would be rare to print so many at once; a small run could be printed to test a market before producing many more identical copies. This is very different from using a press with moveable type (known in China by the 11th century), which would require the printer to anticipate how well a work would sell and make that many pages, then reset the type to create the next page; the possibility of error was much greater on both points.[12] Perhaps for this reason – and because of the large number of characters – Chinese books continued to be printed from woodblocks into the 19th century.

In spite of the probability that many hundreds of impressions of printed books and images were made from each block, the fact that they were inexpensive works on paper meant that the prints were often reused or destroyed. Early examples of printmaking – either texts or images or both – are very rare, so the trove found in the Library Cave presents a truly extraordinary collection that shows the range of prints and their uses.

Certainly, the most famous printed work from the Library Cave is a Chinese translation of the *Diamond Sutra*, often considered the world's earliest complete printed book; it is dated 868. The *Diamond Sutra* is one of the most important texts of Mahāyāna Buddhism; it is a relatively short text that was often committed to memory, since by doing so, one could attain enlightenment.[13] It is a dialog between the Buddha and the monk Subhūti; the conversation itself is depicted in the frontispiece (Fig. 8.1). The Buddha faces toward the beginning of the text, although the first words in this scroll are not from the sutra itself, but are instructions to the devotee to purify the mouth and recite certain mantras before beginning.[14] The densely packed composition shows the Buddha in the center surrounded by monks (identified by their bald heads), while delicate apsaras (angel-like figures) float above. Subhūti himself is shown praying below, with his slippers set neatly by his side, while two not very ferocious lions guard the throne. More fearsome are the guardian figures on either side. In the right corner is the patron whose name is recorded in the colophon at the end of the scroll, which states that the sutra was 'Reverently made for universal free distribution by Wang Jie on behalf of [his] two parents on the fifteenth day of the fourth month of the ninth year of Xiantong [11 May 868].'[15] For Buddhists, the accurate reproduction of the text was important, but perhaps more important was the idea that copying and dispersing sacred texts and images brought merit to the makers and patrons. That merit could be transferred to others including deceased relatives.

Artistically, this is a much more sophisticated image than many other printed images found in the cave; the rich composition as well as the skill it took to carve the curving lines of varied thicknesses suggests that a well-trained artist was at

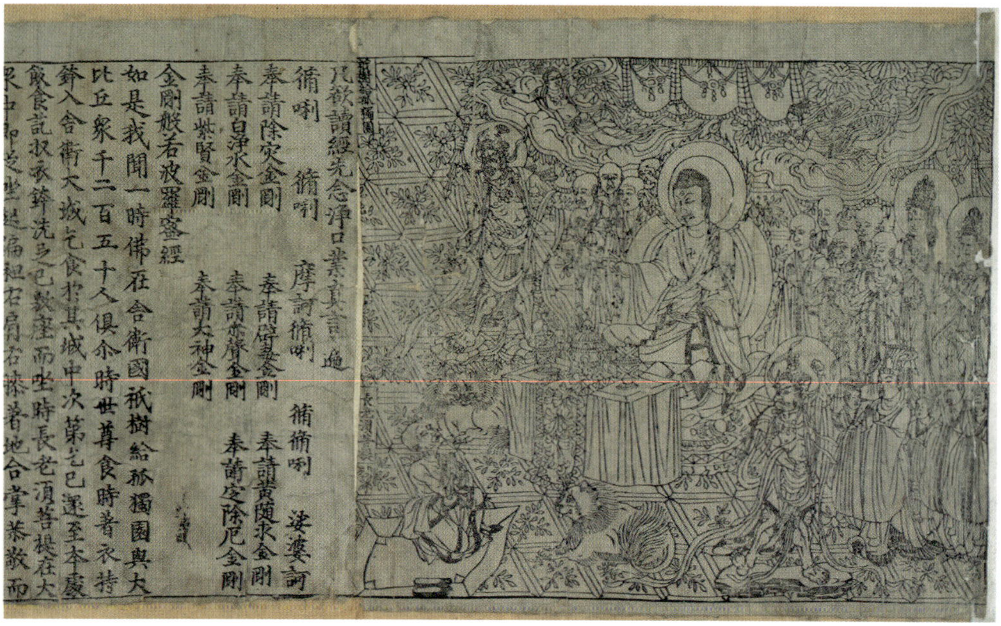

Fig. 8.1. The Diamond Sutra, *frontispiece and text. Woodblock print. Dated 11 May 868. British Library, Or. 8210/P.2. © The British Library Board.*

work. The paper too is very fine and is dyed with three different yellow compounds. Yellow was the sacred, imperial color – it represented solemnity and was thought to be auspicious; but the dyes, made from cork trees, also have insecticidal properties and made the sheets water-repellant. Scholars have concluded that it was probably made in Sichuan province, where the print industry was already well developed; the paper also appears to have been made in southwestern China, near the border of Tibet.[16] Presumably, the owner brought the scroll with him on a pilgrimage to the caves and perhaps left it as an offering. The fact that this is a printed book suggests that it was one of many copies, but only the one example survives. It is not, however, the only copy of the *Diamond Sutra* that was found in the cave; some 500 manuscript copies or fragments survive, including some with colophons that describe why the copy was made: in one case the donor wanted the merit to be given to his ox, another had copies made in thanksgiving for a promotion.[17]

However, not all woodblock prints found in the Library Cave were made primarily to disseminate texts; rather, they are closely tied to ritual practices. Three types of woodblock prints were found in the cave: first, scrolls of repeated Buddha or bodhisattva images with no text at all; second, prayer sheets where an image of a sacred figure is shown above the text; and finally *dhāraṇī*, which give the text of Sanskrit ritual incantations, encircling images of the Buddha or a bodhisattva.

I begin with the repeated images, because they so clearly illustrate the idea expressed in many Buddhist texts, that simply repeating the image of Buddha brings merit, through which believers hoped to escape suffering and rebirth.[18] The repeated motif of the thousand Buddhas is seen in wall paintings throughout the Mogao complex; indeed the first cave is said to have been cut around the year 400 after a monk had a vision of a thousand Buddhas there.[19] Replication in this manner conveys the infinite nature of the Buddha.[20] The scrolls have small, simple images of the Buddha or a bodhisattva, repeated along the length of the scroll (Fig. 8.2). The individual images are very modest in size (about 4–10 cm tall), but the scrolls themselves can be very long; in one case the entire scroll measures more than 26 feet long and contains 2030 impressions of the Buddha.[21] Planning was involved in making these printed scrolls; pages of paper were pasted together first (sometimes the prints go over the seam). The woodblocks are carved in simple outline usually with only one figure; each block would be individually stamped so that it aligned with the others in a vertical column.[22] The scrolls are almost always printed in a single color (black is most common, but there are examples printed in red), and the quality of the printing varies. The paper used on these scrolls is probably locally made; it is not as fine as that used in the *Diamond Sutra* and no examples that I observed are on dyed yellow paper. All these qualities suggest that the scrolls were not made as artistic works for sale or for collecting – instead it is the act of printing that is valued.[23] It is generally thought that these Buddha scrolls were produced on behalf of devotees by monks who lived at Dunhuang. However, the hand-colored example shown in Figure 8.2 provides evidence that a person with some skill finished the scroll. In this scroll, the woodcut impressions are quite faint, so thc outlines were redrawn with ink, and several

Fig. 8.2. Preaching Buddha, *detail of a paper roll in two parts. Woodblock print with added ink and color. Early Tang, second half of 7th century (?). 32.2 cm × 316.5 cm long (combined). British Museum 1919,0101,0.254. © Trustees of the British Museum.*

shades of orange and green were used to fill in the details. Whether the person who finished the roll was a monk or layperson, he probably created it in exchange for an offering.[24] Another Buddha scroll shows clear evidence of devotional use. The fragment of a scroll, now in the British Museum, has a small date inserted after every 21 impressions of Buddha Bhaiṣajyaguru, the Buddha of Medicine.[25] The dates indicate the 8th, 14th, 15th, 23rd, 29th and 30th days of the month, which correspond to the Buddhist fast days. It is likely that this scroll was produced over time, perhaps at intervals corresponding to the dates.[26] This scroll, which has been dated to the late 7th century, was probably kept in a monastery and used to cure illness.

The second type of print found in the Library Cave is the prayer sheet, which has a large image of the Buddha, a guardian figure or a bodhisattva, with texts below; often cartouches on either side of the main figure identify the subject and the patron. This type of sheet is quite common among the woodcuts from Cave 17; scores were found, and there are duplicates among them. Figure 8.3 shows Mañjuśrī, the bodhisattva of supreme wisdom. The inscriptions on either side of the main figure proclaims that the bodhisattva, 'universally urges all to resolve to make offerings to and receive and hold [this icon]'. Below, the devotee is told that this is the 'true countenance' of Mañjuśrī who 'will fulfill all prayers' and 'return [all sentient beings] to eternal bliss'; this is followed by two incantations used to worship him.[27] In one instance four impressions of the Mañjuśrī woodcut were mounted together to form a single unit, as if to multiply the efficacy of the image and the rituals surrounding it (Fig. 8.4). The size of the ensemble may have made it too unwieldy to hold (as instructed in the text). However it was displayed, this ensemble was completed – perhaps decades after the prints were first made – with a small image of Amitābha Buddha, painted in a later style and glued onto the center.[28]

In the next example (Fig. 8.5), the image of the bodhisattva Avalokiteśvara (Guanyin in Chinese) was printed from one block while the text was printed from another, as seen from the slight misalignment of text and image. The image block could be printed separately from the text, and in one instance, nine impressions of the image without the text were joined together, recalling the repeated images found in the scrolls.[29] The ensemble was probably displayed as a kind of banner; other prayer sheets have added strips of color-printed paper at top and bottom, or have tabs at the top, indicating that they may have been hung as funerary banners.[30]

The texts on the Avalokiteśvara prayer sheet are of great interest. The cartouche on the left gives the name of the bodhisattva, calling him 'the Great Compassionate, Great Merciful Savior from Hardship, who Hears the Cries of the World' while on the right we learn that the print was made for Cao Yuanzhong, Imperial Commissioner at Dunhuang from 946 to 975.[31] The lower text gives further details about Cao:

> The Disciple Cao Yuanzhong, Military Commissioner in charge of the Army of Returning to Righteousness, Supervisory Commissioner for the Surveillance of Gua and Sha Districts, and Commissioner for Agriculture and for Western and Other Affairs in the Region, Lord Specially Advanced Grand Mentor and Dynasty Founding Marquis of Qiao Commandery, had

> this printing block carved and offers it in order that the city walls and trenches be secure, and the villages and prefecture towns be peaceful. May the roads to the east and west remain passable, and the trouble-makers in the north and south become obedient.[32]

The prayer goes on to express wishes that diseases disappear and that peace and happiness prevail.[33] It is striking how the patron's identity and military aspirations are recorded on this print. Rather than stating more general desires for merit or freedom from rebirth, the hoped-for effects are more practical: secure walls and trenches, passable roads, and that 'the trouble-makers in the north and south become obedient'.

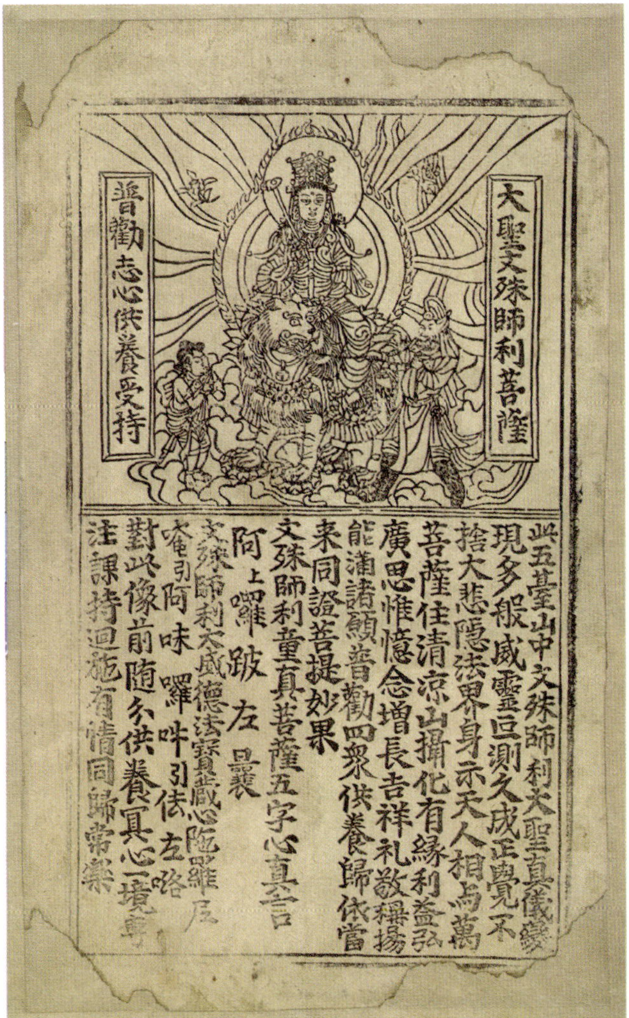

Fig. 8.3. Bodhisattva Mañjuśrī. *Woodblock print. Five Dynasties, 10th century. 27.9 cm × 16.8 cm. British Museum, 1919,0101,0.237. © Trustees of the British Museum.*

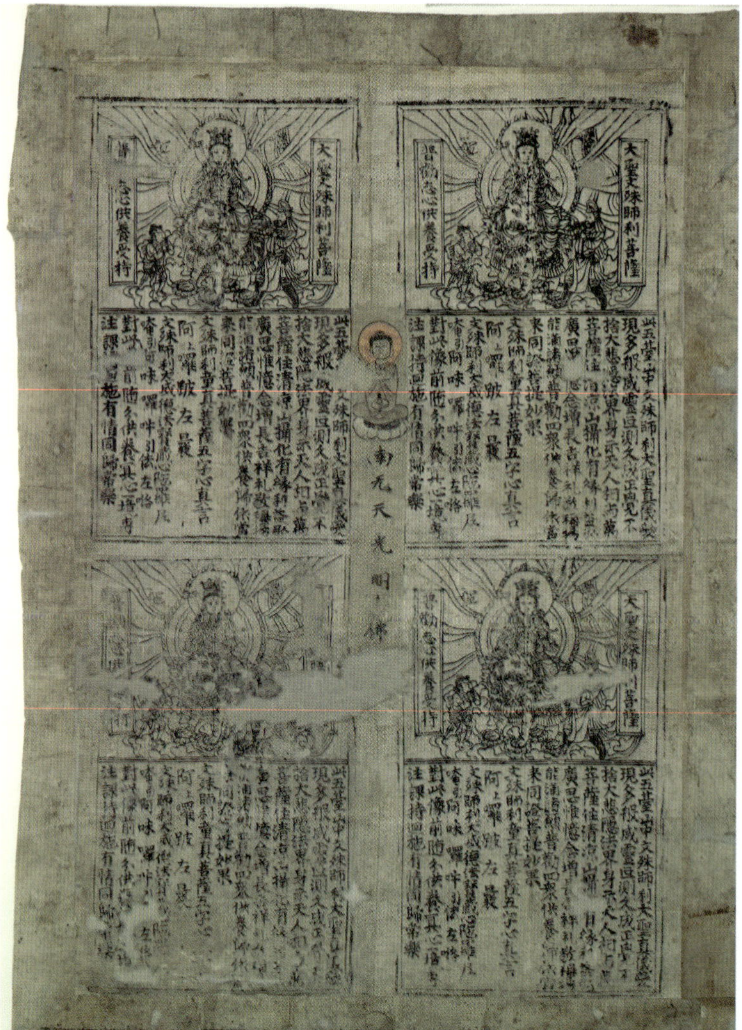

Fig. 8.4. Four impressions of Mañjuśrī prayer sheet, mounted on one sheet, with a small painted figure of Amitābha added later to the central column. Woodblock print with added color on paper. Mid-10th century, Five Dynasties. BM1919,0101,0.239. 63 cm × 46 cm. © Trustees of the British Museum.

This example also provides some unexpected contextual information. At the very end, the inscription states that the block was cut by the workman Lei Yanmei in 947. Another printed *Diamond Sutra* found in the cave (a different one than the one described earlier) ends with a colophon that names the same artisan and is dated two years later. But in the sutra colophon, Lei Yanmei is called a 'woodblock official', who was associated with the frontier commandery government. Sarah Fraser sees these inscriptions as evidence for a Dunhuang academy where professional woodcutters

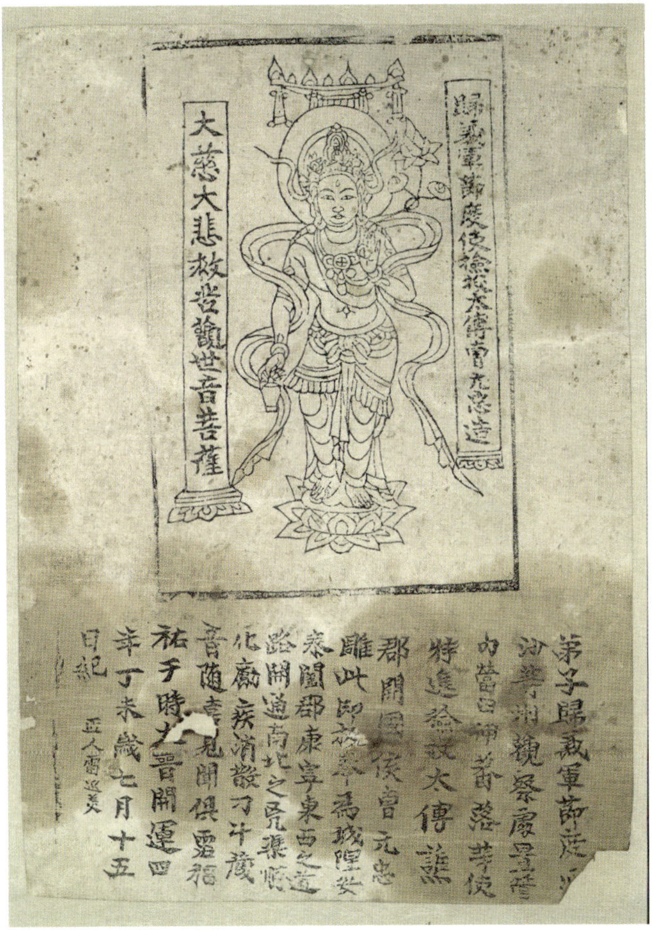

Fig. 8.5. Great Compassionate, Great Merciful Avalokiteśvara Savior from Perils. Dated 947, Five Dynasties. Woodblock print. H: 31.7 cm, W: 20.2 cm. British Museum 1919,0101,0242. © Trustees of the British Museum.

formed complex organizations, with ranks through which they could rise and become so well paid that they could themselves become patrons of art.[34]

The third type of woodcut print is the most complex. These are *dhāraṇīs* or 'charms' which record Sanskrit incantations meant to be voiced by devotees. The words were transliterated rather than translated into Chinese so that the sounds were preserved; believers must have felt that the spell would lose its power if it were translated, so Chinese characters were found that approximated the sound of the Sanskrit spell. The writing itself is ungrammatical and largely incomprehensible in any language; it is meant to move the worshipper beyond seeking the meaning of the words.[35] Such incantations might preface a prayer or a sutra (the *Diamond Sutra* discussed above begins with a *dhāraṇī*, for example).

A *dhāraṇī*, however, is not only a ritual text, whether written or chanted; the power of the spell was often embodied, or transferred to an object or image. Fragments of *dhāraṇī* texts were often placed in small stupas, which were then gathered together in larger stupas; the *dhāraṇīs* and the papers on which they were written were thought of as relics of the Buddha.[36] Other times, wearing the *dhāraṇī* on the body activated its protective power. Statements within the texts of the *dhāraṇīs*, often instruct believers to wear it on their head, arm or around the waist. In 1944, a printed amulet was found in a Tang dynasty burial in Chengdu; the print was contained within a silver armlet worn by the deceased person. Since then the discovery of many other burials has confirmed that the practice of wearing amulets was widespread.[37] Cases for the *dhāraṇīs* range from simple round containers, boxes or tubes that were probably carried in pouches or worn as pendants, to silver armlets. Written or printed on silk or paper, the *dhāraṇīs* within the cases measured between 21 cm square to about 45 cm square. However, not all *dhāraṇīs* were found in cases – some were laid over the corpses in burials, and others give indication of being sewn into clothing. The Chengdu amulet offers further evidence regarding production. The *dhāraṇī* was printed on very thin paper, which was badly deteriorated, but two inscriptions could still be read: 'Dragon Pool Workshop' and 'spell text printed for sale'.[38] These brief inscriptions give firm evidence that there was a popular market for *dhāraṇīs*. Early *dhāraṇīs* sometimes include spaces for the believers to write their own names or draw a picture of whatever they wished to receive. Later examples, like the one shown in Figure 8.7, include the patron's name in the printed text.[39]

The *dhāraṇī* woodcuts found in the Library Cave were preserved with other scrolls and sheets, without the contextualizing information of the archeological sites. However, the two examples I illustrate here do refer to the practice of being worn on the body; they are good examples of the range of *dhāraṇī* woodcuts especially in the mid-10th century. The first (Fig. 8.6) is quite small and simple. The image shows Amitābha Buddha seated on a lotus throne, surrounded by Sanskrit text with Chinese writing along the left side. Even though *dhāraṇīs* are by nature untranslatable, Gergely Hidas has produced a rendering that will give modern readers a sense of the words. The Sanskrit text reads:

> Veneration to the Three Jewels. Veneration to the noble Amitābha, the Tathāgata, the Arhat, the Perfectly Awakened One. Namely, *oṃ* O Immortality, O the One Arisen from Immortality, O Immortality-born, O Immortality-child, O Immortality-perfect, O Immortality-power, O Immortality-valour, O the One Acting by Immortality-valour, O Immortality-sky-fame-maker, O Immortality-kettledrum-sound, O the One who Accomplishes all Aims, O Destroyer of all Defilements originating from [bad] Actions *svāhā*. *oṃ* O Light of the Jewel-lotus that is the Great Seal of the Unfailing Vairocana advance *hūṃ*.[40]

Even reading the English translation, you can almost hear the sound of a monk chanting the words, with repeated O's and litany-like repetitions of the Buddha's name.

The Chinese text that runs along the left side of the print is quite different:

> This Great Vow [spell] of the Infinite Life [Buddha] is enormous and extensive. The Wish-fulfilling [spell] is [like] whatever your heart wishes, it will necessarily follow. The Buddha Eye Mother [spell] is extraordinarily auspicious. The Consecrated Light [spell] can destroy the bad paths [of rebirth]. The dragon spirits [will] protect the place where the Ucchuṣma-*mantra* is put. As for the Verse of the Dharma-body, those who wear it at the waist will be equal to the Buddhas. The four assemblies are universally encouraged to keep and wear this [amulet] to create a karmic basis [for a good future] and it is also avowed that they [will] ascend together to the true and eternal wonderful fruit.

The Chinese text names and explains the effects of the various spells, and then describes how the print should be worn as an amulet at the waist of the devotee. The reference to the 'consecrated light' of the *dhāraṇī* helps to explain why the text surrounds the image, like a halo. The text is not just a record of words to be spoken, but also a representation of the light that surrounds the Buddha and extends in all

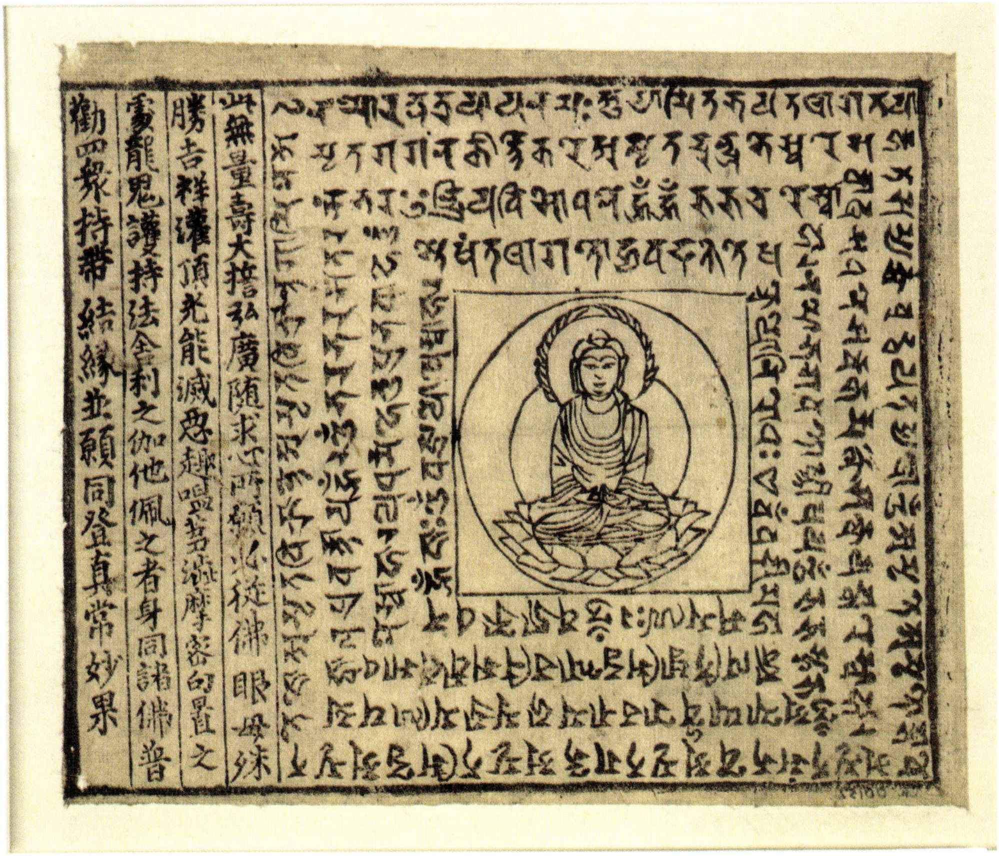

Fig. 8.6. Amitābha, seated within a square of Sanskrit text (dhāraṇī) with Chinese inscription on the left. Woodblock print. Mid-10th century. Five Dynasties. H: 13.7 cm, W:16.7 cm. British Museum, 1919,0101,0.247. © Trustees of the British Museum.

directions through the cosmos.[41] Although earlier manuscript *dhāraṇīs* also had a central image surrounded by text, they more often took on the character of a mandala – a representation of the universe meant to guide ritual practice. This woodcut and others like it were instead probably objects of contemplation.[42]

A much larger and more elaborate example of a *dhāraṇī* is dated the 24th day of the seventh month 980 (Fig. 8.7). In the cartouche to the left of the central image, the donor is named – Li Zhishun – and in the cartouche to the right, the name of the woodblock carver is given as Wang Wenzhao. This *dhāraṇī* is an example of the 'Incantation of Wish Fulfillment', a popular spell that is based upon a late 7th-century translation of the *Mahāpratisarā Dharani Sutra* attributed to the monk Baosiwei. A translation of the text begins:

> *The Great Dharani of Wish Fulfillment.* If one practices this spirit-incantation, one's position will be victorious. If there is one who can copy it out and carry it on his head or on his arm, he can complete all the meritorious acts of utmost purity. He will ever be held in the protective embrace of gods and dragon kings and will be remembered and kept in mind by all buddhas and bodhisattvas. This spirit-incantation can bestow the utmost security and bliss upon all beings.[43]

At the center is the bodhisattva Mahāpratisarā, 'the great accorder of what is sought', who holds eight attributes: pestle, axe, lasso, knife, jewel, wheel, lance, and book. He is surrounded by the *dhāraṇī* written in 19 concentric circles of Sanskrit text, which create an even more dramatic radiance around the bodhisattva. As with the smaller *dhāraṇī* woodcut, the inscription tells the devotee to wear the amulet on the body ('... carry it on his head or on his arm ...') to be protected by its power.[44] The Sanskrit text is so small that that it would be difficult to read; it is unlikely that anyone would have tried to use the print in that way. The skill required to carve the block is impressive, and there is further technical innovation in the use of a second block to overprint the eight Sanskrit seed syllables at the outer edges. Several very similar prints were found in the Library Cave, suggesting that they were produced for wide distribution. It is likely that the prints were made as votive images for lay worshippers, in exchange for offerings to the monastery.[45]

I will conclude with a few thoughts about the value of thinking about these prints, especially for someone who is not a specialist in Chinese or Buddhist studies. Although some of these prints – especially the frontispiece of the *Diamond Sutra* – are sometimes mentioned in histories of printmaking, little attention is given to them. It is worth remembering that these complex prints were made at least 500 years before anything like them appeared in Europe. Even more importantly, these prints and the contextual information surrounding them provide very specific details about how religious prints were used (worn on the body, for example) and how the making of them might itself be part of a ritual (as with the repeated stamping seen in the Buddha scrolls). Is it possible that the earliest makers of religious prints in Europe felt that they too were engaged in an enterprise that would similarly bring a spiritual benefit? Many of the woodcuts discussed here

Fig. 8.7. Mahāpratisāra Bodhisattva *with text of* Da Sui qiu tuoluoni. *Dated 980 CE. Northern Song. H: 41.7 cm, W: 30.3 cm. British Museum, 1919,0101,0.249. © Trustees of the British Museum.*

Fig. 8.8. Christ on the Cross, *hand-colored woodcut with gold-leaf on vellum. German, 1485. 32 cm × 19.8 cm (sheet). Ruth and Jacob Kainen Collection, Gift in Honor of the 50th Anniversary of the National Gallery of Art, 1990.19.1. Courtesy National Gallery of Art, Washington.*

provide 'instructions to the user' telling devotees to wear the image or to recite an incantation. Only recently have scholars of western printmaking noticed similar sorts of instructions or indications of use in the works they study.[46] At Dunhuang, we have an extraordinary time capsule of visual and print culture unlike any seen in the West until well into the 16th century; however, at Dunhuang, as in Europe, what survives is only a sampling of the printed images that were made at the time. Whether we are talking about Chinese or European prints, we must acknowledge that hundreds – perhaps thousands – of such printed images have been lost. Why any prints survive from either area is truly fortuitous, and the circumstances surrounding their survival are worth attending to. Most 15th-century German woodcuts, for example, were preserved only because they were pasted into covers of books held in certain monasteries, and French woodcuts survive because they were kept in lids of traveling boxes, perhaps carried by merchants or pilgrims to serve as portable altars.[47] Furthermore, what is preserved in print collections today is there partly because collectors found them beautiful – Western examples are often colored, sometimes with added gilding, and visually 'enframed' as if they are inexpensive paintings (Fig. 8.8). We should always keep in mind that collecting prints is a culturally specific, meaningful act. The processes of making and using religious prints might be something else entirely.

Notes

1 I wish to thank the organizers of the conference, *The Silk Roads: From Local Realities to Global Narratives*, held at Wake Forest University in March 2019, where a version of this paper was first presented. Funding for my research was provided by Wake Forest's Art Department and the Charlotte C. Weber fund. Susan Whitfield, Ann Farrer, Yu-Ping Luk, and Michele Wang very generously shared their knowledge of the prints, Dunhuang, and the IDP. I am also grateful to my colleagues Jay Ford and Di Luo for their help with questions about Buddhism and medieval China.

2 Whitfield *et al.* 2000, 85; Fraser 2004, 4.

3 Rong Xinjiang 1999, 247–275; Fraser 2004, 4–6.

4 Whitfield *et al.* 2000, 47; Whitfield 2004, 231–257. For the almanacs, see Whitfield 1998, 4–22.

5 On the stencils and sketches, see Fraser 2004, 102–113.

6 Stein 1922, 2, 801–813. For an accessible account of Stein's encounter with Wang Yuanlu and his access to the cave's contents, see Wood and Barnard 2010, 32–39.

7 A catalog of an exhibit at the Bibliothèque Nationale in Paris discusses many examples from the Pelliot collection: Cohen and Monnet 1992.

8 Tsien 1985, 5–6, 136–139; Ledderose 2000, 159–160.

9 *The Annals of the Sui Dynasty*, from *c.* 600 describes how Daoist priests would 'make "seals" [yin] of wood upon which are carved the constellations, the sun, and the moon. Holding their breath, they take them with their hands and make impressions [yin]. There are many whose ailments are cured thereby.' Quoted in Loehr 1968, 2 with n. 6. Stamps were also made on the skin of sick or dying people, and when printed on paper, were used as talismans or burnt and swallowed to cure illness; see Ledderose 2000, 160 with ns. 50–52. On images impressed in sand or silt, see Tsiang 2010, 205.

10 Tsien 1985, 1–2.

11 Loehr 1968, 1; Tsien 1985, 146–151.
12 Blair 2011, 351. On the development of moveable type in China, see Ledderose 2000, 151–152.
13 Van Schaik 2011–2012, 2.
14 Tsiang 2010, 207.
15 Whitfield 2004, 301.
16 Whitfield 2004, 295, 301; Wood and Barnard 2010, 67; Helman-Ważny 2011–2012, 7. A film produced at the British Library by the International Dunhuang Project, and narrated by Susan Whitfield provides an excellent description of the restoration process.
17 Morgan and Walters 2011, 138–140.
18 Shen 2019, 6–8.
19 Whitfield 2004, 272.
20 Shen 2019, 7–8 and chapter 3.
21 Cohen and Monnet 1992, no. 24, Pelliot chinois 3938.
22 Farrer 2007. There is a related practice in Japan as early as the 12th century, where scrolls of repeated stamped Buddhist figures were placed inside carved wooden statues. However, in Japan, each carved block had multiple figures and they were printed in horizontal rows.
23 Tsiang 2010, 205.
24 Farrer 2007; Tsiang 2010, 203; Whitfield in von Spee 2010, 62.
25 Whitfield, in von Spee 2010, 61. British Museum 1919,0101,0.256. The scroll in four parts is about 33 cm tall × 563.5 cm long (about 18.5 feet long).
26 Shen 2019, 80–81.
27 Translations in Copp 2008, 264.
28 British Museum, 1919,0101,0.239.
29 British Museum, 1919,0101,0.244. The ensemble measures 63 cm high × 42 cm wide.
30 Shen 2019, 49–51.
31 On Cao, see Rong 2004, 57–62; and Fraser 2004, 37. Another woodcut commissioned by Cao Yuanzhong is discussed in Cohen and Monnet 1992, 53–57. There are 14 impressions of this prayer sheet to Vaisravana in the Bibliothèque Nationale, Paris.
32 Translation by Tsiang 2010, 216.
33 Rong 2004, 62; Shen 2019, 27–29. Shen emphasizes how carving texts on woodblocks ensured their accuracy and efficacy.
34 Fraser 2004, 28–30, 34. The later sutra is dated 14 June 949 or 3 June 950.
35 Copp 2014, 3–5.
36 Tsiang 2010, 209–212; Copp 2014, xiv–xvi, 33–36.
37 Copp 2014, 61–63. The other burials where similar amulets have been found are listed in Copp 2014, 233–237. Written references show that the practice of wearing enchanted armlets or cords existed in China by the early 5th century, and it derives from a much more ancient Indic practice of wearing enchanted cords; Copp 2014, 79–80.
38 Copp 2014, 59, 74–79 on the cases.
39 Copp 2014, 118.
40 Hidas 2014, 111–112.
41 Copp 2014, 120–122.
42 Copp 2014, 98–119.
43 Translation in Copp 2014, 61. The author summarizes the history of the spell on 64–74.
44 Copp 2014, 61.
45 Fraser 2004, 155–158, 277, n. 27; Whitfield in von Spee 2012, 68.
46 Areford 2010, chapter 2.
47 For examples, see Parshall *et al.* 2005, 171–172, 193–195.

Bibliography

Areford, D.S. 2010. *The Viewer and the Printed Image in Late Medieval Europe*, Farnham, England/ Burlington, VT.

Blair, A. 2011. Afterword: Rethinking Western Printing with Chinese Comparisons, in: *Knowledge and Text Production in an Age of Print: China, 900-1400*, Leiden, 349–360.

Cohen, M. and Monnet, N. 1992. *Impressions de Chine*, Paris.

Copp, P. 2008. Altar, Amulet, Icon Transformations in *Dhāra P* Amulet Culture, 740–980, *Cahiers d'Extrême-Asie* 17, 239–264.

Copp, P. 2014. *The Body Incantatory: Spells and the Ritual Imagination in Medieval Chinese Buddhism*, New York.

Farrer, A. 2007. Early Devotional Prints in China: Technique, Format and Function, unpublished lecture, Budapest.

Fraser, S.E. 2004. *Performing the Visual: The Practice of Buddhist Wall Painting in China and Central Asia, 618-960*, Stanford.

Helman-Ważny, A. 2011–2012. A Preliminary Study of the Paper of the *Diamond Sutra* (Or.8210/P.2), *IDP News* 38, 6–7.

Hidas, G. 2014. Two *Dhāras*, Prints in the Stein Collection at the British Museum, *Bulletin of the School of Oriental and African Studies, University of London*, 77, 105–117.

Ledderose, L. 2000. *Ten Thousand Things: Module and Mass Production in Chinese Art*, Princeton.

Loehr, M. 1968. The Beginnings of Printing in the Far East, in: Loehr, M., *Chinese Landscape Woodcuts from an Imperial Commentary to the Tenth-Century Printed Edition of the Buddhist Canon*, Cambridge, MA, 1–18.

Morgan, J. and Walters, C. 2011. *Journeys on the Silk Road: A Desert Explorer, Buddha's Secret Library, and the Unearthing of the World's Oldest Printed Book*, Guilford, CT.

Rong Xinjiang. 1999–2000. The Nature of the Dunhuang Library Cave and the Reasons for Its Sealing (trans. V. Hansen), *Cahiers d'Extrême-Asie* 11, 247–275.

Rong Xinjiang. 2004. Official Life at Dunhuang in the Tenth Century: The Case of Cao Yuanzhong, in: Whitfield, S. and Sims-Williams, U. (eds), *The Silk Road: Trade, Travel, War and Faith*, London, 57–62.

Shen, H. 2019. *Authentic Replicas: Buddhist Art in Medieval China*, Honolulu.

Stein, A. 1921. *Serindia: Detailed Report of Explorations in Central Asia and Westernmost China,* 3 vols., Oxford.

Tsiang, K.R. 2010. Buddhist Printed Images and Texts of the 8th–10th Centuries: Typologies of Representation and Replication, in: Kapstein, M.T. and van Schaik, S. (eds), *Esoteric Buddhism at Dunhuang: Rites and Teachings for This Life and Beyond*, Boston, 201–252.

Tsien, T. 1985. *Science and Civilisation in China: Vol. 5: Chemistry and Chemical Technology, Part 1: Paper and Printing*, Cambridge.

Van Schaik, S. 2011–2012. The *Diamond Sutra*: History and Transmission, *IDP News* 38, 2–3.

von Spee, C. 2010. *The Printed Image in China from the 8th to the 21st Century*, London.

Whitfield, R., Whitfield, S. and Agnew, N. 2000. *Cave Temples of Mogao: Art and History on the Silk Road*, Los Angeles.

Whitfield, S. 1998. Under the Censor's Eye: Printed Almanacs and Censorship in Ninth-Century China, *The British Library Journal* 24, 4–22.

Whitfield, S. 2004. *The Silk Road: Trade, Travel, War and Faith*, London.

Wood, F. and Barnard, M. 2010. *The Diamond Sutra: The Story of the World's Earliest Dated Printed Book*, London.

Chapter 9

Dome of Heaven: From the Lantern Ceiling to the Chinese Wooden Dome[1]

Di Luo

Alexander Soper identified a specific type of ceiling, which he called the '*laternendecke*' or lantern ceiling, in his article 'The "Dome of Heaven" in Asia' published in 1947.[2] He described the lantern ceiling as highly distinctive of its 'square-and-diamond' shape. Before Soper, the lantern ceiling had been examined by art historians including Albert Grünwedel, Josef Strzygowski, and Albert von Le Coq, whose studies revealed similarities between ceilings found in China's Western Regions, North India, Pakistan, and the West.[3] Based on this body of scholarship, Soper investigated the lantern ceiling in an even broader geographical scope and under the theme of the Dome of Heaven whereby a range of distinctive ceiling types could be brought together and viewed in comparison. Writing in response to Karl Lehmann's earlier article, 'The Dome of Heaven', Soper attempted to expand Lehmann's study of the symbolism of the Western dome by adopting a similar conceptual framework for Asian examples.[4]

Soper's article has since received criticisms and revisions, notably from Oleg Grabar and Nancy Steinhardt, especially regarding the implicit Eurocentric and diffusionist view in his study.[5] Using the Western dome as a point of departure and the Asian dome as a comparison might be problematic methodologically, but this narrative does not necessitate a diffusionist view. Soper never declared in the article that domes in Asia sprang from some Western prototype, but his knowledge of Western examples afforded him a frame of reference within which Chinese and other non-Western domes could be compared. Advancing from the known to the unknown, from the familiar to the unfamiliar, is how the human brain works. Notwithstanding the criticisms, Soper's article remains a seminal piece in the study of Asian architecture. It instills an awareness of the underlying connection between Eastern and Western ceiling forms and technologies, which must not be ignored when studying either tradition. Soper's model of the lantern ceiling, of its highly identifiable square-and-diamond shape and its wide application across Asia, sheds light on the history of Chinese architecture.

Following a west–east trajectory, Soper related the lantern of a Hindu temple at Pandrethan, Kashmir, with the ceilings of the Buddhist cave-temples at Kizil and Mogao in China and the stone-chamber tombs in Korea (Fig. 9.1). The visual similarity between his examples is immediate: the square ceiling contains a second square, which is turned 45 degrees (hence the diamond) and placed on top of the first square. Often, additional squares are added, each rotated 45 degrees and decreased in size, forming a multi-layered ceiling space. The superimposition of the squares can go on forever theoretically; in practice, it helps to contract the ceiling into a smaller opening while increasing its height. However, Soper was not aware, due to the lack of available

Fig. 9.1. Ceiling plan, Hindu temple at Pandrethan, Kashmir, 10th century (drawing by author after Cunningham 1848).

material evidence at the time, that the square-and-diamond shape also permeates the domes of Chinese wooden architecture. The Fengshengsi dome that he discussed is, in fact, a later adaptation of the lantern, though the square-and-diamond form in this case is not immediately identifiable.

A better example of the wooden lantern in China is the Jingtusi (Pure Land Monastery) in Yingxian, Shanxi Province. The ceiling of the Main Hall is divided into nine *zaojing* 藻井, which are sunken panels or coffers in the ceiling (Fig. 9.2). All of the *zaojing* are square-and-diamond in shape. The diamonds in the central row each contain an octagon, while those in the front and the back either look slightly 'squashed' or are transformed into hexagons. No two *zaojing* are identical, but they are all based on the same square-and-diamond layout that can be identified as lanterns and examined under the theme of the Dome of Heaven. The use of the golden dragons and phoenixes, the red sun, the lotus flower, and the miniature pavilions as ornamentation indicates the importance of celestial symbolism, as is appropriate for

Fig. 9.2. Ceiling plan, Main Hall of Jingtusi, Shanxi Province, China, 1124 (drawing by author).

this building. Dated to 1124 in the Jin dynasty, the monastery was established under an imperial decree, and the building was probably converted from the ancestral shrine of a Later Tang emperor.[6]

The Jingtusi lantern, unique though it is, proves to be a highly popular ceiling type in North China from 1124 to the 17th century. It was adopted for imperial structures especially temples and palaces, the most famous example being the Hall of Supreme Harmony (rebuilt 1695–1697) in the Forbidden City in Beijing. The twin lanterns from the Zhihuasi (1443/1444), now at the Philadelphia Museum of Art and the Nelson-Atkins Museum of Art respectively, and a lantern from the Longfusi (1452), also incorporated the distinctive square-and-diamond shape. The charm of the lantern was so widespread and penetrating that its reverberations can still be felt in modern times. The Chiang Kai-shek Memorial Hall in Taiwan, completed in 1980, has a ceiling that is clearly modeled on Ming and Qing examples. At the Metropolitan Museum of Art in New York City, one finds a mock lantern, reminiscent of those plastered lanterns at the 5th-century Mogao Caves, in one of the galleries that host Chinese Buddhist statues.

The Dome of Heaven model helps to illuminate the spiritual meanings of these lanterns, yet it fails to establish a historical relationship between Chinese and non-Chinese lanterns beyond formal resemblance and celestial symbolism. The structure of the square-and-diamond, its technology, source, and associated building practice, remains largely unexplored, but such information is crucial to understanding the significance of the lantern and the modes of transmission of this and other ceiling types across cultures and will provide new perspectives on the nature and history of Chinese and Asian architecture.

The wooden dome is itself a technological conundrum. Unlike a masonry dome, a wooden dome comprises wooden ribs placed across the ceiling space, and is essentially part of the post-and-lintel structure. However, the 'vaulting' of the ceiling is rather at odds with the rectilinear Chinese roof structure and the curvature of the roofline. It almost feels like a foreign structure, an intrusion, to the otherwise largely homogenous timber-frame building. The wooden dome is a fake dome: it does not provide a technical solution to the problem of spanning over a large interior space as real domes do. Why, then, did it prevail in Chinese architecture? Why was the square-and-diamond shape adopted? What made it an enduring pattern?

The current study owes much to Nancy Steinhardt, who has researched and published multiple works on ceilings in and beyond China, addressing many of the core questions presented here. In a book published in 2014, she gives a thoughtful overview of the half-century-long debate on the Dome of Heaven, adding new archaeological evidence to the discussion, which refreshes our view and greatly expands our knowledge of the subject. She maintains that the architecture of Han China (206 BCE–220 CE) was as likely a source as Roman domes, and that Chinese domes, including the lanterns, were recognized as such a powerful symbol of rulership and imperial grandeur that nomads and semi-nomads, who later conquered part of the Han territories, consciously adopted them in their own building practices in emulation of the Han model. The first dated lantern in China, as Steinhardt has correctly identified, is that of the Jinguyuan tomb (7–23 CE) near Luoyang, Henan Province. She suggests that the lanterns found in Goguryeo (4th–6th centuries), Penjikent (8th century), and the Bamiyan Valley (ca. 400–800?) could have been based on the Han rather than Roman models, thus reversing Soper's eastward trajectory of diffusion.[7] Steinhardt's diffusion theory is substantiated by firmly dated examples, but like Soper, the links that she provides between lantern ceilings across Asia heavily rely on formal resemblance and celestial symbolism whereas the difference in building technology is underexamined.

Steinhardt's study further points to the relationship between ceiling and building. Octagonal architecture, from the eight-sided tombs of the Han dynasty to the wooden and brick pagodas and wooden pavilions of later dynasties such as the Liao (947–1125), was charged with cosmic symbolism where an octagonal dome of heaven was often incorporated. The octagon, she suggests, was created to approximate the circle: it was to intermediate between the square earth and the round heaven.[8] This point is expanded by Jun Hu in his examination of the domical structures in China and Nara Japan.[9] It leads to the question if the octagonal room in some way inspired the octagonal dome. This is interesting because the Jingtusi lantern also contains an octagon, and the wooden domes in China are more generally described as 'octagonal' rather than 'square-and-diamond' in shape. But as I will clarify in the following, the octagon, like the diamond, is derived from rotating and superimposing the square. It was a later development and a more elaborate form based on the square-and-diamond.

The octagonal dome is an important characteristic of Liao architecture.[10] The oldest example, the Guanyin Pavilion (984) at the Dulesi in Jixian, Tianjin, is a two-storied wooden pavilion with an octagonal dome above the 16 m tall statue of Bodhisattva Guanyin. The dome is an eight-sided cone resting on a square base, its eight wooden ribs rising from the square base and converging at the apex of the cone. Between the eight ribs are wooden panels covered by a certain lattice pattern, its horizontal and diagonal lines echoing the ribs and the beams in direction. Several other notable Liao structures, namely the Sutra Library (1038) of the Huayansi in Datong and the Yingxian Wooden Pagoda (1056), include octagonal domes over the Buddha statues. In the south, the octagonal dome appears in the Baoguosi Main Hall (1013, Northern Song dynasty) in Ningbo.

The octagonal dome is termed as the *douba zaojing* 鬬八藻井 in the *Yingzao fashi* 營造法式, the building standards promulgated by the Northern Song court in 1103, which includes a section on the technical specifications of the dome. The text prescribes that the dome should be a tripartite structure composed of a square base (*fangjing* 方井) roughly 8 feet per side, an octagonal frame (*bajiaojing* 八角井) in the middle, about 6 feet across, and lastly an eight-ribbed cupola (*douba*) of 4 feet across at the top. The total height is about 5 feet.[11] However, the octagonal dome was only sparingly used during the 10th and 11th centuries, and as Steinhardt has observed, it rarely appeared in Northern Song structures.[12] In fact, some of the grandest Liao examples, including the Fengguosi Buddha Hall (1019), and other large-scale Song and Jin structures, the Jinci Sage Mother Hall (1038–1087) and the Chongfusi Amitabha Hall (1143), for instance, used no ceilings but opted for an 'exposed roof structure (*cheshang mingzao* 徹上明造)'.

On the other hand, the lantern ceilings of the Jingtusi, Zhihuasi, and Longfusi display more divergence from, rather than adherence to, the octagonal dome. The ribs do not conjoin at the center but rotate around it, forming a diamond pattern placed across the square base. In the case of the two Ming examples, the ribs intersect, generating a complex web of squares, diamonds, octagons, and triangles around the central round plate where a dragon coils up its serpentine body, its head suspended from the center of the ceiling and its eyes looking straight into the ground. The lantern does not assume a conical form because of the lack of vertical ribs and is shallower than the octagonal dome. It belongs to a building system that differs fundamentally from the post-and-lintel structure.

The Stone Lantern

The lantern emerged no later than the 1st century CE in China. Han-dynasty tombs yield plenty of visual evidence suggesting that it was a prevalent type of ceiling used in funerary, and probably residential, structures across China at the time, from the north to the east coast and the southwest. By contrast, the other type, the octagonal dome, was rarely if ever found.

The Jinguyuan tomb mentioned above contains a pair of lanterns in the rear chamber. The lanterns are not actual structures but are painted in red, blue, and black directly on the plastered stone ceiling. Each contains a square, a diamond, and a square from the outside to the inside, creating a square-diamond-square (SDS) sequence of superimposition. The squares and diamonds are double-lined, suggesting that these represent wooden ribs used to construct the lantern. The rear chamber, though built of stone bricks, mimics a timber-framed house with a large gable roof supported on beams and columns.[13]

The celestial symbolism of the Jinguyuan lanterns is consistent with that of Han-dynasty funerary art.[14] The south lantern encloses a red sun with a black bird, its four outer triangles covered by scrolls of clouds evocative of the *qi*-energy depicted in earlier Han tombs such as on the Mawangdui silk banners. On the other end of the roof ridge, the north lantern, though much of the painting has deteriorated, encloses a moon inhabited by probably a toad, its outer triangles similarly filled with scrolls of clouds. Painted in the space between the lanterns are spirit animals (dragons, phoenix, serpents, tiger) roaming in the clouds, passing through the holes of eight *bi* 璧 jade disks that are, interestingly, arrayed in a diamond formation. The ceiling is much simpler in structure than the Jingtusi lantern, but the celestial symbolism is in full force: it is a 'spyglass' leading the searching eye, and perhaps the ascending *hun* 魂 soul of the deceased, toward the celestial dome.

The SDS lantern and the sun-and-moon motif resurfaced in Han cliff tombs in Qijiang, Sichuan Province. Bailingpo Tomb 1 (117 CE) is a large multi-chambered tomb where the lantern appears in at least four chambers.[15] Songlinzui Tomb 1, Fentaizui Tomb 1, and Zijingwan Tomb 1, to name but a few examples, have single-lantern square chambers and/or double-lantern rectangular chambers. Moving forward to the 3rd century, Dahuting Tomb 2 in Mixian, Henan Province, switches from the gable roof to the barrel vault where the SDS lantern still lingers. The lotus flower is introduced to the visual vocabulary of ceiling decoration and is interspersed with the lantern while another new element, the lattice, now fills the center of each lantern, replacing the sun-and-moon motif.[16]

The SDS lantern, the lotus, and the lattice make their way to another Eastern Han dynasty tomb, Yinan Tomb 1, a masonry tomb in Shandong Province contemporaneous with Dahuting Tomb 2 (Fig. 9.3). This is a key example of the stone lantern in China. The vaulting technique is distinct: no gable roof, no barrel vault, but a corbeled (*diese* 疊澀) ceiling made of multiple courses of stone, each course suspended slightly from the one below, forming a tapering sunken space in the ceiling. From the outside, the roof looks like a group of stepped pyramids with varying heights. Four SDS lanterns appear in the four front chambers of the tomb. Here they are not painted or carved, but are built following the corbeling method and are part of the load-bearing roof. Two of the lanterns contain, respectively, a lotus flower and a lattice carved on the capstone. The rest of the tomb is covered by a corbeled ceiling with carved lanterns. This creates a rather intriguing coexistence of the built lantern and the carved lantern, one functional and the other decorative. The inclusion of the lattice suggests this is yet another case where underground architecture consciously simulates wooden domestic buildings, but the lantern, unlike any other example from the Han tombs, has been translated into a structurally authentic masonry lantern integral to the corbel vault.[17]

The lotus is customarily identified as a Buddhist motif imported from India and Central Asia to China during the Han dynasty. The interpretations of the lotus in the ceiling by Eastern Han and later scholar-officials, however, appear largely secular, or at least non-Buddhist. In the *fu* 賦 rhapsody, a highly stylized and prestigious literary genre of the Han dynasty practiced exclusively by high court officials, the 'inverted lotus-flower planted in the ceiling' becomes almost a trope associated with majestic wooden buildings. Zhang Heng's 張衡 (78–139) 'Western Metropolis Rhapsody' (*Xijing fu* 西京賦), Wang Yanshou's 王延壽 (ca. 140–165) 'Rhapsody on the Numinous Hall of the Lu State' (*Lu Lingguangdian fu* 魯靈光殿賦),' and He Yan's 何晏 (ca. 195–249) 'Rhapsody on the Jingfu Hall (*Jingfudian fu* 景福殿賦),' for instance, all mention the lotus planted in the *zaojing* or *fangjing* of the halls.[18] The Numinous Hall, a structure attributed to Prince Liu Yu 劉餘 (?–128 BCE) of the Western Han dynasty, was in ruins when Wang composed the rhapsody, which suggests that the lotus-in-ceiling had become an important and already highly symbolic architectural motif by then. None of the literati associates the lotus with Buddhism, and indeed its appearance in the Numinous Hall predates the official introduction of Buddhism to China in

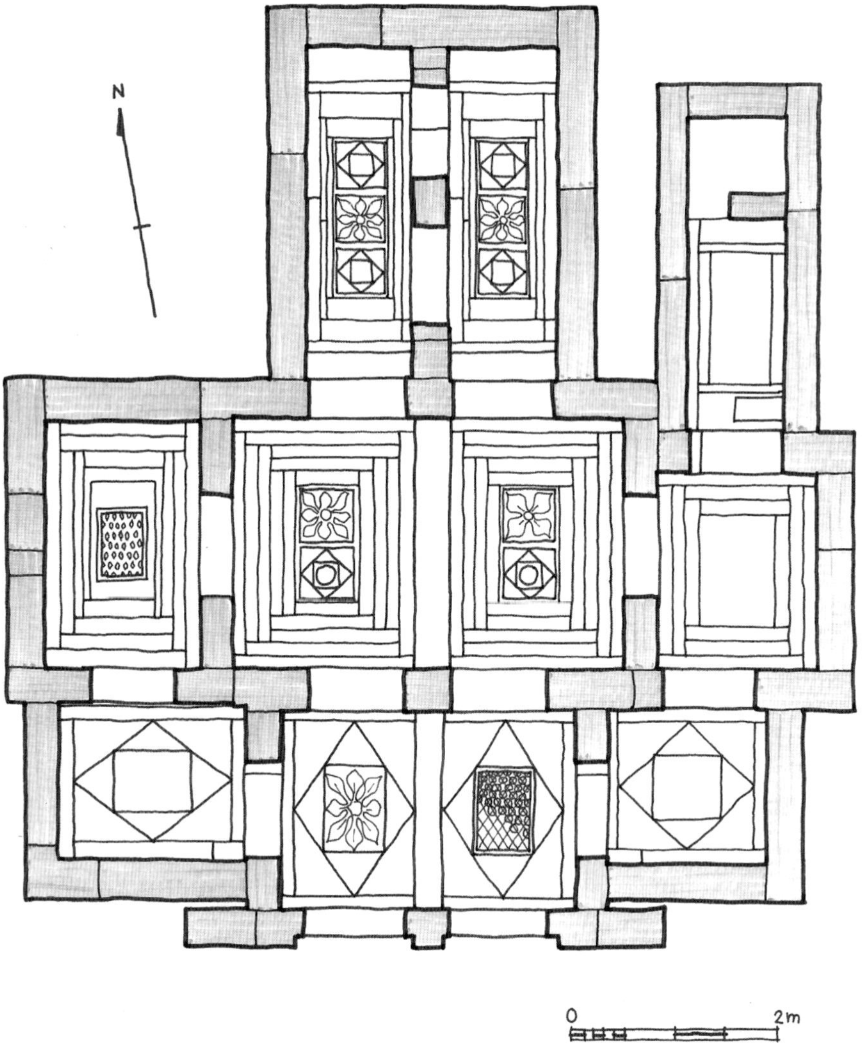

Fig. 9.3. Ceiling plan, Yinan Tomb 1, Shandong Province, China, 3rd century (drawing by author after Zeng, Jiang, and Li 1956).

67 CE. Instead, the poets and later annotators all emphasize that the lotus is an aquatic plant and that the ceiling has a strong affinity to water or rainwater (the term *zaojing* literally means 'water-weed well'), which they claim to help 'subjugate fire', the archenemy of all wooden buildings.[19]

Also notable is the consistent reference to the ceiling as '*jing* (well)' in the rhapsodies. The term might have more to do with the apparent association with a stable source of freshwater in common households, but the typical structure of a well, two pairs of bars placed perpendicular to each other, as shown in the character *jing* 井, is known as *jinggan* 井干 and was reportedly used for towers as tall as several hundred feet in the Han.[20] A modern parallel of the *jinggan* structure is the log cabin. Be it a well dug into the earth or a tower thrusting into the sky, the *jing*, and likely the early *zaojing* as well, maintains a shaft-like structure with an accessible interior.

Viewed in light of a continuous building tradition shared between China's east coast, northeast China, and the Korean Peninsula, the Yinan lantern affords a base point for the stone lantern to be examined as a pan-Asian ceiling type across modern national borders. Steinhardt has already investigated the striking similarity between the Yinan tomb and many Goguryeo tombs of the 4th to 6th centuries.[21] This similarity, as she observes, extends from the overall layout of the tomb, the content and meaning of the mural and tomb occupant's portrait, to the use of the highly distinctive corbel vault with the SDS lantern. The earliest dated Goguryeo tomb, Anak Tomb 3 (357) in South Hwanghae, North Korea, features such a lantern built in each of its five chambers. The same lantern appears in the Twin Pillar Tomb and the Tomb of the Celestial Kings and the Earthly Spirits, two examples mentioned by Soper, and other Goguryeo tombs including Anak Tomb 1, Ohoe Tomb 5, Jinpha-ri Tomb 7, and Kangso Great Tomb, to name but a few, where a lotus or dragon is often found painted on the capstone of the lantern.[22] Indeed, it is in the Goguryeo tombs, rather than Han tombs, that scholars have been so far able to discover the majority of the load-bearing stone lanterns in East Asia, which they refer to as '*mojiao zaojing* 抹角藻井' (Kr. *malgakjojeong* 말각조정, Jp. *sankaku jososhiki tenjo* 三角持送式天井).[23]

The occupant of Anak Tomb 3 was a certain Dong Shou 冬壽. The epitaph found upon excavation has provided a sketch of this person's life and allowed for the accurate dating of the tomb. Dong hailed from the State of Former Yan (337–370), a political regime of the Xianbei ethnic group in northeast China. He later fled to Goguryeo where he received land and official titles in the former Lelang Commandery of the Han dynasty. This is one of the many historical accounts whereby the diffusion theory of architectural forms can be substantiated. Steinhardt has already presented and carefully examined the evidence for the interrelationship between Han and Goguryeo architecture. Here, instead of giving a summary of her examination, I invite the reader to consider the validity of the diffusion theory beyond the Han-Goguryeo exchange. What if the same phenomenon (the similarity in the ceiling, or in any part of architecture for that matter) occurred over a geographical distance so great that no textual or other material evidence has yet been found in support of the connection? We do not run to hasty conclusions, but do we dare to entertain even the slightest possibility that two examples, separated by half a globe, might have sprung out of the same source or architectural prototype that exposes, perhaps, the 'psychic unity of mankind'?

This is dangerous business; diffusionist thinking is extremely vulnerable to attacks and easily slips into the pitfalls of 'heliocentrism' and racism.[24] It plagued Soper and those who came after him, but the question, as well as challenges and traps, remains: whether or not the multiple 'dots' of material evidence across Eurasia, dispersed as they are, could be threaded and linked successfully with a compelling narrative deploying a non-biased historical view and supported by comprehensive archaeological and written evidence. The Silk Road is usually invoked to explain the movement of things and peoples across long ranges, but it is not always successful when immovable objects, such as tombs and monuments, are involved. Pictorial representation, celestial symbolism, ritual practice, and building technology, on the other hand, are intangible properties of architecture traveling along the trade routes and can be more confidently rationalized by the Silk Road model.

The next examples, most of which are from outside China, present a challenge for the diffusion theory and the Silk Road model. The challenge lies mainly in the great geographical distance between the comparable examples and the leap of faith to acknowledge exchange between builders of different cultures, with or without intermediaries. The same challenge, meanwhile, leads us to new understandings of Asian architecture.

The world's oldest surviving stone lanterns are found in Thrace and Anatolia, where, according to Nikola Theodossiev, 16 examples have been excavated and reported so far. All of them are 'lantern-roofed tombs' dated between the 6th and the 1st century BCE, predating the Han and Goguryeo lanterns.[25] The ceiling configuration ranges from the two-tiered SD to the three-tiered SDS, four-tiered SDSD, and five-, six-, and seven-tiered lanterns.

The tumulus at Gordion (capital of ancient Phrygia, in Ankara, Turkey), dated 2nd to the 1st century BCE of the Hellenistic period, is a typical example with detailed excavation reports available in English.[26] The tomb includes an antechamber and a rear chamber each covered by a seven-tiered stone lantern which is square, octagon, square, octagon, square, diamond, and square (SOSOSDS) in sequence (Fig. 9.4). Viewed from the exterior, the roof resembles a stepped pyramid made of courses of stone, similar to the Yinan lantern. The multi-layered interlocking squares, diamonds, and octagons is a common feature that defines this type of 'Hellenistic' tombs.[27]

The same vaulting technique is found in southwest Turkey, in Gumuskesen, where a monument dated to the 2nd century CE still stands today. Simply referred to as the 'monument at Mylasa (Mylassa)' by architectural historians of the late 19th and early 20th centuries, this marble structure is the earliest lantern-roofed building above ground to date (Fig. 9.5). A burial monument with a lower level where the sarcophagi were kept, its upper level is an open pavilion, a cubic space with a lantern roof supported by 12 Corinthian columns and piers, perhaps a shrine where offerings would be made. Like the Gordion tomb, the roof is a stepped pyramid made of five stone courses. The stones appear unadorned outside, but inside, they are carved with intricate moldings and patterns, perhaps to mimic a wooden ceiling. The diamond in the center of the SOSDS lantern is echoed by 12 smaller diamonds carved into the ceiling. Josef Durm's illustration of the monument indicates that there were originally more stone courses on top forming a pyramidal roof (now lost). He includes a plan

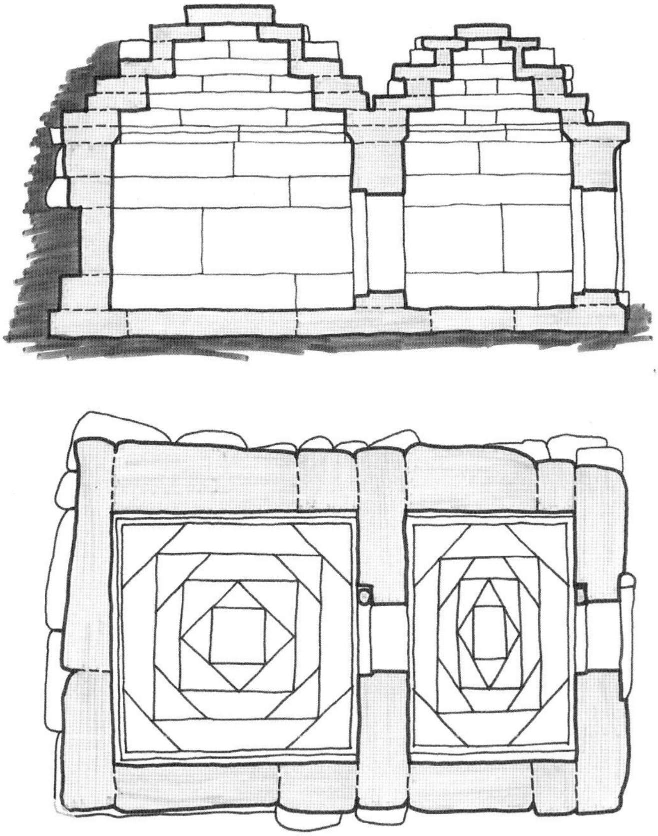

Fig. 9.4. Section and ceiling plan, Tumulus at Gordion, Ankara, Turkey, 2nd–1st century BCE (drawing by author after Young 1956, fig. 4).

of the Armenian wooden lantern ceiling as a comparison, but he does not explain the connection.[28]

Why does the diamond appear again and again? What is the technological advantage of it in lieu of, say, a ceiling with beams and joists placed in parallel (or at right angles) to the wall? The diagonal member, forming the four sides of the diamond, not only provides braces for the square room against lateral forces, such as wind, which can skew or distort the square, but it is also advantageous in saving material and labor. For a post-and-lintel structure, the roof is often where the largest quantity of building materials is required and where the load is the greatest. The problem of every roof is to transfer the load from the horizontal members of the roof structure to the vertical supports while reaching the maximum span of the beams and providing as much covered space as possible. Had the beams of the Mylasa monument laid in parallel to the lintels, the longest beam required would be three times the width of the bay (the distance between two adjacent columns).

Fig. 9.5. Ceiling plan, temple at Mylasa, Gumuskesen, Turkey, 2nd century CE (Borra 1750).

When laid diagonally, by contrast, the longest beam needed is two times that length (this occurs at the second square of the lantern), and because of the shorter span they could be made thinner and lighter. In cases where a large space with no interior columns is desired, or when long and thick beams are lacking or difficult to work with, builders might choose to 'break apart' the roof into a superimposed, multi-layered lantern as a solution. Hence the diagonal works for multiple purposes: resisting lateral forces, reducing load, saving material and labor, increasing roof span and height, and maximizing the column-less space under the roof. In addition, the lantern creates a pyramidal roof, which pleases the eye and repels rainwater and snow more efficiently than a flat roof.

Most lanterns from Anatolia predate the Han tombs. Does this mean that Han tomb builders learned this technique from West Asia and modeled their constructions on Anatolian or other non-Chinese examples? Or could it be that they developed the technique independently? The similarity in form, the square and diamond, and the mechanism of spanning the roof using diagonally placed beams is too great to be dismissed as a mere coincidence; both Han and Anatolian lanterns attest to a prevalent, perhaps universal, building method devised and practiced by the earliest human societies in the world. It is understood that Han tombs marked a watershed in the history of tomb building in China. Wu Hung suggests that foreign influences played a role in the transition from vertical-pit burials (*shuxuemu* 竪穴墓), composed of earthen chambers and wooden coffins, to horizontal-pit burials (*hengxuemu* 横穴墓) either dug into the cliff of mountains or approached via a ramp down to an underground stone 'house'.[29] Stone bricks had never been used to build tombs before the Han; they came to be used probably due to the Han expansion of the territory westward, whereby a closer contact with the Central Asian kingdoms, notably the Parthian (247 BCE–224 CE) and the Kushan Empires (ca. 1st–3rd centuries CE), was established.

The Mylasa monument is often compared to classical architecture in ancient Greece and Rome. The Greco-Roman influence is evident in the fluted Corinthian columns with obvious entasis, but the lantern roof, though often discussed in a Hellenistic or pre-Hellenistic context (since the majority of the lantern-roofed tombs in Anatolia and Thrace date from this period), appeared in the architectures of Asia more often and more widely than in Europe. The city of Mylasa, nowadays Gumuskesen, was a port city of the Roman Empire when the monument was built. But before Rome, it had been ruled by Mausolus (r. 377–353 BCE), a *satrap* (governor) of the Achaemenid Empire (ca. 550–330 BCE) and ruler of Caria (west Anatolia). The Achaemenid Empire, also known as the First Persian Empire, expanded across Eurasia from the Balkans to the Indus Valley. The wars between the Persians and the Greeks explain the exchange and syncretism in art and architecture, and the monument was very likely a hybrid structure born out of multiple architectural traditions. The lantern roof, as demonstrated by archaeological evidence, had long since been used in the local tomb construction of Anatolia.

Mylasa used to be the seat of Caria before Mausolus moved the capital to Halicarnassus, where his mausoleum, hailed as one of the Seven Wonders of the Ancient World, was built between 353 and 350 BCE. The original structure is long lost, but reconstructions show a pavilion with a pyramidal roof, a model which the Mylasa monument (and a few modern structures around the world, such as the Soldiers and Sailors Memorial Hall in Pittsburgh, built in 1910) is based on. The mausoleum was designed by two Greek architects, who must have acquired some knowledge of the local building tradition while trying to blend in elements of Greek architecture. A study by René Ginouves and Anne-Marie Guimier-Sorbets has proved that in the 4th century BCE, the Anatolian lantern was adopted across a wide geographical range, mainly as a solution to spanning over large rooms. An example is the West Wing of the Palace of Vergina in Greece, a wooden structure of three square rooms each measuring 16 by 17 m where no interior columns were used, hence a lantern was likely adopted.[30] The ruined House of Dionysos in Pella, Greece, which also dates from the same period, contains a floor mosaic that looks like the projection of a seven-tiered SDSDSDS lantern, the type that would later resurface in the Gordion tomb.[31]

At the other end of Eurasia the stone lantern enjoyed a wide distribution over the Indian subcontinent. For Soper, the temple at Pandrethan (10th century) in Kashmir serves as a key example to bridge the Asian dome with European ones (see Fig. 9.1). The ceiling contains four single and four pairs of confronted figures performing a kind of ritual dance, holding ribbons and in a vigorous bodily movement, all circling around a lotus flower in the center. With the dancers and the lotus, the ceiling does fall into Soper's category of the Dome of Heaven where iconography and symbolism, rather than structure, were the mason's major concerns.

It is Strzygowski who first drew the connection between Pandrethan and Mylasa, taking note of the similar square-and-diamond ceiling in spite of a distance of roughly 3,000 miles between the two locations.[32] Unlike Mylasa, however, the Pandrethan lantern is an SDS lantern, which is more comparable to the Han and Goguryeo examples where the lotus became the centerpiece. But in contrast to the East Asian practice, curiously, lanterns in India never appeared in funerary structures (predominantly stupas) but were applied almost exclusively to temple buildings, in particular the *mandapa* (porch) and the *garbhagriha* (sanctuary), the two most important spaces of a temple. The Baroli Temple (10th century) and the Ranakpur Jain temple (15th century) in Rajasthan, for example, both incorporated the SDS lantern, yet the lanterns were adopted not as a technological solution but mainly as an ornate architectonic framework for the iconographic program.

James Fergusson's survey of Indian architecture demonstrates that the lantern, despite its religious undertone, was foremost an ingenious roofing technique. Fergusson's diagrams show how a simple SD lantern advances to a five-tiered lantern using a growing number of stone blocks, resulting in an increase in both ceiling height and room square footage.[33] To cover a four-pillared square room, for instance, one could use a single stone slab, but rotating the stone by 45 degrees will double the square footage. Adding additional tiers of stone also yields a greater square footage, from the SD lantern (five stones, two tiers, twice the area) to the SDS (nine stones, three tiers, quadrupled area; Pandrathen), the SOSD (13 stones, four tiers), and the SOSDS lanterns (17 stones, five tiers; Mylasa). That is to say, a large room does not necessarily require larger stones, but stones of a moderate size, as if a module, can be used generally to cover spacious halls as well as intimate chambers. This technique is yet again at odds with early temple construction in India, from the Bhaja Buddhist cave (1st century BCE) and the Karle Cave (100–120 CE) to the Ajanta Cave 26 (5th century) where a ribbed barrel vault covers the oblong, horse-shoe shaped *chaitya* hall.

The Wooden Lantern

It is in India that perhaps the oldest surviving wooden lantern of the world is found. This is the SDS lantern in the *mandapa* of the Lakshana Devi temple in Bharmour, Himachal Pradesh, a structure dating back to the late 7th century when Meru Varman, ruler of the Hindu kingdom of Champa, settled there as his capital and dedicated this temple to the Hindu goddess Durga (locally known as Lakshana Devi).[34] The lantern is carved with ornate floral patterns, including a lotus in the center and roundels of leafy branches at the side. Dancing/flying deities occupy the triangular corners of the lantern. The structure and iconography are highly comparable to the Pandrethan lantern, a resemblance which can be explained by geographic proximity.

The rest of the existing wooden lanterns outside China come from a much later period, and their locations seem to be concentrated in the mountainous regions and river valleys near the Himalayas where the Lakshana Devi lantern is found. At the Baltit Fort and the Altit Fort in Hunza Valley, Pakistan, the reception rooms are equipped with a wooden SDSDSD lantern over a central fireplace. Unlike all previous examples, the lantern there is both ceiling and roof; moreover, it is an open roof, which lets in natural light and allows the free flow of air. The open lantern is generally found in the rural houses of the Pamir Mountains in modern Pakistan and Afghanistan. Aurel Stein described such a rural house he visited during one of his expeditions to Serindia (Xinjiang) in the early 20th century.[35] He was quite amazed at the coziness of the humble house, and his photograph showed a room with a hearth illuminated by natural light pouring through a large SDSDSD lantern, the same type used at Baltit and Altit.

Wooden lanterns before the 7th century are completely missing, even though all the stone lanterns from Han, Goguryeo, and Anatolia are said to have been modeled on wooden ones. This is, of course, due to the perishable nature of the building material, but more problematic is the fact that little to no material evidence, not even stone lanterns, exists between Mylasa and Pandrethan, the two ends of Eurasia, to validate any probable route of transmission of this roofing technology. Nonetheless, reconstruction drawings, ethnographic studies, and approaches to the global history of building technology indicate that the wooden lantern was, in fact, widely adopted in the dwellings of Central Asia, North Asia, and South Caucasus.

The Old Nisa Castle in Turkmenistan, the palace of the Parthian Empire dated around the 2nd century, probably had a lantern ceiling. Reconstruction drawings provided by excavators show a majestic square hall where four clustered columns in the center of the room support an SDSDS lantern.[36] Light penetrates the roof through the lantern, but the lantern does not open directly to the sky but is surmounted by an octagonal turret. The turret, sometimes referred to as a lantern rather confusingly, shelters the opening from rain and direct sunlight, and meanwhile admits breezes and diffused light through the side. This feature was crucial to dealing with excessive heat and cooling off the building in summer. Vernacular architecture in modern Central Asia, Iran, and Arab countries is typically equipped with a wind-catcher (*bâdgir.*), a turret-like roof structure similar in design and purpose.

Another comparable reconstruction from Central Asia, an 8th-century Sogdian mansion in Penjikent (modern Samarkand, Uzbekistan), again features an open wooden SDSDS lantern for residential use.[37] The reconstruction is probably based on both historical and ethnographic observations of the local building practice, with a comparative study of the traditional dwellings of nearby regions such as the Pamiri house. Though scholars used later materials to study earlier instances, the general continuity of the local building tradition has largely ensured the reliability of such a deduction. The Juma Mosque in Khiva, Uzbekistan, built in the late 18th century, uses multiple SDSD lanterns in the wooden hypostyle hall. Though these lanterns belong to a religious building and are not open to the sky, they provide clues for the search of early wooden lanterns in the same area.

Rolf Stein has noted how building an open roof was a common practice shared among peoples of Central Asia, North Asia, and Siberia. Prehistoric pit-dwellers would dig a hole into the ground, enclose it with solid walls on all sides, and leave a single entrance at the top of their subterranean house. The hole in the roof was a conduit for rainwater, wind, sunlight, and moonlight and a vent for the smoke of the cooking fire to escape. It was the door, the window, the skylight, and the chimney all at once. From the Eskimo winter house to the Mongolian/Turkish tent and the Tibetan mountain residence, the open roof above the fireplace was ingrained in these cultures as the heart of the home. Though the materials are gathered by modern ethnographers via direct observations during fieldwork, Stein claims, and I agree with him, that they suggest a very ancient practice that has persisted through multiple technological advances in history. It started as a 'hole in the roof' in the earliest and humblest huts and developed into various types of ceiling including the wooden lantern. In China, the opening in the roof also transformed into the 'well of heaven (*tianjing* 天井)' in traditional courtyard houses.[38]

The hole in the roof has had diverse structures and manifestations around the world, and the lantern is but one of the many options. A similar design is found in the houses of Pueblo Americans, in Colorado and New Mexico where subterranean and semi-subterranean communal houses were made of wooden logs. The logs were placed horizontally underground and stacked up to enclose the room as if it were a well or a tunnel to the mines. The room was often square or octagonal, growing increasingly smaller toward the top where it culminated in a small hole. Like the lantern ceiling, each layer of logs is placed across (not parallel to) the layer below, forming a series of 'rotating' squares and octagons. The earliest examples, known as *kivas*, were built during the 12th century as communal houses for the men in the village. The *hogan* of the Navajos also had a comparable structure. These similarities suggest that the application of the lantern ceiling has far exceeded the scope of the present Silk Road model and it demands a broader framework of investigation and an alternative historical narrative.

Rolf Stein informs us that certain religious practices in Asia did involve the (actual or symbolic) opening of the roof, such as some funeral when the soul of the dead would be called upon to ascend to the sky through a ladder to the roof hole. The ascent and descent through the hole and the climbing of the ladder (symbolic of the world tree) have evolved into different practices in various cultures. The *bi* jade disks in 2nd-century China were sewn to the burial suits of royal princes and princesses so that their celestial souls could escape through the 'heavenly gate (*tianmen* 天門)' on top of the skull. The image of the world tree and the Shamanistic practice of tree worshipping still linger today despite the christianization of Christmas. That Santa Claus always comes down from the chimney suggests a link (however flimsy) to this ancient practice.

The archetypal image of the primitive hut, therefore, is an image of the 'womb house'. The enclosed space is warm, dark, and intimate, and the only source of light comes from the hole, the only exit, and crawling out the hole in the roof is tantamount to rebirth from mother earth's womb. The womb house must have originated in multiple places in the world because of the shared experience of humans' own birth, the first rite of passage in life, a long and perilous journey to this world. From the lantern to the oculus, from the Chinese courtyard house to the Roman compluvium, the open roof, or the 'hypaethral' structure as Ananda Coomaraswamy calls it, continued to be the hallmark of the womb house. The circular roof plate in Indian temples, according to Coomaraswamy, also stands for the original opening in the roof. Though the roof was sealed in later development, the image and significance of the hole remain alive.[39] The rationale behind the wooden lantern, therefore, goes beyond concerns of economy and efficiency, but the opening of the roof necessitates that ceiling beams must stay clear of the center. They must not be placed as if the spokes of a wheel to converge at the center, as in the case of the Chinese octagonal dome, but have to 'rotate' around the center to form a kind of interwoven pattern like a bird's nest.

Viewed in the context of global wooden architecture, the lantern pertains to what Galyna Shevtsova terms the 'blockhouse' system. A blockhouse is similar to a log cabin. It is built by stacking up horizontal timbers and uses no vertical supports. According to Shevtsova, the blockhouse is commonly found in areas where there is no strong wind or other lateral forces, where winter is harsh and trees are abundant. In other words, its distribution is primarily determined by local geographical conditions. Mongolia, Siberia, Manchuria, North China, North Korea, northern Central Asia, and northern Europe meet these criteria and are where blockhouses abounded.[40] Wooden churches in Norway and Switzerland, chapels in Uzbekistan and Ukraine, the *jinggan* towers recorded in Chinese historical documents, and the Shosoin Repository in Nara, Japan, are such examples. The way logs are stacked seamlessly to form an impenetrable enclosure is highly evocative of the shipbuilding technology whereby planks are bound together to create a watertight vessel. This is distinct from the post-and-lintel system, a more 'porous' structure, which can deal better with hot and humid summers and is more suitable for regions with heavy wind and frequent earthquakes.

Shevtsova refers to the traditional Georgian house called the *darbaza*, which has a pyramidal roof (*gvirgvini*) with an opening in the center. There were a variety of roof structures: the SDSDS lantern, the superimposed squares, and the honeycomb, some of which are still in use in certain areas today. The traditional Armenian house *hazarsban* used a similar roofing technique, as noted by Durm, Styswoski, and Le Coq. In fact, Le Coq has compared the lanterns of the Armenian *hazarsban* and the Kizil caves in China to support his point on the Western influence on Kuchean art.

From House to Temple

Wooden houses do not last long, but their forms were copied in more durable materials and translated into sacred spaces such as temples and shrines. The distinctive square-and-diamond shape of the wooden lantern was, therefore, immortalized and disseminated. During the process, the shape became standardized while the practical function of the lantern was forgotten. Yet the pattern itself, transcending all mundane concerns and considerations, has been elevated to a sacred symbol, a pure iconic form detached from the original roof structure and loaded with theological and liturgical significance.

Back in China, the Mogao Caves in Dunhuang are some of the earliest instances of using the lantern in a Buddhist rock-cut temple. An example is Mogao Cave 285 (534–556, Western Wei dynasty) where an SDS lantern is painted in the center of the truncated pyramidal roof (*fudouding* 覆斗頂). The lantern occupies less than a tenth of the total ceiling area and is completely flat. The form is highly reminiscent of the stone lanterns of the Han tombs, but the multicolored triangles jutting from the border of the lantern, representing the double-layered fringe of a canopy, and the arcade with seated figures along the bottom of the ceiling, suggest that this is a consecrated lantern appropriated for religious use. In Mogao Cave 254 (Northern Wei dynasty, 386–534), the ceiling of the main chamber is divided into 14 squares each containing a painted SDS lantern (Fig. 9.6).[41] There is also a miniature arcade with seated figures at the bottom of the ceiling as in the first example. The combination of the 'lantern and arcade' would re-emerge in later wooden temples including the Jingtusi and the Zhihuasi (see below).

The two major cave-temple types in China, the *chaitya* (stupa-pillar cave; Mogao 254) and the *vihara* (meditation cave; Mogao 285), were both imported from India, but cave temples in India, as noted earlier, did not use the lantern at all. Those who built the Mogao Caves must have learned of the lantern somewhere else, most likely from the same source of the Han tombs. They might have never seen or built a real lantern, a multi-layered structure, in either wood or stone, but were simply copying the shape from some visual sources. This is because the lanterns they made are all flat and painted; there is no indication of volume or depth, and the general lack of diversity in representation suggests either ignorance of, or disinterest in, the lantern technique. Tombs of the Northern Dynasties (4th–6th centuries), though contemporaneous with

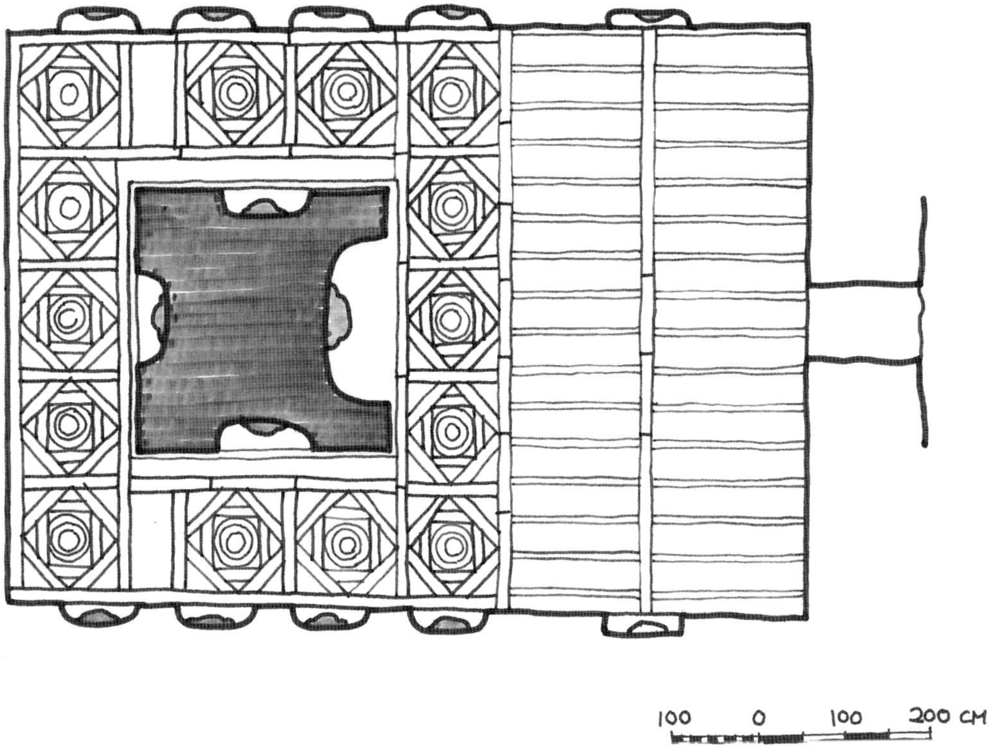

Fig. 9.6. Ceiling plan, Mogao Cave 285, Dunhuang, China, 386–534 (drawing by author after Tonko bunbutsu kenkyujo 1980–1982, v. 1, 260).

the Mogao Caves and comparable in structure and layout, never used lanterns. In fact, the SDS lantern that characterized the Han tombs has altogether disappeared in funerary structures in China since the end of the 3rd century.

The lantern did not last long at the Mogao Caves. Since the Sui dynasty, the square-and-diamond shape has been largely replaced by non-architectural elements. It never reached the Chinese heartlands into Yungang or Longmen where the earliest caves were excavated during the Northern Wei period, contemporaneous with the lantern-roofed caves at Mogao. Indeed, the SDS lantern was short-lived in Chinese Buddhist architecture and has never become a prevalent feature east of Dunhuang. To the west of Dunhuang, however, it was not an isolated instance, as similar lanterns are found at Bamiyan and Kucha, two important oasis towns on the trans-Himalayan routes of the Silk Road.

Thirty-four lantern-roofed caves in the Bamiyan Valley of Afghanistan have been identified and recorded.[42] Despite the Taliban's bombing of the two colossal Buddha statues in 2001, and despite the history of iconoclasm in the area since the 14th century during the Islamification of Asia, many of the rock-cut caves, some monumental in scale, are intact as far as the architectural features are concerned. The unique features of the ceilings of these caves were noted in the 1930s when French archaeologists first surveyed the site.[43] Cave V, one of several caves surrounding the feet of the 150-foot-tall Buddha, has a SDS lantern that imitates a wooden construction. A more recent comprehensive survey by Japanese archaeologists reveals that the lantern is one of the most frequently applied ceiling types at Bamiyan.[44] These include the SDS and SDSD lanterns, which greatly vary in size, form, and detail. A standardized pattern for the Bamiyan lantern does not exist.

A comparison between the Bamiyan and the Mogao lanterns suggests that the SDS lantern was known to the builders of both areas as a ceiling type closely associated with Buddhism. How this association came to be is not yet clear, but images found in the lantern suggest that celestial symbolism was at work: the open roof of the 'womb house' was here appropriated and reinterpreted as a portal to Buddhist heavens. At Bamiyan Cave Na, the only cave where the original painting on the ceiling still remains, the SDS lantern is transformed into a Buddhist mandala: the Cosmic Buddha (Vairocana) sits in the center and is encircled by, from center to perimeter, 20, 4, 32, and finally 4 buddhas, which are arranged in four separate layers of the square-and-diamond frame (Fig. 9.7). There are a total of 61 buddhas.

The 'lantern and arcade' has also made its way to Bamiyan, adding to the celestial symbolism of the ceiling. Cave V, mentioned above, includes 16 miniature archways, four on each side, encircling and supporting the lantern. The trefoil-shaped archway suggests that there might have been seated Buddhist figures in the arcade originally like the two Mogao caves, though none of those figures survived, and that the arcade might be considered a representation of the 'heavenly palace'. The figures-in-the-arcade motif, which Benjamin Rowland calls the *homme-arcade*, was used for the famous Bimaran Reliquary (ca. 1st century), a gold casket bearing the image of a standing Buddha excavated from a stupa in Jalalabad, Afghanistan, about 200 miles east of Bamiyan.[45] The same motif, though with some variation, appeared in the Karle Cave, the Takht-i-bahi Monastery, and the Gandharan stupas, all of which were probable sources of the Bamiyan cave architecture.

The dating of the Bamiyan caves has always remained tentative. It is unknown who built these caves. The scale of such a grand project means it must have required extraordinary commandeering and mobilization of labor and resources to complete. Early travelers' records, including those by the Chinese monk Xuanzang (ca. 602–664), have helped to date the two colossal Buddhas to no later than 800, while some of the caves could date to as early as 600, perhaps even earlier, due to comparisons in style and iconography. Roofing techniques, on the other hand, received little consideration in the process of dating. The similarity between the Bamiyan and the Mogao lanterns adds to the puzzle: though the SDS pattern seems to have been largely shared between the two, each incorporated it in a distinct way. Mogao suggests an ignorance of the lantern technique while Bamiyan displays the grasping of the technique as demonstrated by the articulation and clarity of the structure. In other words, workers at the Bamiyan

Fig. 9.7. Ceiling plan, Bamiyan Cave Na, Afghanistan (drawing by author after Iwade and Kubodera 2013, 167).

site seem to have had some experience building the wooden lantern, which they were able to mimic realistically in stone. If a transmission route is to be proposed between Mogao and Bamiyan, it is highly unlikely that Bamiyan learned of the lantern ceiling from Mogao, even though the Bamiyan lantern might have dated from a later period.

Yet another piece of the puzzle is Kucha. Kizil Cave 167, tentatively dated to around 7th to 12th centuries based on the analysis of the painting style (due to the lack of written records), is a single-chamber cave covered by a full-scale, seven-tiered lantern.[46] The lantern is not painted on the ceiling but hewn from the living rock. Even though scholars like to compare Kizil in general with Mogao and Bamiyan, emphasizing the exchange between these sites and the overall dissemination of Buddhist art along the northern route through the Taklamakan, the ceiling structure of Cave 167 differs from all Mogao and Bamiyan examples. It is one of the few SDSDSDS lanterns in the world, almost as if a distant reverberation of the Gordion tomb and a close relative of the Pamiri wooden lanterns. The patterns painted on the ceiling are also unique. A group of small intersecting circles (*qiuwen* 毬文) and round multi-petaled flowers interlaced with entwined or hooked threads (*suowen* 鎖文) fill the triangles of the lantern, which suggests the use of textiles or wooden latticework in the ceiling.

The closest examples to the Kizil lantern are from the Buddhist temple at Alchi in Himachal Pradesh, India, bordering Tibet in an area known as Ladakh. There one finds the oldest surviving wooden lanterns of Buddhist architecture. The Alchi temple is a complex of several *chortens* (passage-door stupas; Ch. *guomenta* 過門塔) and image halls built of wood and plastered earth. The Great Chorten, a structure of five spires on top of a cubic base, has an eight-tiered SOSDSDSD lantern under the central spire (Fig. 9.8). Inside the Great Chorten is a miniature *chorten* equipped with an SDSD lantern. At the same monastery, the twin-spired Kakani Chorten (ceremonial gateway) similarly holds miniature *chortens*, and the lantern is applied to all *chortens*, large and small. None of these lanterns, though made of wood and structurally functional, serves as chimneys or skylights as the Parmiri lantern, but they are incorporated in the overall pictorial program of the interior to create a sanctified space.

The main chamber of the Great Chorten is covered by tiny Buddha figures painted in small circles, often called the 'Thousand Buddha' motif, which also appears on the vertical risers of the lantern. The horizontal surface of the lantern, on the other hand, is covered by figures and animal/floral motifs set against a background of geometric shapes, recalling the *qiuwen* and *suowen* of the Kizil lantern. These patterns differ from tier to tier: pairs of confronting four-armed figures on the lowest tier, roundels, hooked lines, and stripes on the second and third tiers. The fourth tier seems to show a typical façade of a Tibetan monastery often represented by a T-shaped element in a Tibetan mandala marking the four entrances to the central altar. The painting shares certain elements with the Bamiyan mandala but is a more complex design. Gerald Kozicz's point that it is a Vajradhatu (Diamond World) Mandala is interesting, considering how the multiple layers of the lantern might have corresponded to the different sections of the mandala, and how the Buddhist appropriation of the lantern must have given the structure a new meaning.[47]

Roger Goepper has related the geometric designs on the Alchi lanterns to textile patterns.[48] A study by Christiane Papa-Kalantari has verified this point. She stresses the 'textile-conscious' culture of Tibet and the fact that traditional wooden architecture in the West Himalayas, including Alchi in Ladakh, has involved the use of various fabrics, baldachins, silk banners, banderoles, and *thangkas*, in temple buildings. Textiles circulated and traded along the Silk Road, bearing typical Central Asian motifs such as floral medallions from Sogdiana and Iran, were used for altars and became depicted in ceilings.[49]

The Alchi *chortens* are the only known cases where the lantern was used for a stupa. They are unique also because of the nesting of the main stupa and the miniature inner

Fig. 9.8. Section showing structure of lantern ceilings, Great Chorten at Alchi, Himachal Pradesh, India, 12th–13th century (drawing by author after Kozicz 2011).

stupa. Studying the inscriptions found at Alchi, Goepper identifies the Great Chorten as the 'Auspicious Stupa with Many Doors'.[50] The inner stupa of the Great Chorten contains portraits of four eminent Buddhist sages and teachers painted on the walls, including Rinchen Zanpo (958–1055), who translated Sanskrit Buddhist sutras into Tibetan.[51] Goepper's analysis of the written records confirms the establishment of the monastery by a certain Tshul khrims 'od, an active practitioner and promoter of Buddhism in Ladakh in the late 12th and early 13th centuries, when Alchi became a major Buddhist center of the Guge Kingdom in Western Tibet and played a role in preserving and continuing Buddhism after the Islamification of Asia.[52] After the fall of Gandharan Buddhism, Kashmir inherited much of its legacy and remained a stronghold of Buddhism into the 10th century, its influence radiating into the West Himalayas and the Pamirs. This might explain how the Bamiyan mandala entered Alchi.

There and Back Again

The Yuan dynasty (1271–1368) brought Xinjiang, Tibet, and North and South China under a central unified rule after centuries of political division. The Mongol rulers of the Yuan were patrons of many religions including Tibetan Buddhism, and through them, Buddhist art of Tibet and Nepal enjoyed a great influx in North China. The White Pagoda of the Wutai Mountains in Shanxi was built by the Nepali architect Anige (1245–1306) and was the model for the other white pagodas, including the one next to the Forbidden City in Beijing. The five-spired *chorten* from Alchi came to be known as the 'Diamond-throne Pagoda (*jingang baozuo ta* 金剛寶座塔)' and was adopted for Buddhist monasteries of the Yuan, Ming, and Qing dynasties, such as the Wutasi (Five-pagoda Monastery) and the Biyunsi in Beijing.

Tibetan Buddhist art also infiltrated the traditional wooden buildings of China. The mandala, for instance, was incorporated into the Zhihuasi, built in 1443/1444 by Wang Zhen 王振, the most powerful eunuch at the Ming emperor Yingzong's court. Unlike Bamiyan and Alchi, the mandalas are here painted on the checkerboard ceiling and are paired with Sanskrit seed syllables (Ch. *zhongzizi* 種子字), *dharanis*, and mantras executed in calligraphy. The halls and pavilions were envisioned as an integral ritual space for Tibetan Buddhism.[53] In this light, the twin lanterns of Zhihuasi, briefly noted at the beginning of this essay, should bear traces of Tibetan influence. But the lanterns we see differ greatly from the Alchi example, not only because the mandala is replaced by carved dragons and phoenixes, but the lantern itself has also changed in structure. The beams are not stacked to make a multi-layered ceiling space but are intersected at the same height, framing almost a flat plane. The intricate pattern thus formed is probably derived from an SOSD lantern, though more prominent is perhaps the square-and-diamond harking back to the Han times. The central dragon is surrounded by eight smaller dragons in the inner octagon, flying deities and flower bouquets in the middle, and four phoenixes (reminiscent of Kizil) in the outer triangles. An arcade of miniature wooden pavilions, seven per side and 32 in total (including four corner towers), forms the border of the ceiling. It hosts an assembly of Buddhist figures and floats above the clouds, and is referred to as the 'Heavenly Palace (*tiangong louge* 天宮樓閣)' in Chinese architecture, evidently derived from the *homme-arcade* that prevailed in the Mogao and Bamiyan lanterns.

The Zhihuasi lantern, therefore, is a hybrid structure based on Han traditions and absorbing multiple influences from outside China. The ceiling structure has transformed from what is essentially a corbeled vault (as in the cases of Yinan, Gordion, Mylasa, Pandrethan, and Alchi) into a wooden dome, a process in which the earliest lantern was translated and reconfigured by the Chinese timber-frame system. An often neglected fact is how the timber-frame was itself compatible with the lantern form. Post-and-lintel structures often add diagonal beams to brace the corners of the timber frame. The *Yingzao fashi* records the use of *mojiaofu* 抹角栿 (chamfering/corner beams) in the roof. The Offering Pavilion of the Master Dou Shrine (1267) in Taiyuan, Shanxi Province, for instance, uses four diagonal beams in the square roof, creating a square-and-diamond base leading to a square-and-octagon upper part or an SDSO lantern. Another key example is the theater stage of the Niuwangmiao (1283) in Shanxi where the roof is made of multiple tiers of wooden beams forming an SOSDSO lantern. Chen Mingda has pointed out that the Chinese post-and-lintel system, especially the *tailiang* 抬梁 (lit. elevating the beams) framework, was born out of the *jinggan* technique, which prioritized the piling up and joining of horizontal beams without adding vertical supports.[54] One might argue that adding the diagonal beams was for practical considerations, but large wooden halls in Chinese history, the famous Foguangsi included, have been erected without using the diagonal beams, and the aesthetic, even spiritual, attributes of the diamond and the diagonal cannot be dismissed when their purpose is considered.

Conclusion

The distribution of the lantern ceiling in Asia is amazingly broad, covering an area encompassing China, Korea, India, Pakistan, Afghanistan, Uzbekistan, Turkmenistan, Armenia, Georgia, Turkey, and probably reaching Greece. A hypothetical trajectory of the transmission of the lantern can be proposed based on the Silk Road model, along the main arteries across Eurasia, driven by the incentives to travel, conquer, and trade. The earliest lanterns, all of which were made of stone in a funerary environment, were likely modeled on wooden houses, primitive, semi-subterranean wooden huts with a hearth and a chimney/skylight in the middle of the ceiling, bringing light and water into the room. Later, the technological and practical advantages of the lantern were 'forgotten' while religious and ritual concerns overrode, and in the rise and dissemination of Buddhist art and architecture, the lantern was sanctified as it became associated with the Buddhist notions of heaven and enlightenment. It was also largely due to this sanctification, however, that the particular square-and-diamond shape of the original lantern was kept and started to radiate from its birthplace, emitting to other building customs and adapted in various forms at Bamiyan, Mogao, Kizil, Alchi, and Beijing, taking on new cosmological patterns (the mandala) and new theological and ritual significance. Meanwhile, what is perhaps the oldest form of the lantern, the open lantern of the primitive hut, is still alive in rural Pamiri houses.

Eurasian empires, the Achaemenid Empire, the Hellenistic World, and the Mongol Empire, set the stage for cross-cultural transmissions. But the nomads and semi-nomads and the kingdoms they established, Northern Wei for instance, certainly played a role. The steppes in the northern part of Asia, from the Black Sea to the Urals, from Russia to Mongolia, North China, and Korea, were homes to early nomads whose movement and migration might have brought the lantern across regions. The trading of silk and textiles to and from China could have enhanced mutual awareness of architectural traditions, and silk fabrics, used as canopies in religious structures, might have even carried the square-and-diamond shape to be copied by other builders.

The Dome of Heaven in China, and more broadly in Buddhist Asia, is characterized by the lantern ceiling as the basic structural framework. This is not only because of the celestial symbolism, from Han and from India, associated with the structure in history, but the lantern itself harks back to early human experiences of looking through the roof's hole into the sky, of climbing through the hole, of gazing at the smoke of the fire wafting and exiting the hole. It is the archetypal image of the cosmic house common to all peoples and cultures that gained a solid ground in Asian architecture.

Notes

1 I would like to thank Yaohua Shi, Jeffrey Lerner, Chanchal Dadlani, Qiong Zhang, Andrew Gurstelle, Tansen Sen, Hyunhee Park, Judith Lerner, Aleksandr Naymark, Saba Samee, Joe Alchermes, and Anna Vallye for reading and/or providing helpful comments on this article at various stages of its development. Thanks also go to Susan Whitfield, who shares my interest in the lantern ceiling and has encouraged me to pursue this project.
2 Soper 1947, 225–248.
3 Grünwedel 1912; Strzygowski 1918; von Le Coq 1925.
4 Lehmann 1945, 1–27.
5 Grabar 1990, 15–21; Steinhardt 2014, 271–281.
6 Steinhardt 2003, 76–119.
7 Steinhardt 2011, 27–71.
8 Steinhardt 2014, 281–83. For the octagon in Liao architecture, see Steinhardt 1997.
9 Hu 2014.
10 Steinhardt 1997.
11 Li 1954, vol. 1, 165–169.
12 Steinhardt 1997.
13 Huang 1996, 105–120.
14 For an overview of the Han funerary art, see, for instance, Tseng 2011 and Wu 2010.
15 Zhong *et al.* 2005, 14–35; Sichuansheng wenwu kaogu yanjiuyuan *et al.* 2007; Steinhardt 2019, 44–46.
16 Henansheng wenwu yanjiusuo 1993.
17 Zeng, Jiang, and Li 1956.
18 See translation in Knechtges 1982–1996.
19 Some of these comments are recorded in Li 1954, vol. 1, 35–36.
20 The Jingganlou 井干樓 of the Han emperor Wudi (r. 141–87 BCE), for instance, reached 50 *zhang* 丈 (more than 100 m, or 328 feet) according to *Records of the Grand Historian* (*Shiji* 史記). An annotation by Xue Zong 薛綜 (d. 243) on the 'Western Metropolis Rhapsody' explains the *zaojing* as 'made by intersecting timbers into a square as if making a *jinggan*-structure'. According to the Qing etymologist Duan Yucai 段玉裁 (1735–1815), '*Gan* is the wooden railing of the well; its shape is either quadrilateral or octagonal.'
21 Li 2014; Steinhardt 2014.

22 A good image source for the Goguryeo tombs is Kim 2004.
23 Choi 2017, 195–202.
24 Nancy Steinhardt explains 'heliocentrism' as an idea that propagates 'the origin of many cultures in one, or at best, the origin of many cultures in several' (Steinhardt 2014, 274).
25 Theodossiev 2006. See also Fedak 1990.
26 Young 1956, 249–266.
27 Ginouvès and Guimier-Sorbets 1994, 311–321.
28 Durm 1910, 187.
29 Wu 1986; 1997; 2010.
30 Ginouvès and Guimier-Sorbets 1994, 317–319.
31 Ginouvès and Guimier-Sorbets 1994, 319–321. In China, the correspondence between ceiling and floor design is recorded in the *Yingzao fashi*.
32 Strzygowski 1918, 619–625.
33 Fergusson 1910, vol. 1, 314–315.
34 Bernier 1997, 139–142.
35 Stein 1921, vol. 1, 47–49 (photo of interior on p. 49); vol. 3, pl. I.
36 Пилипко 1996.
37 Беленицкий 1973, 14–15.
38 Stein 1990. See also Stein 1957, 163–186.
39 Coomaraswamy 1997, 192–242.
40 Shevtsova 2017, 93–108.
41 A virtual tour of the interior of this cave is available at: https://www.e-dunhuang.com/cave/10.0001/0001.0001.0254.
42 Steinhardt 2014, 274.
43 Hackin and Carl 1933.
44 Iwade and Kubodera 2013, 282–302.
45 Rowland 1946, 44–47.
46 Early documentation of this cave includes Grünwedel 1912, fig. 1; and von Le Coq 1925, 31–33, 99–104 (figures). The plans and descriptions of all Kizil Caves are available in Xinjiang Qiuci shiku yanjiusuo 2000.
47 Kozicz 2011, 48–56. See also Goepper 1993, 136–137; Klimburg-Salter 2001, 323–348.
48 Goepper 1993.
49 Papa-Kalantari 2007, 149–178.
50 Goepper 1993, 116.
51 Goepper 1993, 137–139.
52 Goepper 1993, 115–116.
53 Chen and Zhang 2017, 51–62.
54 Chen 1990.

Bibliography

Bernier, R.M. 1997. *Himalayan Architecture*, Madison.

Borra, G.B. 1750. *Plan of a Temple at Mylasa and Detail of a Coffered Ceiling from a Tomb*, Yale Center for British Art, Paul Mellon Collection.

Chen Jie 陳捷 and Zhang Xin 張昕. 2017. Han-Zang jiaorong hua jingtu: Zhihuasi shensheng kongjian de yiyi suzao 漢藏交融化净土: 智化寺神聖空間的意義塑造, *Meishu yanjiu* 美術研究, No. 1, 51–62.

Chen Mingda 陳明達. 1990. *Zhongguo gudai mujiegou jianzhu jishu: Zhanguo-Beisong* 中國古代木結構建築技術: 戰國-北宋, Beijing.

Choi, S. 2017. Bindirme Tavan: Bir Adlandirma Sorunsali, *Ankara University Journal of Social Sciences* 8, issue 2, 195–202.
Coomaraswamy, A.K. 1997. The Symbolism of the Dome, in: Coomaraswamy, A.K., *The Door in the Sky: Coomaraswamy on Myth and Meaning*, Princeton, 192–242.
Cunningham, A. 1848. *An Essay on the Arian Order of Architecture: As Exhibited in the Temples of Kashmir*, Calcutta.
Durm, J. 1910. *Die Baukunst der Griechen*, Leipzig.
Fedak, J. 1990. *Monumental Tombs of the Hellenistic Age: A Study of Selected Tombs from the Pre-Classical to the Early Imperial Era*, Toronto.
Fergusson, J. 1910. *History of Indian and Eastern Architecture*, 2 vols., London.
Ginouvès, R. and Guimier-Sorbets, A.-M. 1994. Voûte 'Galate' et Charpente Macedoniénne, *Revue Archéologique*, No. 2, 311–321.
Goepper, R. 1993. The 'Great Stūpa' at Alchi, *Artibus Asiae* 53, 111–143.
Grabar, O. 1990. From Dome of Heaven to Pleasure Dome, *Journal of the Society of Architectural Historians* 49, 15–21.
Grünwedel, A. 1912. *Altbuddhistische Kultstätten in Chinesisch-Turkistan*, Berlin.
Hackin, J. and Carl, J. 1933. *Nouvelles Recherches Archeologiques a Bamiyan*, Paris.
Henansheng wenwu yanjiusuo 河南省文物研究所 (Henan Institute of Cultural Relics) (ed.). *Mixian Dahuting Hanmu* 密县打虎亭漢墓. 1993. Beijing.
Higuchi, T. and Barnes, G. 1995. Bamiyan: Buddhist Cave Temples in Afghanistan, *World Archaeology* 27, 282–302.
Hu, J. 2014. *Embracing the Circle: Domical Buildings in East Asian Architecture ca. 200–750*, PhD dissertation, Princeton.
Huang Milan 黄明蘭 and Guo Yinqiang 郭引强. 1996. *Luoyang Hanmu bihua* 洛阳漢墓壁画, Beijing.
Iwade, Mayu, and Shigeru Kubodera. 2013. *Structure, Design and Technique of the Bamiyan Buddhist Caves*, London.
Kim, R. 2004. *Koguryo Tomb Murals: World Cultural Heritage*, Seoul.
Klimburg-Salter, D. 2001. Tucci Archive Report, 4: Dung Dkar/Phyi Dbang, West Tibet, and the Influence of Tangut Buddhist Art, *East and West* 51, 323–248.
Knechtges, D. (trans.). 1982–1996. *Wen xuan*, 3 vols., Princeton.
Kozicz, G. 2011. Die Laternendecke: Von den Höhlentempeln Bamiyans zu den Stupas von Alchi, in: Kostka, R. (ed.), *Wakhan: Talschaft zwischen Pamir und Hindukusch*, Graz, 48–56.
Lehmann, K. 1945. The Dome of Heaven, *The Art Bulletin* 27, 1–27.
Li, C. 2014. From Han to Koguryo: The Spread of Stone Chamber Tombs in Northeast Asia (1st–7th centuries), in: Lopes, R.O. (ed.), *Face to Face: The Transcendence of the Arts in China and Beyond – Historical Perspectives*, Lisbon.
Li Jie 李誡. 1103. *Yingzao fashi* 營造法式. Reprint 1954. 4 vols., Shanghai.
Papa-Kalantari, C. 2007. From Textile Dress to Vault of Heaven: Some Observations on the Function and Symbolism of Ceiling Decorations in the Western Himalaya Temples of Nako/Himachal Pradesh, India, in: Hardy, A. (ed.), *South Asian Archaeology: The Temple in South Asia. Proceedings of the Eighteenth International Conference of the European Association of South Asian Archaeologists, London, 2005*, 149–178.
Rowland, B. 1946. Gandhāra and Early Christian Art: The Homme-Arcade and the Date of the Bīmarān Reliquary, *The Art Bulletin* 28, 44–47.
Shevtsova, G. 2017. Beam-pillar and Blockhouse Wooden Construction Systems in the World: The Areas of Domination and Mixing Zones, *Traverse: Kyoto University Architectural Journal* 18, 93–108.
Sichuansheng wenwu kaogu yanjiuyuan 四川省文物考古研究院 (Sichuan Provincial Institute of Cultural Relics and Archaeology), Santaixian wenwu guanlisuo 三台縣文物管理所 (Santai

County Cultural Relics Administration), Mianyangshi Bowuguan 綿陽市博物館 (Mianyang Museum) (ed.). *Santai Qijiang Yamu* 三台郪江崖墓. 2007. Beijing.
Soper, A.C. 1947. The 'Dome of Heaven' in Asia, *The Art Bulletin* 29, 225–248.
Stein, M.A. 1921. *Serindia: Detailed Report of Explorations in Central Asia and Westernmost China*, 5 vols., Oxford.
Stein, R.A. 1957. Architecture et Pensée Religieuse en Extrême-orient, *Arts Asiatiques* 4, 163–86.
Stein, R.A. 1990. *The World in Miniature: Container Gardens and Dwellings in Far Eastern Religious Thought*, Brooks, P. (trans.), Stanford.
Steinhardt, N.S. 1997. *Liao Architecture*, Honolulu.
Steinhardt, N.S. 2003. A Jin Hall at Jingtusi: Architecture in Search of Identity, *Ars Orientalis* 33, 76–119.
Steinhardt, N.S. 2011. The Sixth Century in East Asian Architecture, *Ars Orientalis* 41, 27–71.
Steinhardt, N.S. 2014. *Chinese Architecture in an Age of Turmoil, 200–600*, Honolulu.
Steinhardt, N.S. 2019. *Chinese Architecture: A History*, Princeton.
Strzygowski, J. 1918. *Die Baukunst der Armenier und Europa*, 2 vols., Wien.
Theodossiev, N. 2006. The Lantern-Roofed Tombs in Thrace and Anatolia, *Proceedings of the 10th International Congress of Thracology*, Athens.
Tonko bunbutsu kenkyujo (ed.). 1980–1982. *Tonko bakukokutsu*, 5 vols., Tokyo.
Tseng, L.L. 2011. *Picturing Heaven in Early China*, Cambridge.
von Le Coq, A. 1925. *Bilderatlas zur Kunst und Kulturgeschichte Mittel-Asiens*, Berlin.
Wu, H. 1986. Buddhist Elements in Early Chinese Art (2nd and 3rd Centuries AD), *Artibus Asiae* 47, 263–352.
Wu, H. 1997. *Monumentality in Early Chinese Art and Architecture*, Stanford.
Wu, H. 2010. *The Art of the Yellow Springs: Understanding Chinese Tombs*, Honolulu.
Xinjiang Qiuci shiku yanjiusuo 新疆龜茲石窟研究所 (Xinjiang Kucha Grottoes Research Institute) (ed.). *Kezier shiku neirong zonglu* 克孜尔石窟内容總録. 2000. Urumqi.
Young, R.S. 1956. The Campaign of 1955 at Gordion: Preliminary Report, *American Journal of Archaeology* 60, 249–266.
Zeng Zhaoyu 曾昭燏, Jiang Baogeng 蔣寶庚, and Li Zhongyi 黎忠義 1956. *Yinan guhuaxiang shimu fajue baogao* 沂南古畫像石墓發掘報告, Beijing.
Zhong Zhi 鐘治, Zhou Kehua 周科華, and Li Sheng 李生. 2005. Sichuan Santai Qijiang yamuqun Bailinpo yihaomu fajue jianbao 四川三台郪江崖墓群柏林坡1號墓發掘簡報, *Wenwu* 文物, No. 9, 14–35.
Беленицкий А.М. 1973. *Монументальное искусство Пенджикента: Живопись, скульптура*, Москва.
Пилипко, В.Н. 1996. *Старая Ниса: здание с квадратным залом*, Москва.

Chapter 10

'Malacca' – From Fabled Port to Muddy Lagoon: A Cautionary Tale of Ecological Disaster[1]

Margaret Sarkissian

This article focuses on one Malaysian coastal city, Melaka.[2] In its 15th- to 16th-century heyday Melaka was a central node in what has become known as the 'Maritime Silk Road'. Today it is a medium-sized city of around 500,000 inhabitants and the eponymous capital of Malaysia's third smallest state.[3] Despite only ranking 14th among Malaysian cities by population, its historical allure is second to none.[4] Marketed as Malaysia's *Bandaraya Bersejarah* (Historical City), 'Malacca' is the focal point of a tourist industry that has become one of the state's primary sources of income.[5] Since the turn of the 21st century, developers have used the aura and patina of history as a catalyst for massive building and land reclamation projects along the coast. As a result, Melaka has reemerged as a node in a new transnational maritime project, China's so-called 'Belt-and-Road Initiative' (BRI). The cautionary tale of my title is rooted in my conviction that such projects, built upon historical mirages, are not sustainable, will cause irrevocable environmental damage, and will ultimately destroy the heritage they purport to celebrate.

The Fabled Port

That the very name 'Malacca' evokes a powerful aura in the region is due originally to geographical serendipity. This was crucial at a time when long-distance trade was dependent on seasonal winds that blew six months in one direction and six months in the other. Although historians are unable to pinpoint precisely when the city was founded, they agree that it stands as the oldest city on the Straits of Malacca (Selat Melaka in Malay). Their 'best guess' is that Melaka became a cosmopolitan entrepôt and center of the entire South China Sea spice trade 'sometime around 1400'.[6] A fine natural harbor, located at a narrow point in the straits through which traders had to pass in order to sail between India and China, made it the ideal place to meet and exchange goods. Arabs, Persians, and traders from India began to arrive each May and remained until January, while Chinese merchants, coming from the opposite direction, arrived in December and remained until late June, when the wind direction changed and they could begin their voyage home.[7]

Geographical good fortune certainly explains Melaka's location and rapid expansion but it does not adequately account for the aura that developed. To understand that, we need to explore the space where history and myth converge. I am not alone in believing this. While attempting to summarize two full days of conference papers on Melaka's early history, the eminent Southeast Asian historian Anthony Milner concluded: 'Myth has become social fact.'[8] He has a point. 'Malacca' as it exists today is a discursive space in which myths swirl and intertwine to form crystals of 'social fact', which grow and are juxtaposed in the present in ever more creative ways. To illustrate this, I will introduce three central (semi-)historical figures and suggest that, regardless of historical veracity, their myths cumulatively enhance Melaka's aura and continue to resonate into the present.

Parameswara

Parameswara is popularly described as a Palembang-born prince who fled from Srivijayan Sumatra in the late 14th century. There are many variants of his myth, of course, but the basic story – known to every Melakan schoolchild – is that while Parameswara was out hunting one day, he saw his dogs chase a small mousedeer up a hill. When the mousedeer reached the top, it turned to face its pursuers. The dogs were so startled by the mousedeer's unexpected bravery that they tumbled down the hill and into the river. Parameswara chose this hill, the place where the weak stood up to the strong, as an auspicious site for his settlement, which he named 'Malacca' after a large shady tree that stood upon the hill. This myth is memorialized in Melaka's coat of arms: two mousedeer stand rampant on either side of a shield bearing the image of a tree on a hill, bordered at the top by a row of five Malay swords. The two deer's hind legs are entwined in a ribbon that renders the motto *bersatu teguh* ('strength in unity') in Malay on the left and in Jawi on the right.[9] Above the shield is an Islamic crescent moon and star, which together index another strand in the myth: that Parameswara was influenced by the many Persian and Arab traders who came to the new port, converted to Islam, and took the name Iskandar Shah. Whether this is true, or a conflation of stories about later rulers, is irrelevant. It is the myth of Parameswara that explains the location, pinpoints the beginning of the Malay Sultanate and the adoption of Islam as state religion, and cements Sang Kancil, the clever mousedeer, as hero of countless children's folktales.

Cheng Ho (Zheng He)

Zheng He – Cheng Ho as he is known in Melaka – was a Muslim Chinese admiral sent with massive fleets by the Ming emperor Yongle to the South Seas and Indian

Ocean region six times between 1405 and 1421.[10] Here we see the seeds of regional engagement: Parameswara joined the Ming orbit, paying tribute to China in return for protection against piracy, while Cheng Ho established a trading depot in Melaka. In a powerful example of semantic snowballing, Admiral Cheng Ho has been associated with a well dug near the foot of Bukit Cina (a nearby hill),[11] a temple located near that well, and – decades after his actual death – escorting Hang Li Po, an unconfirmed 'Ming princess' sent to marry Mansur Shah (1459–1477), himself both the sixth sultan in Parameswara's line and the person said to have gifted Bukit Cina to the Chinese of Melaka. The myth of Cheng Ho cements the beginning of the long-standing relationship with China, a relationship doubtless facilitated locally by his Muslim identity.

Afonso de Albuquerque

By the time our third figure, Afonso de Albuquerque, the Portuguese Viceroy of India, arrived on the scene a century later, Melaka was the busiest, most cosmopolitan port in the region, crowded with goods, agents, and traders from all over the world. As Tomé Pires noted in 1512, 'in the port of Malacca very often 84 languages have been found spoken, every one distinct, as the inhabitants of Malacca affirm.'[12] Melaka was by then well-established as a crucial node in international trade – as witnessed by Pires's testimony that 'whoever is lord of Malacca has his hand on the throat of Venice'.[13] These two statements, often quoted, have become part of 'Malacca's' lore: the city was such a prize that Albuquerque laid siege to and captured it in 1511, built a surrounding wall and fortress (called *A Famosa*) to protect the city, and attempted to populate it with a resident community loyal to Portugal. Though only vestiges of the fort's wall and one gate, the Santiago Gate, remain, a small community survives to this day and forms one of Malaysia's more colorful minorities. Still loyal to their 'Portuguese' heritage, resolutely Catholic, and speaking a creolized form of Portuguese called *Kristang* or Melaka-Portuguese, about 1200 descendants live along the coast in an area called Ujong Pasir, not far from the ruins of Albuquerque's fort. Their village, which became known as the Portuguese Settlement (Kampung Portugis in Malay), was established in the 1930s during British times as a haven for the poorest of the local Portuguese fishing community.[14]

(Re-)Constructing 'Malacca'

Melaka's decline from its commanding position as the hub of the international sea trade was long and slow, fueled by the monopolistic practices of Dutch traders followed by a century plus of British colonial rule and strategic neglect in favor of Penang and Singapore. By 1879, Victorian traveler Isabella Bird described Melaka as 'a veritable Sleepy Hollow'.[15] Despite this, the mystique of 'Malacca' never entirely dissipated. Why else would Malaya's first leader, Tengku Abdul Rahman, drive to Melaka on 20 February 1956, to announce Malaya's independence from Great Britain?

For a brief moment, 'Malacca' was back on the map as the Tengku tapped into auspicious memories of past glory as – with his back to the sea – he faced Albuquerque's Santiago Gate, raised his right fist, and declared independence three times, 'Merdeka, Merdeka, Merdeka!'[16]

This salute to Melaka's past only momentarily disturbed its slumber. It was not until 1990, sparked by the federal government's first 'Visit Malaysia Year', that Melaka finally experienced a full-fledged renaissance, complete with a new slogan that rekindled nostalgia for the past: 'Malacca, *Bandaraya Bersejarah*: The Place Where It All Began.' Like ghosts returning from another time and place, Parameswara, Cheng Ho, and Afonso de Albuquerque reappeared, revived perhaps by the cosmopolitan sound of multiple languages, this time spoken by tourists, not traders. A 1990 TV advertisement for Visit Malaysia Year began with mysterious music emanating from a swirling gray mist:

> Slowly a nebulous image formed: as calm gray sea melted into cold gray sky, a ghostly apparition floated eerily toward us through thinning fog. As the mist cleared, the shadows coalesced into a ship, a huge wooden ship, European in style, with white canvas sails hanging wraith-like from proud yard-arms. Though the image remained artistically blurred, the reference was crystal clear to everyone sitting around the flickering television set: the year was 1511 and we were witnessing Afonso de Albuquerque's flagship, the *Flor de la Mar*, sail silently into the fabled port of Malacca. An instant later, magically transported to the present, we were bombarded by a kaleidoscopic array of current attractions flashing above a single command: 'Visit Malacca, Malaysia's Historic City'.[17]

If Albuquerque's presence was subliminal in the TV ad, it soon became palpable as the government began building a full-sized replica of the *Flor de la Mar*, as the site of its new Maritime Museum. Completed in 1994 and officially opened by then (and now-again) Prime Minister, Tun Dr Mahathir Mohammad, Albuquerque's ship – moored (in concrete) at the river mouth where the old traders' warehouses used to be – remains the most visited museum in Melaka.

Tying past to present, Albuquerque's descendants, the residents of the Portuguese Settlement, became the face of 1990s 'Malacca' as the State Economic Development Corporation promoted cultural performance as the crowning jewel of its tourist package. Every Saturday night the cultural show they sponsored at the Portuguese Square was packed with tourists who ate local Portuguese food and drank at restaurants surrounding the stage. Each week the SEDC showcased the city's cultural diversity – Chinese, Indian, Malay dance groups – but the grand finale always featured one of the Portuguese cultural dance troupes. Dressed in imitation of mid-20th-century Portuguese *rancho folclóricos*, they performed a heteroglot mix that included Portuguese dance imported in waves between 1952 and the 1980s, locally composed and choreographed songs and dances in similar style, and their own indigenous dance called *branyo*. While dances of the first two categories connected the community with its mythical Portuguese heritage (though an anachronistically modern representation thereof), the latter, a social

Fig. 10.1. Jerry Alcantra (musician) with Sherine Gallyot (dancer). Sherine is wearing local traditional dress known as kebaya kumpridu*, 2017. Photo M. Sarkissian.*

dance that accompanied song-duel quatrains called *mata kantiga*, was truly local. Sung in Kristang, their own Portuguese creole language, *mata kantiga* verses and *branyo* rhythm and melodies unified generations on the dance floor at weddings and community celebrations, as well as multicultural audiences at cultural shows. Switch the language from Kristang to Malay, keeping everything else the same, and the dance becomes *joget*, Malaysia's national social dance. The ironies are multiple. Modern Portuguese folkloric dances become a symbol of Historical 'Malacca', musicians and dancers become fictive ancestors of their own parents, and the repertoire that is uniquely theirs – and ties them to their own soil – is distinctly hybrid (Malaysian-Portuguese) and associated with local sarongs and *kebayas* (Fig. 10.1).

Modern 'Malacca'

Almost 30 years later Parameswara, Cheng Ho, and Afonso de Albuquerque remain pervasive – and crazily juxtaposed – in present-day Melaka. They are memorialized in everything from street and business names to colorful wall murals and commercial tourist kitsch.[18] Albuquerque's ship-shaped museum is a short walk from the Sultan's Palace (an imagined 'replica' harking back to Parameswara's Melaka Sultanate), itself located a stone's throw from Albuquerque's Santiago Gate. Just across a short bridge is the Cheng Ho Cultural Museum, a private enterprise founded in 2006 by a local Chinese businessman on what he claimed to be the very site of the admiral's depot, 'proof' of which he demonstrated by displaying 'Ming relics' conveniently unearthed during the renovation process. It is hardly surprising that in 2008 UNESCO declared Melaka a World Heritage City.

Mythification is a continuing process. Sometime since 2015, a large polished black granite stone appeared at the end of a narrow alleyway, facing the old Dutch buildings across the river. Etched with Cheng Ho's (imagined) likeness, the stone supposedly marks the place where 'Admiral Cheng Ho arrived in Malacca in 1405'.[19] The alleyway has been renamed (rather optimistically in view of its size) the 'Malaysia-Chinese Friendship Botanical Square'. Each tree (three so far) in the otherwise paved area has its own smaller granite marker that memorializes the Chinese dignitaries and local Malay politicians who planted it. That the dignitaries were Mr Meng Jianzhu, Secretary of the Committee of Political and Legislative Affairs of the CPC Committee (10 February 2015), Mr Zhu Xiaodan, Governor of Guandong Province (19 September 2015), and His Excellency Huang Huikang, the Ambassador to Malaysia (26 March 2016) clearly demonstrates the increasing importance of what we might call 'Cheng-Ho diplomacy'. There is, of course, absolutely no historical basis for this being the place where Cheng Ho stepped ashore. The past and the present exist in a dizzying post-modern mélange.

Modern 'Malacca' has become a victim of its own success, and that has led to rapid urban development. The very foundation of that success – its maritime romance – has meant that development has been unfairly weighted towards coastal reclamation. One early casualty of this was Assumption Chapel at the end of Praya Lane.[20] Until the early 1980s the chapel sat at the edge of the water with a statue of the Virgin Mary looking out over the seawall, standing watch over local fishermen. Today the statue faces the back of the church; Mary has turned her back on the half mile of urban blight built on reclaimed land that now separates her from the sea.

Building a New 'Malacca' ...

Reclamation projects have proliferated exponentially over the last two decades to the point that it is now development for its own sake. The entire subdivisions of Melaka Raya and Kota Laksmana are built on reclaimed land. Rows of shophouses stand empty and unfinished. A glut of hotel rooms ensures intense competition for Chinese tour busses that stop in Melaka overnight en route between Singapore and Kuala Lumpur. This brings little real benefit to the local economy; hoteliers hate it, but have become dependent on it to fill their otherwise empty rooms. At the other end of the economic scale, luxury condominiums are being marketed to wealthy Singaporean, Chinese, and Taiwanese investors. Gleaming towers sprout up as fast as land can be reclaimed, each stealing the sea view of yesterday's prime address. One of the newest, SilverScape Residences, a 45-storey high-end condominium built on reclaimed land, looms over the Portuguese Settlement and blocks the sunset. It is currently the tallest building in Melaka, visible all over the city, and part of 'Hatten City, Phase 1'. Phase 1 alone has 'an estimated gross development value (GDV) of RM2.05 billion [and] comprises Elements Mall, SilverScape Residences, Hatten Place, and Double Tree by Hilton Hotel Melaka'. When Phase 2 is completed the combined project will have an aggregate gross floor area of 9.6 million square feet.[21] These figures are as bedazzling as the buildings themselves. No one seems to care, however, that this is square footage of empty space. After two years, the only human activity in Elements Mall can be seen in the vicinity of the cinema on the 11th floor. Apartments, bought as investments by foreigners, remain unoccupied unless the owners lease them to a property company that promises to manage them as short-term rentals at a guaranteed rate for a six-year period. What will happen after that is uncharted territory. For now, though, the hyperactive short-term rental market is further killing the hotel business.

Melaka's allure has always been tied to the sea and to trade. Parameswara came by sea from Sumatra, Cheng Ho from across the South China Sea, and Albuquerque sailed halfway around the world to trade with Melaka. With old 'Malacca' ever further inland, 21st-century developers – with not-so-hidden connections to China – are now building 'Melaka Gateway', a new 'Malacca' in the sea.

According to its website:

> Melaka Gateway is a premier mixed development project initiated by KAJ Development Sdn. Bhd (KAJD), which [will] comprise of three (3) reclaimed and one (1) natural islands totalling 1,366 acres, set in a strategic and idyllic area of Straits of Malacca, Malaysia.[22]

Promotional material drips with nostalgic references to 'Malacca's' proud heritage as an emporium and its connection to the sea.[23] There is even a 'Melaka Gateway Song', the lyrics – in English – of which reference the sea, history, and new adventure (see Appendix A). The (now former) Chief Minister of Malacca, Idris bin Haron who was responsible for authorizing the project in 2014, clearly understands the power of

Fig. 10.2. The 45-storey SilverScape Residencies and beginnings of the Melaka Gateway development as seen from the far side of the Portuguese Settlement, showing trash buildup due to lack of water flow. 2018. Photo by Margaret Sarkissian.

nostalgia – and the myth of Cheng Ho – as a lure for Chinese developers and is keen to take advantage of the money they are so willing to invest:

> 'If you ask why Beijing is interested in Melaka, I think the historic legacy is also one of the reasons. Admiral Zheng He – his presence in Melaka in the 1400s is still remembered,' Idris Haron, Melaka's chief minister, said in an interview at his official residence.
>
> Zheng He, the Chinese explorer, remains revered for expanding China's maritime trade to places as far as Africa. Melaka's alliance with China at the time made it a key hub for the lucrative trade of spices.
>
> The state of Melaka is seen as the cradle of the contemporary Malaysian nation with its sultanate stretching back to the 15th century. Zheng, a Muslim from Yunnan province, is said to have been instrumental in converting Melaka's first ruler Parameswara to Islam.
>
> The subsequent invasion of the province by the Portuguese, followed by 19th- and 20th-century colonial ruler Britain's preference for Singapore as its regional entrepôt, led to the demise of the Melaka port. 'I was asked if I was ready to revive the deep-sea port that once made Melaka the Venice of the East, with nearly 80 languages spoken at the river mouth of Melaka. I said we will happily accept, with pleasure,' Idris said.[24]

Although the local developer, KAJ Development, describes itself as 'a wholly privatized Malaysian company',[25] the project is widely advertised as part of the BRI.[26] Giant billboards outside KAJ's headquarters on Pulau Melaka (itself a failed island project dating from the mid-2000s)[27] make clear that the building work is being done by China

State Construction's Third Engineering Bureau, heavily financed by China's state-owned Power China International Group with investment partners from Shenzhen and Shandong. The billboards – and the website – promise a 'Heritage Promenade', a massive Ferris wheel, a 70-storey hotel rivaling Dubai's Burj al-Arab, a container port, a Free Economic Trade zone, a private marina for luxury yachts, and a terminal for Royal Caribbean mega cruise ships. The 'Heritage' and entertainment precincts will provide a Disney-like, luxury experience – an imagined 'Malacca' – for wealthy tourists, while the marina and terminal will, ironically, block access to the sea for locals.

Melaka Gateway is a deeply flawed project that ignores social and environmental impact studies in favor of quick profit and under-the-table incentives to local and national politicians. The 2018 general election and transfer of power from entrenched interests to the new Pakatan Harapan, Party of Hope (led, ironically, by 93-year-old former PM Tun Dr Mahathir Mohamad), led to hopes that sanity might prevail. Despite canceling some of the BRI projects, notably the East Coast Railway Project, Dr Mahathir was wooed heavily by China and on 3 January 2019, Malacca's new Chief Minister Adly Zahari announced that the Gateway Project would continue. Reclamation, which proceeded around the clock after the election, had gone too far to be stopped.[28]

It is no coincidence that while this project's future was up in the air, the Chinese Embassy of Malaysia released a song that celebrated the long connection between Malaysia and China across the sea (see Appendix B).[29] Released on 28 January 2019, *Bahu Kiri* ('Left Shoulder') is sung in Mandarin and Malay by two 19-year-old Malaysian girls, one Chinese (Jeryl Lee Pei Ling), the other Malay (Siti Nurmaysitah Mohd. Yusoff). A *South China Morning Post* correspondent interprets the song as describing 'the two countries as 'brothers through it all' who 'nothing can stop' from bonding'.[30] The video is full of images of Asia for the Asians: artistic brushstrokes revealing images of the Great Wall, Kuala Lumpur's Petronas Towers, Dr Mahathir and Xi Jinping, Chinese festivals, and pandas. A spinning globe in the hands of Lee segues into a shot of the two girls sitting side-by-side on a boardwalk, looking out to the open sea. Their story, played out in rural rice *padi* fields and beside the sea, is juxtaposed with scenes in which three children interact playfully at school (a Chinese boy and girl helping a Malay girl with her homework), in a bucolic field, and by the water's edge. The song ends with the two singers sitting side-by-side on a pier, looking out at a calm and peaceful sea: 'Although there are setbacks, you calm my heart. My loyal and honest companion throughout our lifetime. My right shoulder will protect you from the wind and rain. My left shoulder will be where you lean on.'[31] In *Bahu Kiri* the sea is empty, calm, and peaceful, a natural place full of longing and strong connection; it is the antithesis of the soaring buildings and non-stop hyperactivity promised by Melaka Gateway.

... Or Creating an Environmental Disaster?

Parameswara and Cheng Ho's ghosts might approve of this romance, but what about the ghost of Albuquerque? It is his descendants, the residents of the Portuguese

Settlement who are affected most by the Melaka Gateway project, for it is being built immediately in front of their village – the very village that was created to give the 'poor Portuguese fisherfolk' a home by the sea. Fishing has been in decline as a source of income for the community since land reclamation projects began in the early 1980s to the extent that today few can make a living by fishing alone. The environmental impact is clear. Changing currents have destroyed the once bountiful coastal shrimp breeding grounds. Net fishing now requires taking small boats deeper into the Straits of Melaka and dodging huge container ships that ply the busy international shipping lanes. Silting between the coast and the growing island complex is already so bad that Settlement fishermen now have to wade waste deep in mud for clams. With no way of flushing the water between the coast and the new islands, the narrowing channel will inevitably become a garbage dump; a process that has already started (Fig. 10.2).

But what of the social impact? Reclamation affects fishing communities all over Malaysia,[32] but it affects the Portuguese Settlement more than most, for it is not only their livelihood that is threatened, but also their unique culture. As former community leader Richard Hendricks explained, 'Where there are Portuguese around the world, there is the sea. There is no separating the Portuguese from the sea.'[33] This is exactly what is at stake. Settlement fishermen are being divided by the developer's promises of lump sums as 'compensation', pitting fishing families against each other and against non-fishing families, thus threatening the cultural stability of the whole community. Whether active fishermen or not, all Settlement residents have a profound spiritual connection to the sea. Government officials may say that the sea belongs to the State, not to the Settlement, but Gerard Danker, a leading fisherman, responds angrily, 'The sea does not belong to the State; God made the sea and He gave it to the fishermen.'[34]

As the Portuguese Settlement becomes little more than a muddy lagoon, residents will inevitably be forced from their land and away from the sea. Once their land is gone, they will have nowhere to go, nowhere to call home. *Bahu Kiri* writes them out of the picture completely. There are no cute Malaysian-Portuguese schoolchildren in the video though they, too, are Asians of long standing. They also look longingly out to sea, but not out of nostalgia. As surely as Melaka Gateway will block the UNESCO-certified world heritage city's *feng shui* and its connection to the sea, it will also sever the ties of the last remaining Malaysian-Portuguese community to the proud seafaring tradition of their ancestors. Instead Albuquerque's descendants will look out at a massive wall of glass and metal where an ersatz 'Malacca' will provide non-stop entertainment and high-tech distractions for 21st-century sojourners who no longer depend on monsoon winds to traverse China's new Maritime Silk Road. Perhaps 84 languages will be heard once again in this Disney-like playground for the super-rich, but what myths will they have to talk about when the aura of old 'Malacca' has been erased? Or, will the project fail leaving a monumental mess to be reclaimed slowly by rising tides and encroaching mangroves (Fig. 10.3).

Fig. 10.3. View from the Portuguese Settlement. March 2019. Photo Colin Goh. Used with permission.

Appendix A: Melaka Gateway Song[35]

As the tide swept to shore
I stand spellbound at its wake
It was a sight to adore
I gaze breathlessly at the Strait
She shimmers so magically
The great waterways of the brave
Setting the keystones of history
Melaka emerges from the waves
The bricks of history pave the way
Let's set on a new adventure
Peer through Melaka Gateway
The Past presents the Future.

Appendix B: *Bahu Kiri*, 'Left Shoulder'[36]

The original version of this song, premiered on 28 January 2019, was performed in Mandarin with Malay subtitles. The following is a translation of the Malay subtitles.

A second version of the song premiered on 30 April 2019. This second version, sung in Malay with Mandarin subtitles, uses the same video but has been renamed *Sahabat Sejati* ('Eternal Friends'). The Malay text has replaced specific references to the South China Sea and the two shoulders with more indirect expression idealized through a perfect friendship.[37] The differences between the two texts are significant, but the subject for another paper.

Cahaya mentari menembusi awan yang bergumpal-gumpal
Gunung-ganang yang tinggi sama-sama kita medaki
Hanya sekilas toleh senyumanmu yang mesra menyentuh hatiku
Sahabat sejati tidak akan dipisah jarak dan waktu

The sun shines through the thick clouds
As we climb high mountains together
A glance of your friendly smile touches my heart
True friends that will never be separated by time or distance

Menyaksikan Laut Cina Selatan sebagai satu sanjungan yang memilukan
Kita berkongsi segaris sisiran pantai yang menjadi jalinan perpaduan
Pasti ada rintangan, namun kau menenteramkan hatiku
Teman sepanjang hayat yang setia dan jujur

Watching the South China Sea from either end
Our shared coastline is a bond of unity
There will be obstacles, but you calm my heart
My lifelong companion, loyal and sincere

Hari menjelang siang, namun hujan belum reda
Kita tetap duduk bersampingan
Ikatan persahabatan tidak akan putus
Ombak datang dan bawaku pergi
Bahu kananku menghalang angin dan hujan untukmu
Bahu kiriku untukmu berpaut

It's close to noon, yet the rain has not yet abated
Seated side by side
A bond of friendship that will never break
The waves come and take me away
My right shoulder protects you from the wind and rain
My left shoulder is for you to lean on

Notes

1 I would like to thank Colin Goh for taking many of the photographs that inspired the oral version of this paper but sadly could not be included in the published version, Datuk Dionysius Sharma and Christine May Yong for advising on matters of translation, Frank Citino for help with formatting figures, and Jerry Dennerline for reading and commenting on drafts of this paper. I am also grateful to Jeff Lerner and his team, including Elizabeth Clendinning, for inviting

me to the stimulating conference, *The Silk Roads: From Local Realities to Global Narratives*, held at Wake Forest University (27–30 March 2019).

2 The city's name has been spelled differently over time: Malaca (in Portuguese), Malacca (in English), and Melaka (in Malay). I will use the contemporary Malaysian preferred spelling, Melaka, except when I am clearly referring to the historically imagined mythical city, which I will represent as 'Malacca'. I will not change, however, spellings in quotations.

3 According to the 2010 census, the population of the state of Melaka was 821,110, while that of Melaka Tengah (Central Melaka, one of three subdistricts) was 503,127 and Bandar Melaka (Melaka City), 484,855. Melaka Tengah encompasses Bandar Melaka and surrounding neighborhoods, including Ujung Pasir, where the Portuguese Settlement/Kampung Portugis is located. (See JPM 2010.)

4 JPM 2010.

5 A 2015 study states that tourism was 'the most important economic sector in the state, contributing 46.6% of GDP' (445). Amir *et al.* 2015, 443–452.

6 Lowey-Ball 2019.

7 Hussin 2007, 3–5.

8 Milner 2019.

9 Jawi is a form of Arabic script formerly used for writing Malay and related languages in the region. It continued to be taught in primary schools until the mid-1970s as part of the Bahasa Malaysia language curriculum (personal communication, Datuk D. Sharma, 15 October 2019). It is now only maintained in more conservative Malay areas such as Kelantan and Trengganu.

10 See Wade 2005; Dreyer 2007.

11 Bukit Cina: Malay, 'China Hill'.

12 Cortesão 1944, 269.

13 Cortesão 1944, 287.

14 For a study of this community and a history of the creation of its village, see Sarkissian 2000, 36–49.

15 Isabella Bird arrived in Melaka on 20 January 1879, and recorded her first impressions, 'Very still, hot, tropical, sleepy, and dreamy, Malacca looks, a town "out of the running", utterly antiquated, mainly un-English, a veritable Sleepy Hollow' (Bird 1990, 125).

16 Newsreel footage of this historical moment is available on YouTube: *20 Februari 1956 Pengistiyaran tarikh Merdeka di Melaka*. The Tengku's car passes the Santiago Gate at 0:04. The speech is given from the *padang* (field) opposite the gate, although with the crowd it is impossible to see the actual gate. Today this *padang* is now a large shopping mall.

17 Sarkissian 2000, 8.

18 The giant murals, painted on the walls of buildings that abut the Melaka River, are only visible to those who ride river cruise boats or walk along the recently created river boardwalk, and thus rarely seen by locals.

19 This alley runs from Lorong Hang Jebat (still known by its old name, First Cross Street) to the riverbank.

20 Praya Lane in Bandar Hilir (halfway between the Santiago Gate and the Portuguese Settlement) was a mixed-race community with a significant number of Portuguese families. The poorest members of this community, almost all fisherfolk, were among the first beneficiaries of homes in the new Portuguese Settlement in the mid-1930s. A few families continued to live near Assumption Chapel, despite the land reclamation around them, until developers bought the land and razed all the homes.

21 NST Property 2018.

22 Melaka Gateway 2015b.

23 Melaka Gateway 2015d. Ironically, the company's motto, which appears on all of its promotional

literature, claims: 'The Past presents the Future.'
24 Jaipragas 2017.
25 As stated on the website, Melaka Gateway 2015a.
26 'Originally termed the "One Belt One Road" initiative and launched by [President] Xi [Jinping] in 2013, the BRI is a massive attempt to re-develop and modernise the old overland Silk Road and its maritime equivalent. Its end goal is to reconnect China to Central Asia and Central Europe on the one hand, and to South East Asia, the Indian subcontinent and East Africa, on the other' (Loh 2018).
27 Aside from a beautiful mosque that appears to float on the water, Pulau Melaka (Melaka Island) is full of empty shophouses. The small channel between the coastline (itself reclaimed land) and the island has almost completely silted up; it will soon be possible to walk to the island when the tide is out.
28 Yee and Amarthalingam 2019.
29 China Embassy Malaysia 2019.
30 Sukumaran 2019.
31 Lai 2019.
32 Fishing communities elsewhere in Malaysia, especially in Penang, have organized themselves to challenge the developers more effectively than has the Malacca Portuguese community. The Penang case is highlighted in a recent environmental study published on the website of Sahabat Alam Malaysia, an organization dedicated to working towards environmental justice and member of Friends of the Earth International, see Sahabat Alam Malaysia 2019.
33 Augustin 2017.
34 Gerard Danker, personal communication, June 2017.
35 An audio file of this song can be found at Melaka Gateway 2016c. There is no attribution to the song's composer or lyricist. The website only states: 'It was firstly composed as a poem to beautifully present the very essence of Melaka's past history and Melaka Gateway to the world. It was then recorded at a recording studio in Abbey Road, United Kingdom.'
36 *Bahu Kiri* is available on YouTube, see China Embassy Malaysia 2019.
37 *Sahabat Sejati*, is also available on YouTube, see Lee and Masyitah 2019. I am indebted to Christine May Yong for insights into differences between the two texts.

Bibliography

20 Februari 1956 Pengistiyaran tarikh Merdeka di Melaka [20 February 1956 Anniversary of Independence in Melaka]. 2013. https://youtu.be/3fHCMe-TWkM (accessed 19 October 2019).

Amir, S., Osman, M.M. and Ibrahim, S.B.M. 2015. Sustaining Local Community Economy through Tourism: Melaka UNESCO World Heritage City, *Procedia Environmental Sciences* 28, 443–452.

Augustin, R. 2017. Melaka Gateway raises heritage concerns for Portuguese Settlement, *Free Malaysia Today*, September 7. https://www.freemalaysiatoday.com/category/nation/2017/09/07/melaka-gateway-raises-heritage-concerns-for-portuguese-settlement/ (accessed 19 October 2019).

Bird, I. 1990 (1883). *The Golden Chersonese: The Malayan Travels of a Victorian Lady*, Singapore.

China Embassy Malaysia. 2019. *Bahu Kiri* (45th Diplomatic Ties Anniversary). https://www.youtube.com/watch?v=6hFUTiTiDFo (accessed 17 October 2019).

City Population. Melaka Tengah. n.d. https://www.citypopulation.de/en/malaysia/admin/melaka/0403__melaka_tengah/ (accessed 19 October 2019).

City Population. Melaka State. n.d. https://www.citypopulation.de/php/malaysia-admin.php?adm1id=04 (accessed 19 October 2019).

Cortesão, A. (ed.) 1944. *The Summa Oriental of Tomé Pires: An Account of the East, from the Red Sea to Japan, Written in Malacca and India in 1512–1515*, London.

Dreyer, E.L. 2007. *Zheng He: China and the Oceans in the Early Ming Dynasty, 1405–1433*, New York.

Hussin, N. 2007. *Trade and Society in the Straits of Melaka: Dutch Melaka and English Penang, 1780–1830.* NIAS Monographs v.100, Singapore.
Jaipragas, B. 2017. Can China really deliver Malaysia's Singapore Slayer? *This Week in Asia*, April 15. https://www.scmp.com/week-asia/politics/article/2087402/can-china-really-deliver-malaysias-singapore-slayer (accessed 15 October 2019).
JPM 2010. Jabatan Perangkaan Malaysia/Department of Statistics, Malaysia. 2010. *Banci Penduduk Dan Perumahan Malaysia/Population and Housing Census of Malaysia: Taburan Penduduk Dan Ciri-Ciri Asas Demografi/Population Distribution and Basic Demographic Characteristics.* https://web.archive.org/web/20150301154300/http://www.statistics.gov.my/portal/download_Population/files/census2010/Taburan_Penduduk_dan_Ciri-ciri_Asas_Demografi.pdf (accessed 15 May 2010).
Lai, A. 2019. Chinese Embassy Composes Song to Celebrate Malaysia-China Ties, *The Star Online*, 28 January. https://www.thestar.com.my/news/nation/2019/01/28/chinese-embassy-composes-song-to-celebrate-malaysia-china-ties/ (accessed 18 October 2019).
Lee, J. and Masyitah, M. 2019. *Sahabat Sejati.* https://www.youtube.com/watch?v=6FM-eMB0sRk&feature=youtu.be (accessed 23 October 2019).
Loh, F. 2018. BRI, Spike in Chinese Investments: What Are Implications for Malaysia's Politics, Sovereignty? *Aliran*, 3 May. https://aliran.com/aliran-csi/aliran-csi-2017/bri-spike-chinese-investments-malaysia-implications-malaysias-politics-sovereignty/ (accessed 17 October 2019).
Lowey-Ball, S. 2019. Oral summary comments, *Melaka-in-Fact COnference: Melaka in the Long-Fifteenth Century*, August 2–4, Ramada Plaza, Melaka.
Melaka Gateway. 2016a. About Us. http://melakagateway.com/kaj-development-sdn-bhd/ (accessed 17 October 2019).
Melaka Gateway. 2016b. Introduction. http://melakagateway.com/introduction/ (accessed 17 October 2019).
Melaka Gateway. 2016c. Logo and Song. http://melakagateway.com/our-logo/ (accessed 17 October 2019).
Melaka Gateway. 2016d. Why Melaka? https://melakagateway.com/why-melaka/ (accessed 15 October 2019).
Milner, A. 2019. Oral summary, *Melaka-in-Fact Conference: Melaka in the Long-Fifteenth Century*, August 2–4, Ramada Plaza, Melaka.
NST Property. 2018. Banking on Melaka booming tourism, *New Straits Times*, June 7, https://www.nst.com.my/property/2018/06/377580/banking-melaka-booming-tourism (accessed 15 October 2019).
Sahabat Alam Malaysia. 2019. *Impacts of Coastal Reclamation in Malaysia.* https://assets.nationbuilder.com/foemalaysia/pages/24/attachments/original/1551161113/190226_Impacts_of_Coastal_Reclamation_in_Malaysia-compressed.pdf?1551161113 (accessed 18 October 2019).
Sarkissian, M. 2000. *D'Albuquerque's Children: Performing Tradition in Malaysia's Portuguese Settlement*, Chicago.
Sukumaran, T. 2019. 'You are here for me for life': Beijing releases song marking 45 years of diplomatic ties with Malaysia, *South China Morning Post*, 28 January. https://www.scmp.com/news/asia/southeast-asia/article/2183975/you-are-here-me-life-beijing-releases-song-marking-45-years (accessed 17 October 2019).
Wade, G. 2005. The Zheng He Voyages: A Reassessment, *Journal of the Malaysian Branch of the Royal Asiatic Society* 78.1 (288), 37–58.
Yee F.K. and Amarthalingam, S. 2019. RM30b Melaka Gateway Project still on; talks now on changing island's status, *theedgemarkets.com*, July 29. https://www.theedgemarkets.com/article/rm30b-melaka-gateway-project-still-talks-now-changing-islands-land-status (accessed 18 October 2019).

Chapter 11

21st-century Trading Routes in Mongolia: Changing Pastoral Soundscapes and Lifeways

Jennifer C. Post

The soundscapes and soundways of pastoralism in Mongolia today may be connected to acoustic worlds in steppe lands and deserts that contributed in significant ways to the economy, ecology, and culture of the old trading routes. We cannot fully reconstruct them, but we know they were not characterized by the sounds of trucks rumbling by settlements, mining explosions that picrcc thc air and shake the land, motorcycles in grasslands weaving inbetween grazing animals, and songs blasting from jeep radios. Yet pastoralists' soundways – the ways they express their relationships to sound and its circulation – during periods we have isolated as critical times of conquest and trade, surely played a significant role in their lifeways as they tended to their daily needs and contributed to both ecological and economic production.[1] Soundscapes of Mongolian mobile pastoralists are linked to local ecosystems and ways of life, and are impacted by changes brought by social, political, and ecological events. The lifeways of these semi-nomadic pastoralists, historically and in current practice, have been widely supported by the natural environment, by human-animal and human-plant relationships that have played roles in the maintenance of pastoral practices for hundreds of years.[2] There is a connection between sensory experience and ecological knowledge, as well as links between other sound practices, such as music, and lifeways that provide social, spiritual, and practical support for pastoralists. For example, in the Altai Mountain region in western Mongolia, the practice of listening and sound production, including music-making, among the local Kazakh herders, engages them directly with the different environments where they grow and care for their livestock. Interacting in their acoustic communities – their socially shared sonic spaces – they access a continuous set of sounds of insects, birds, wild and domestic animals that communicate critical information about environmental health.[3] These herders indicate that animals such as deer, ibex, and argali offer comfort about the health of the land; they say sounds of different kinds of winds gently move grasses and leaves, or rush through rocky crevices and blow sand and soil, scaring animals and herders alike; sounds of wolves in the night prompt some herders to then vocalize in a particular style in order to scare the predators away from their livestock; songs about place reinforce identities and remind listeners of values that support pastoralists' hard work and encourage them to continue their lifeways, even during difficult times.[4] Today, the biodiversity challenges and new mobility opportunities generate a swirl of movement; lands once dedicated to supporting livestock and local families now also include miners, tourists, and representatives of organizations engaged in surveying or evaluating local practices, managing or adapting land use.[5] The herders' soundscapes and lifeways have been changed by forces that they cannot control, and this has impacted every facet of their lives.

The cycle of relationships celebrated in Silk Roads studies between those who moved as a way of life in support of their livestock, and those who are identified with sedentary ways of life, contributed to and influenced various forms of cultural production and exchange.[6] These cycles continue in Mongolia today, although the power structures and sites for exchange have now changed partly because the land has been redesigned. Opportunities for those who rely on mobility for their economic stability have diminished considerably in recent years. Lands once used largely by pastoralists have been appropriated for mining, roadways, and railways, projects driven not only by Mongolian government concerns about national growth, but by international interests in the country's rich resources and the easy access between China and Russia the country provides.

The difficulties for pastoralists are made more challenging due to widespread climate-related issues.[7] Their ecological and socio-economic spaces are affected by drought and by *dzud*, a period of bitter cold, snow and/or ice that kills thousands of livestock, as well as unpredictable weather that puts herders on edge daily.[8] Pederson compares the early 21st-century environmental difficulties to similar climatic events that led to socio-political instability in Mongolia during the 12th century and to the rise of Chinggis Khan.[9] Climate-related events, especially, have had a devastating impact on lifeways, and pastoralists are increasingly forced to abandon grasslands that have become unproductive, thereby separating extended families due to limited resources in each location.[10] The impact affects pastoral lifeways directly because families together support local biodiversity with both physical and cultural labor.

In this article, I consider the impact of new trade networks crossing Mongolian pastoral lands, especially China's Belt and Road Initiative (BRI), on the sounds, soundscapes, and lifeways of local herders. Sounds in these settings continue to characterize economic patterns, cultural scenes, and environmental changes. In the 21st century, new roads, new and planned rail services, and tourism growth all improve economic opportunities for Mongolians, their Chinese and Russian neighbors, as well as other nations, but the intrusions, coupled with changing values related to the land and recent climate change events, impact pastoralists' effectiveness in sustainable land management.[11] This contributes to land degradation and to loss of food sources, affecting both human and ecological well-being.[12]

Trade Networks: The Belt and Road Initiative

The Belt and Road Initiative (BRI) is linked to China's early efforts to engage economically with neighboring Central Asian states after the breakup of the Soviet Union. Plans in the early 1990s were linked to China's Central Asian diplomacy efforts with a goal to promote and maintain trade and other economic relationships with Central Asia.[13] The Trans-Eurasian Railroad from China through Central Asia was opened in 1990, and the cities of Almaty in Kazakhstan and Urumqi in Xinjiang Uyghur Autonomous Region in China were linked by rail in 1992. Borders between China and Mongolia were opened for business in that period as well. In that year, China promoted events that focused attention on the Silk Roads to stage national and international trade and tourism growth.[14] In 1994, Prime Minister Li Peng traveled to Central Asian countries to promote the Silk Roads with an interest in developing economic pathways into these regions for oil production, and more.[15] In a speech in Uzbekistan Li Peng referenced the long history of relationships between these regions, stating, 'China attaches great importance to its relations with the Central Asian countries and with them, we are willing to work to build a new Silk Road, with a perspective towards the future and the next century.'[16]

The Silk Road imaginary adopted by China in connection with their development plans used imagery from Central Asian trading route cities, such as Samarkand and Bukhara, that continue to be used in the 21st century.[17] The Chinese government capitalized on the Silk Roads concept by developing and expanding tourist sites from Xi'an to Kashgar. Millward suggests that 'China represents Xinjiang and the Silk Road as a stage on which China plays the leading historical role'.[18] A recent example of the effort to engage an international audience with the Silk Roads concept visually, sonically, and economically occurred during a 2014 Silk Road Cultural Journey organized by a local tea company and the Shaanxi government. With the support of 100 camels, a series of horse-drawn carriages, and more than 100 travelers wearing clothing identified with the Silk Roads era, the group traveled from Shaanxi to Kazakhstan along the Silk Roads route giving performances along the way.[19]

China's Xi Jinping introduced the One Belt One Road (OBOR) Initiative in 2013 to support regional and global collaboration. The initiative combined the 'Silk Road Economic Belt' that encompassed overland routes between China and Central Asia with the '21st-Century Maritime Silk Road'. With this plan the Chinese government expressed the intent to create a road from China to Europe through Central Asia and across the Indian Ocean, and to link their economic corridors to Africa and Oceania. In all, they designed six OBOR Initiative corridors in the new millennium: 1) China-Mongolia-Russia Economic Corridor; 2) New Eurasia Land Bridge (Second Eurasia Land Bridge) Economic Corridor; 3) China-Central Asia-West Asia Economic Corridor; 4)

China-Indochina Peninsula Economic Corridor; 5) China-Pakistan Economic Corridor; 6) BCIM Economic Corridor.[20] Each corridor represents unique social and political relationships between China and the world region, mediated by China's interests. The China-Mongolia-Russia Economic Corridor plan had two goals: to gain access to Mongolian resources and to establish more convenient approaches to Russia for trade.

China-Mongolia-Russia Economic Corridor

In Mongolia, the collapse of Soviet-backed socialism in the early 1990s encouraged them to focus on (re)building the country's economic strength as an independent nation. This included plans to develop road networks to link centralized urban sites, contracts with companies for mining rights, and other activities.[21] When Mongolia began their 'Millennium Road' plan in 2001 – a road and railway development project – they made limited progress initially because they needed foreign investment for its completion. New projects launched in 2013 and 2014 to link China and Russia through Mongolia included roads, railways, gas and oil pipelines. These efforts connected Mongolia even more fully and formally with China's Belt and Road Initiative.[22]

It is not surprising that the Mongolian government was drawn to the developing Belt and Road Initiative (BRI), especially the China-Mongolia-Russia Corridor, which links conveniently to the China-Russia influenced Mongolian programs to expand highway and rail access to all the provinces – supported largely by mining company money from outside the country.[23] The relationship among the three countries was cemented with a summit in Tajikistan in 2014 during which Mongolia's then president (Elbegdorj) sought engagement with the expanding connections between China and Russia.[24] In 2016 presidents Putin, Xi, and Elbegdorj signed a trilateral plan to create the China-Mongolia-Russia Economic Corridor (CMREC) with a focus on linking transportation networks and establishing and maintaining economic cooperation.[25] Activities in connection with the China-Mongolia-Russia Economic Corridor and other related projects that can be linked to the BRI, demonstrate a growing coordination between China and Russia as they relate to Mongolia. With the growth of the projects, Mongolia moved into 'a new era of economic opportunity through cooperation with its two giant neighbors'.[26]

The impact of the projects in Mongolia are apparent in government and industry produced maps that show roads, railways, and mines representing the expanding industries resulting from the cooperation among the three countries. Maps also indicate the inevitably shrinking pastoral land. For example, a 2009 road map from the Ministry of Roads, Transport and Tourism, presented by SMEC International in Australia, a firm involved in engineering and infrastructure projects indicates the status of paved roads in Mongolia. Included are the AH-3 linking markets in eastern Russia and China through Mongolia; the AH-4 in Mongolia linking eastern Russian markets with Xinjiang Uyghur Autonomous Region in China and with Pakistan; the

AH-32, Mongolia's main east–west corridor linking at the western end with the AH-4, and at the eastern end with the AH-31 to access the Korean peninsula.[27] Contractors for many of these projects are Chinese and funding for Mongolia includes substantial loans from Chinese and other international banks. Maps also highlight road and rail networks linking the huge Oyu Tolgoi gold-copper mine in the South Gobi Desert near the Chinese border, along with Tavan Tolgoi coking coal mine and its nearby Ukhaa Khudag open coal mine to China. A new rail link from Tavan Tolgoi to China in 2021 will have the capacity to deliver 30 million tons of coal a year to China.[28] Rail lines and roads in the north pass through Erdenet copper mine, and to coal and silver mines in the west of the country into Russia. Maps of mining licenses also indicate the breadth of the plans for Mongolia. A widely distributed map shows existing and planned railway routes, with an apparent interest to connect mines; the Ministry of Road, Transport and Urban Development's rail map uses Aspire Mining (Australia) data to indicate coal projects as well as other mines.[29] There are well over 3500 active mining licenses, although during the last decade the Mongolian government has suspended and revoked many licenses as they identify violations and recognize the detrimental effects the projects have on local environments.[30] A 2016 map of mining licenses from the Mongolian Mineral Resources and Petroleum Authority indicates widespread activity continues.[31]

The activities surrounding these economically driven plans, though, indicate that they are likely to push Mongolia to a point where the grasslands that pastoralists have strategically and successfully maintained for generations will no longer be able to support essential food security for that country and beyond.[32] The routes are built with benefits for industries, not pastoral practices and peoples.

Mobile Pastoralists in Modern Mongolia

It is easy to minimize the significance of the shrinking world of mobile pastoralists in Mongolia when such big economically-focused projects are being developed to benefit a growing urban population. Yet pastoralists continue to play a role in land conservation and to contribute significantly to the country's economy. They use rangelands to their advantage while agriculturalists struggle to deal with the same increasingly nutrient-poor lands.[33] One important strategy for managing grassland resources includes seasonal migration to allow grasses to regenerate. Pastoralists' mobile ways of life depend on the social experience of sharing cultural and ecological knowledge, and efforts to provide economic support to manage the unstable rangelands. Pastoralists decide where to move their livestock based on the health of the land, livestock dietary needs in each season, and access to water.[34] Their ability to adapt to change has allowed them to maintain ways of life in varied circumstances. In fact, pastoralism is linked to food security for large populations especially because of this ability to manage variability.[35] Rangeland stewardship is increasingly recognized as critical to the future of peoples and lands in many locations.[36] Furthermore, relationships between nomadic and settled peoples contributed significantly to the innovation and adaptation that helped to develop some of the tangible and intangible cultural materials and practices valued today. They continue to be valued in Mongolia not only in rural locations, but in urban contexts as well.

During a period when global climate change is disturbing grasslands and impacting lifeways across the globe, there is little evidence the new economic corridors crossing through Mongolia will support local residents' ability to sustain the social-ecological systems critical to the maintenance of local biodiversity. The changes include drought, flooding, diminished soil quality, insect infestations, generally unpredictable weather patterns, and more frequent appearance of *dzud*.[37] In addition to grasslands and domestic animals, Mongolia supports populations of rare flora and fauna in nearly 100 protected areas, including in transboundary regions such as the Altai-Sayan Ecoregion (ASER) shared with Russia, Kazakhstan, and China, and the Amur-Heilong Ecoregion shared with China and Russia.[38] The new and planned economic corridors passing through fragile ecological systems, and paved roads and rail lines crossing grazing lands impact both human and nonhuman residents. They bring industries, from mining to tourism, and they draw pastoralists out of the countryside as well to travel to cities in Mongolia and beyond.

Sounds and Soundscapes of Pastoralism

Recent scholarship in the sciences and social sciences reveals how important sounds and soundscapes are to our understanding of patterns of biological, social, psychological, and cultural behaviors.[39] Geographer Revill reminds us that more attention needs to be paid to the 'processes and practices by which sound actually makes space', shaping and transforming experiences connected not only to space but to practice.[40] Duffy suggests that 'sound is conceived as neither "conscious" nor "nonconscious" but both; neither "cultural" nor "material" but the two concurrently; neither physiologically nor psychologically but both at once; neither "biological" nor "social" but both simultaneously.'[41] Since sound is linked to environment for pastoralists, studying its meanings and use can reveal the impact of projects such as the BRI that are being developed economically yet are poorly conceived socially and ecologically, a theme that appears in other BRI projects as well.[42] In rural Mongolia, pastoralists' sounds and soundscapes exist in their acoustic communities made up of multi-species assemblages.[43] Sounds in these spaces help herders evaluate, communicate, organize, and mobilize and thus sound is animated in social relationships among human and nonhuman actors in each space. Sound practices, from listening to sound-production to sharing songs in social settings, are critically linked to local ecological knowledge and thus play key roles in the conservation of biodiversity.

Listening. For herders, listening is a strategy for evaluating the quality and needs of both animals and the land. They engage with sounds and soundscapes in each more-than-human social setting, in relationships with domestic animals, grasses and trees, the wind, family members and neighbors to evaluate each acoustic scene and the health and character of individual species and landscapes. They also listen to identify dangers and for signs to help them make decisions about seasonal moves. In a yurt, the felt walls allow family members to hear the sounds of their animals throughout the day and night. Herders managing livestock listen to winds blowing through the grasses, sounds of insects, and various animal sounds as they evaluate the quality of the land and anticipate weather events. Families may time their seasonal moves in connection with the sounds and activities of different species of birds. For example, in Bayan-Ölgii Kazakh herders may schedule moves in relation to the sound of the demoiselle crane (*tyrna*) or the bar headed goose (*aqbas qaz*). Kazakh and Tuvan herders learn of a coming cold period from the red-billed chough (*shauqargha*) that hangs around wetlands and settles in rocky crevices. When groups of these birds gather and fly up singing it is a sign of approaching cold weather for which herders must prepare their animals and families.[44] Altai Tugelbai at his summer settlement (*zhaylau*) referred to the chough as a 'snow predictor'.[45] For herders settled during the winter near a lake, the particular sounds of ice cracking help them determine the progress toward spring when they gather their animals to head for spring birthing sites.

Production. Herders throughout Mongolia also regularly produce sounds to communicate with their animals and to influence animal behavior. They recognize through practice that vocalizing specific sounds, including songs, calms yaks, cows, and horses during milking and gives directions while herding or traveling.[46] The bonds between domestic animals and humans are reinforced through sensory experience, especially sound and touch. A woman milking a yak may lean in, sometimes placing her head close to the animal's side (Fig. 11.1). Vocalizing or singing in such a setting activates a relationship mediated by specific sounds believed to play a role in calming or communicating with animals. For some, songs that express sentiments significant to humans also contain sounds meaningful to animals.[47] Herders also sing songs while riding. It is not unusual to hear songs echoing from rocks or across the steppes as riders travel or tend their animals, the horse's gait encourages particular songs and rhythms.[48]

Sounds while tending animals can also include the music of flutes crafted from local reeds[49] and tossed away when they break down. In western Mongolia, Kazakh and Tuvan herders recall learning to make and play the end-blown-flute when they were young. Olonbayar said he learned from his father to look for *quray*, a local reed, yet sites for growing these plants have diminished in recent years. Some herders said they were encouraged to emulate the sounds of the camel as they were learning to play the *quray* flute.[50] Other Kazakh musicians also indicate there is a mimetic relationship between the sound of the camel and an ideal sound quality for the more permanent end-blown flute (*sybyzghy*) made of wood or bamboo.[51] Furthermore, music played on this instrument can also describe a landscape in sound.[52] Levin and Süzükei identify this as the outcome of timbre-centered listening for rural

Fig. 11.1. Singing while milking in Khatgal, Khovsgol province (J. Post, 2004).

Tuvans who engage in spiritual, acoustic, and visual relationships with the land as herders and hunters. They argue that 'listening carefully to the sounds of the natural world is an essential prerequisite to becoming a musician'. Thus, listening timbrally offers both knowledge about the environment and provides inspiration for music-making.[53]

Socializing. Sounds are shared socially as songs, instrumental tunes, and recitations in informal and formal daily and seasonal family and community events. Song lyrics reinforce relationships with the land and express well-being. They address the values and needs of local communities in relation to their locations, histories, and character; reference livestock, wildlife, and the quality and availability of resources; and express beliefs about social relationships and responsibilities, health, safety, and well-being. Some songs include references to sounding sources that pastoralists enjoy and often rely on. For many Mongolian residents, the *urtyn duu* song form offers general descriptive information about landforms, livestock, wildlife, and social life, acting as a shared mode of expression, musically and lyrically, that binds people to the land.[54] The *terme* genre maintained by Kazakhs in the western provinces also provides descriptive information.[55] When Altai Tügelbai sings about his summer encampment in Bayan-Ölgii's Dayan Lake region, he shares local knowledge about place, resources, and practices. Such songs link ecological knowledge and sense of place, they express local identities and reinforce shared values, including musical values: Altai sings 'with the voice of my grandfather', he says, indicating he is using his grandfather's melody and style of singing. In the multiverse song with characteristics common to other songs of this genre, Altai links his song to *location and place* (he describes his residential community (*auyl*) near the Altai mountains); *characteristics of place* (he references the valley, trees, lake, birds); shows *family and community ties* (he references family relationships); identifies *resources for livestock and for humans* (water and moist grasses for animals, berries for herder families and wildlife); shows interest in *livestock numbers and diversity* (thousands of livestock scattered on the land); expresses concerns about *safety and security for all* (grazing safely away from wolves), including *human and animal well-being* (animals are comfortable and humans are happy in this place).

My *auyl* is on the edge of the Altai mountain	*Location and place*
Its beautiful valley is covered with pine trees	*Characteristics of place*
Dayan Lake is at the bottom of a hill	
Swans and geese live on the shore	
My *zhaylau* is Oiqaraghai at Daiyn lake	*Family and community ties*
It is my homeland where I have lived since childhood	
My people, my relatives, my siblings	
They are settled comfortably in their *zhaylau*	
There is a caravan road that descends to Oiqaraghai	*Characteristics of place*
If you continue on it leads you to Altai	
Örmegeiti lies down between them	
There are so many riches for all the livestock	*Resources for livestock*
Beldi Buyra begins when you walk from there	
My young people live their lives honestly	
They graze their five kinds of livestock safely	*Safety and security for livestock*
Even though with wolves it is difficult	

It [Beldi Buyra] unites with Qaraköl	
The Altai mountains continue from here	
There are moist grasses for all the livestock	*Resources for livestock*
It is my pride and joy	

Herders sing when grazing their animals	
Thousands of animals are scattered all over the land	*Livestock numbers and diversity*
Rich in water, the grass is moist	*Resources for livestock*
My people are happy here	*Human well-being*

The pine trees of Sarybuyra grow everywhere	
For *zhaylau* it is comfortable for animals[56]	*Animal well-being*
The children of herders tend their animals	*Family and community ties*
The two-year-old horses they ride suit them	

Between Tumbaötkel and Tasötkel	
Their shores are covered with a green carpet	*Resources for livestock*
Strawberry, blackberry and gooseberry	*Resources for humans*
The children of the *auyl* will pick them	

There is no other place like Sünggiti and Zhamaty	
It is as beautiful as the girl who dresses up to go to *toi*	
My Qolagash is the home for Uang tribe	*Family and community ties*
It is like the colorful wing of a ring necked pheasant[57]	

Altai mountains, your peaks are covered in fog	*Location and place*
The top always remains cloudy	*Characteristics of place*
The summer stays chilly, the winter is warm	
Anybody who sees it can be attracted to its *zhaylau*	

Engaging with the landscape through sound is connected to spiritual values as well. This is expressed clearly by Mongol herders whose praise songs (*magtaal*) and recitations that accompany specific rituals create soundscapes that embody values and offer lyrics that connect all participants directly to the land and its resources. In a recitation during the ritual *dallaga* (to summon good fortune) for the camel, cited by Chuulunbaatar, Khalkha Mongols reference the valued seasonal characteristics, followed by the repeated word *xurai* [*khurai*] which means to gather.

While the white tundra swans are arriving,
while the snow-covered mountains are melting,

while the yellow-white camel is having its young,
we ask for timeless great prosperity.
xurai, xurai, xurai
While the swans are arriving,
while the snow is melting,
while the old camel is having its young,
we ask for prosperity rich in blessing.
xurai, xurai, xurai[58]

Lindskog suggests that 'the soundscape produced through recitation as well as the perceived power of the words themselves' has a kind of magical power. Such texts may relate to a Buddhist ritual canon, but also contain symbolic content; 'the summoning of the masters and spirits of the land and waters of a particular homeland (*nutag*) generally plays a pivotal part in the prayer itself'.[59]

The experience of listening to and producing sounds and songs is deeply tied to the daily life experience of rural herders in Mongolia. These are activities that are often only meaningful in the unique contexts that characterize pastoral lifeways, a way of life that promotes the health and continuing growth of livestock. The well-being of livestock relies on the health of the land. Furthermore, this requires mobility as well as cooperation in family and community groups.

Changing Lifeways: Transforming Spaces and Sonic Practices

It is clear that the physical footprint of new economic corridors in Mongolia have not been developed with the sounds and soundscapes of pastoralists in mind. The result is that the roads, railways, mining sites linked to new economic models have had an environmentally and socially transformative role for rural residents, both human and nonhuman. For example, roads and railways have created fragmentation and loss of habitat and even barriers for wildlife. Studies led by Ito, Dorsey, and Olson document the impact of railways on the Mongolian gazelle and other ungulates.[60] The gazelles do not cross rails even though they are able to do so, creating a new barrier for them during an already difficult period environmentally. Parris's research on the impact of road noise on ecological communities indicates that the acoustical environment during construction and with traffic sounds impacts animals in several key ways. The sounds make it 'more difficult for animals to hear each other, their predators, and their prey' and creates stress.[61] Even subtle changes can have an impact on landscapes (Fig. 11.2). Herders' responses to such sounds are similar, especially when significant sounds are masked, diminish, and disappear. Furthermore, their music changes as the new economic corridors influence their mobility.

It is difficult for herders to continue to employ the same sounds they learned to use to communicate, evaluate, and share knowledge in the new acoustic environments changed by the economic corridors in which they find themselves. During a period when climate change requires adaptive strategies that draw on local knowledge,

Fig. 11.2. Road to a Chinese mining site through grazing land in Shegirtei, Bayan-Ölgii province (J. Post, 2018).

increasing vehicle traffic, in some regions, heavy trucks, trains, and planes, sounds tied to mining, including dynamite explosions, cover or mask sounds herders have been taught to use to communicate with their livestock and to evaluate land and animal health. The heavy equipment that travels along the roads once used largely by herders, while benefiting some investors economically, has proven to be detrimental to the resource-rich lands.[62] The mining activities impact water, soil, and other local resources herders and other inhabitants rely on for their health and well-being. The effects on auditory spaces and social dynamics are expressed in adapted approaches to the use of sound in daily life, with direct impact on effective pasture-related practices. Some of the sound-related impacts are linked to relationships to resources, vehicle use, and social gatherings.

When sounds and song events shrink in size and frequency, the ecological cycle and social systems supporting biodiversity may be interrupted as well. Similarly, when the characteristics of places change, the sense of place and sense of well-being developed through connections to resources are difficult for pastoralists to maintain. Herders in western Mongolia express discomfort with the transformations of acoustic spaces; most frequently, they reference the unpredictability of weather and other events that impact their mobility and affect their access to resources, they note the changes in the quality of resources, and they recognize the loss of wildlife through the absence of sound. Fijn noted the impact of changes not regulated by ecological knowledge carried by pastoralists as she discussed the importance of the ability for herders to hear their animals through the felt walls of their yurts (*gers*) to alert them if there are any difficulties to which they need to attend. When they were moved to Soviet-style concrete buildings during the Soviet era, 'herders found they could not hear the herd animals well enough and would frequently lose their animals to rustlers and wolf attack.'[63]

The new economic corridors being developed are used by herders as well. There is increasing automobility in Mongolia; herders more frequently leave their rural communities to travel to residential centers for supplies and to attend celebrations. This practice also increases traffic along the newly built roads as many more people attend events and travel from greater distances to these sites than would go to a neighborhood event. In contrast to the relatively small events in rural communities, celebrations in towns attract larger groups who travel long distances to attend. With access to electricity in the towns, the music is often amplified and there is a greater focus on pan-Mongolian identity and a waning interest in local relationships to place with location-specific information in songs. In Bayan-Ölgii province, where Kazakh herders' *toi* (celebration) play a significant role in lifecycle events, new '*toi* palaces' in district centers have had a direct impact on song repertoires and local musical values. Older singers complain that attendees gathered in large spaces are not interested in the local styles and repertoires. Musicians who provide amplified musical entertainment for the events travel on newly paved roads to reach the palaces and often have no other social connections to the communities.

Conclusion

For pastoralists whose acoustic spaces have been impacted by the new economic corridors, the sounds and songs linked to daily and seasonal activities are poorly supported in these repurposed spaces. As new road construction and road sounds, mining and other new industrial activity, movement of tourists through rural lands, new mobilities for herders, and decreasing wildlife change ecological sites in Mongolia, listening patterns are altered and the character of place and local lifeways change as well. The rural population of Mongolia has diminished considerably since the Mongolian post-soviet efforts to use roads and railways, mining and tourism, to bolster its economy, supported considerably by Chinese government and its ever-expanding Belt and Road Initiative along with Russian contributions.[64] In 2000 the rural population in Mongolia was 43% and by 2018 it was reduced to 32%.[65] This represents an overall decrease in rural residents from 1960, when the rural population was 65%, of 33%. As the country loses its pastoralists, they lose sight of the stewardship they have provided over the generations, which has played an important role in sustaining the land. Mongolia is also losing sounds and soundscapes in acoustic communities that characterize the ecosystems and human-nonhuman relationships in each of the Mongolian steppe lands. These, in turn, have played significant roles in the development and production of cultural products that emerged not only from nomadic peoples, but from the exchange of ideas between nomadic and settled peoples over time.

Traveling into the countryside from Ölgii city toward the Chinese border in the western part of the province of Bayan-Ölgii, the rocky dirt roads running parallel to the paving of the Mongolian AH4 road from China to Russia reveal Chinese methods for road building that contrast with the values shared by pastoralists more concerned about the long term future of the land. Local herders and Ölgii city dwellers acknowledge the benefits of easier travel provided by paved roads, and the ease of stress on the grasslands that have been covered by multiple dirt roads as more vehicles are used and drivers seek tracks that avoid flooding, mud, or soft sand. Yet they also comment on the impact of the road projects on livestock and herding. Some express distress about the landscape, especially the deep holes carved in the land where dirt for the road was ripped away by Chinese workers rather than more carefully extracted as Mongolian herders might do to show their respect for the land.[66]

During the last few years, since longer portions of the road have opened for general use, there has been considerable growth of Chinese industries along this Mongolian road and along the road to the north of Ölgii toward Russia. Chinese camps along the southern road have been set up where they are now openly digging for oil. In more remote areas, they are engaged in exploratory mining. It is possible that there are also plans for a cement factory, in China the 'over-polluting industry' has been limited by the government, and so they are now finding sites across their borders where sound, sight, pollution, and smell is of no consequence to the Chinese.[67] These projects are also part of the BRI – not the carefully negotiated China-Mongolia-Russia corridor that links such well-funded projects as the Oyu Tolgoi mine to China and the Erdenet Mining Corporation to Russia, but one in a more remote part of the country, reaching deeply into pastoral lands. It may reflect what happened in the Gobi more than a decade ago, where economic development is marked by mines that punctuate the land in the provinces of Ömnögovi (the location of the massive Oyu Tolghoi mining establishment), Dornogovi, Dundgovi, Övörkhangai, Bayankhongor, and Govi-Altai. Herders in these regions say access to water has diminished, the soil has been impacted due to the force of mining explosions and the rumbling of large trucks.[68] New insects have arrived and beneficial plants – forage for their animals – have moderated due to the mining activities as well as climate change. Finding new grazing land is not only difficult, it divides families as people search for smaller parcels of land to use for their animals, putting aside family connections for the sake of their survival.

Considering the old Silk Roads once more, we cannot know all that was lost during these periods of economic and cultural opportunity, but in a new Silk Roads era, directed largely by a nation concerned with economic growth and not cultural production, Mongolians might consider focusing more on what is being lost in the process. Projects with little concern for essential human and ecological needs now crisscross lands populated with some of the rarest resources including peoples whose resourcefulness have maintained the land for generations. While national and international efforts are being made to address safeguarding habitats during this period of growth, engaging with local communities where local ecological knowledge has helped pastoralists conserve lands for generations will improve the outcomes. In these acoustic communities, the soundways of human and nonhuman residents hold valuable environmental knowledge for all Mongolians.[69] It is important to remember that cultural production and exchange along the historical roads provided evidence of adaptation and resilience that emerged especially because of dynamic relationships between settled and nomadic peoples. Where is the opportunity for cultural growth in these 21st-century plans? Acoustic spaces containing valued sounds – of birds singing and calling, grasses blowing, ibex battering horns in the spring, and social gatherings filled with local songs – are being transformed by the sounds of economic progress. In fact, in some cases they are simply being silenced.

Notes

1 Rath 2003.
2 Dunne 2015.
3 Post and Pijanowski 2018; Post 2019.
4 Post and Pijanowski 2018.
5 Endicott 2012.
6 Barfield 1989; Khazanov 1994; Beckwith 2009; Honeychurch 2015.
7 A recent UNDP report compiled by Trankmann (2018) notes that 'Mongolia is one of the countries most vulnerable to global warming. In the last 70 years, its temperatures have climbed three times faster than the world's average, rising more than 2 degrees centigrade and surpassing the global threshold.'
8 Marin 2010; Rao *et al.* 2015.
9 Pederson 2014. Other studies indicate drought and other climatic events in the Central Asian and Mongolian steppes also led to socio-political change in earlier periods (Di Cosmo *et al.* 2017; Frachetti *et al.* 2017).
10 Fernández-Giménez *et al.* 2017; Soma and Schlecht 2018.
11 Marin 2008.
12 Janes 2010; Barcus 2018.
13 Batbayar 2002; Karrar 2019.
14 *Northwest to Revive the Silk Road* 1992.
15 Karrar 2019.
16 Li Peng 1994.
17 Millward 2005; Karrar 2019.
18 Millward 2009, 64.
19 Agencies 2014.
20 Aoyama 2016.
21 Sneath 2006; Diener 2011; Suzuki 2013; Jackson 2015a.
22 Otgonsuren 2015; Kenderdine 2017.
23 Soni 2018.
24 Soni 2018; Campi 2014.
25 Tracy *et al.* 2017; Soni 2018.
26 Soni 2018, 106.
27 Ministry of Road, Transport, Construction and Urban Development 2009.

28 Stanway 2018.
29 Birney 2018.
30 See for example, Rovnick 2010; Baljmaa 2019; SouthGobi Resources 2019.
31 Mineral Resources and Petroleum Authority of Mongolia 2017.
32 Krätli *et al.* 2013.
33 Krätli *et al.* 2013; Otgonbayar *et al.* 2017.
34 Fernández-Giménez 2000.
35 Krätli *et al.* 2013.
36 Squires 2012.
37 Fernández-Giménez *et al.* 2015.
38 Reading *et al.* 2015.
39 Rice 2003; Duffy *et al.* 2016; Moscoso *et al.* 2018; Post and Pijanowski 2018.
40 Revill 2016, 240.
41 Duffy *et al.* 2016, 49.
42 Tracy *et al.* 2017; Foggin 2018.
43 Fijn 2011; Ogden *et al.* 2013.
44 Post 2019b.
45 Interview with Altai Tugelbai at Sary Buyra near Dayan Lake, Bayan-Ölgii province, July 2015.
46 Fijn 2011; Yoon 2019.
47 Fijn 2011.
48 See Levin and Süzükei 2007.
49 Such as *Angelica dahurica* and *Pleurospermum uralense*.
50 Interview with Olonbayar, Tavan Bogd National Park, Bayan-Ölgii province, July 2017.
51 Interview with Qozhaqapan, Tolbo, Bayan-Ölgii province, June 2015.
52 Interview with Enbek, Ölgii city, Bayan-Ölgii province, June 2015; See also Post 2017.
53 Levin and Süzükei 2019.
54 Pegg 2001; Yoon 2019.
55 Post 2007.
56 In Kazakh, the use of the word comfortable here refers to resource rich land.
57 *Phasianus colchicus*, known as *qyrghauyl* in Kazakh.
58 Chuluunbaatar 2012, 102.
59 Heissig 1980; Lindskog 2016, 12.
60 Ito *et al.* 2005; 2013; 2017; Dorsey *et al.* 2015; Olson and van der Ree 2015.
61 Parris 2015, 151.
62 Dorsey *et al.* 2015.
63 Fijn 2011, 62–63.
64 Batbayar 2002; Enkhbayar 2002.
65 Worldbank 2018.
66 Interview, Säkhen Bugibai, Bayan-Ölgii province, June 2014.
67 van der Kley 2016; Goh and Gordeyeva 2019.
68 Jackson 2015b.
69 BirdLife Asia 2009.

Bibliography

Agencies. 2014. Silk Road Cultural Journey Launched in NW China, *China Daily*, September 21. www.chinadaily.com.cn/culture/2014-09/21/content_18635089_2.htm (accessed 7 May 2020).

Aoyama, R. 2016. One Belt, One Road: China's New Global Strategy, *Journal of Contemporary East Asia Studies* 5, 3–22.

Baljmaa, T. 2019. Mining License of South Gobi Coal Trans Company Revoked. 2019, *Montsame Mongolian News Agency*, August 13. www.montsame.mn/en/read/197668 (accessed 8 May 2020).
Barcus, H.R. 2018. Contested Space, Contested Livelihoods: A Review of Mongolia's Pastureland Management and Land-Tenure Reform, *Geographical Review* 108, 138–157.
Barfield, T. 1989. *The Perilous Frontier: Nomadic Empires and China*, Oxford.
Batbayar, T. 2002. Geopolitics and Mongolia's Search for Post-Soviet Identity, *Eurasian Geography and Economics* 43, 323–335.
Batsaikhan, N., Buuveibaatar, B., Chimed, B., Enkhtuya, O., Galbrakh, D., Ganbaatar, O., Lkhagvasuren, B., Nandintsetseg, D., Berger, J., Calabrese, J.M., Edwards, A.E., Fagan, W.F., Fuller, T.K., Heiner, M., Ito, T.Y., Kaczensky, P., Leimgruber, P., Lushchekina, A., Milner-Gulland, E.J., Mueller, T., Murray, M.G., Olson, K.A., Reading, R., Schaller, G.B., Stubbe, A., Stubbe, M., Walzer, C., Von Wehrden, H. and Whitten, T. 2014. Conserving the World's Finest Grassland Amidst Ambitious National Development, *Conservation Biology* 28, 1736–1739.
Beckwith, C.I. 2009. *Empires of the Silk Road: A History of Central Eurasia from the Bronze Age to the Present*, Princeton/Oxford.
BirdLife Asia. 2009. *Safeguarding Important Areas of Natural Habitat alongside Economic Development. Mongolia Discussion Papers. East Asia and Pacific Region Sustainable Development Department, World Bank*, Washington, DC.
Birney, M. 2018. Russian Rail Option Opens Up for Aspire Coal, *The West Australian*, 17 June. www.thewest.com.au/business/public-companies/russian-rail-option-opens-up-for-aspire-coal-ng-b88869302z (accessed 8 May 2020).
Bulag, U.E. 2014. A World Community of Neighbours in the Making: Resource Cosmopolitics and Mongolia's 'Third Neighbour' Diplomacy, in: Saxer, M. and Zhang, J. (eds), *The Art of Neighbouring: Making Relations Across China's Borders*, Amsterdam, 121–144.
Chuluunbaatar, O. 2012. The Camel and its Symbolism in the Daily Life of the Mongols with Particular Reference to their Folk Songs, in: Knoll, E.-M. and Burger, P. (eds), *Camels in Asia and North Africa: Interdisciplinary Perspectives on Their Past and Present Significance*, Vienna, 95–105.
Di Cosmo, N., Oppenheimer, C. and Büntgen, U. 2017. Interplay of Environmental and Socio-political Factors in the Downfall of the Eastern Turk Empire in 630 CE, *Climatic Change* 145, 383–395.
Diener, A.C. 2011. Will New Mobilities Beget New (Im)Mobilities? Prospects for Change Resulting from Mongolia's Trans-State Highway, *Engineering Earth: The Impacts of Megaengineering Projects*, Springer, 627–641.
Diener, A.C. and Batjav, B. 2019. Axial Development in Mongolia: Intended and Unintended Effects of New Roads, *Mobilities* 14, 1–17.
Dorsey, B., Olsson, M. and Rew, L.J. 2015. Ecological Effects of Railways on Wildlife, in: van der Ree, R., Smith, D.J. and Grilo, C. (eds), *Handbook of Road Ecology*, Oxford, 219–227.
Duffy, M., Waitt, G. and Harada, T. 2016. Making Sense of Sound: Visceral Sonic Mapping as a Research Tool, *Emotion, Space and Society* 20, 49–57.
Dunne, J. 2015. When the Cows Come Home: A Consideration of the Sensorial Engagement between Pastoralists and Their Cattle, in: Pellini, J.R. (ed.), *Coming to Senses: Topics in Sensory Archaeology*, Newcastle upon Tyne, 21–37.
Endicott, E. 2012. *A History of Land Use in Mongolia the Thirteenth Century to the Present*, New York.
Enkhbayar, N. 2002. Mongolia in the Age of Globalization, *Mongol Messenger* 20, 5.
Fernández-Giménez M.E. 2000. The Role of Mongolian Nomadic Pastoralists' Ecological Knowledge in Rangeland Management, *Ecological Applications* 10, 1318–1326.
Fernández-Giménez, M.E., Batkhishig, B., Batbuyan, B. and Ulambayar, T. 2015. Lessons from the Dzud: Community-based Rangeland Management Increases the Adaptive Capacity of Mongolian Herders to Winter Disasters, *World Development* 68, 48–65.

Fernández-Giménez M.E., Venable, N.H., Angerer, J., Fassnacht, S.R., Reid, R.S. and Khishigbayar, J. 2017. Exploring Linked Ecological and Cultural Tipping Points in Mongolia, *Anthropocene* 17, 46–69.
Fijn, N. 2011. *Living with Herds: Human-Animal Coexistence in Mongolia*, New York.
Foggin, J.M. 2018. Environmental Conservation in the Tibetan Plateau Region: Lessons for China's Belt and Road Initiative in the Mountains of Central Asia, *Land* 7, 52.
Frachetti, M.D., Smith, C.E., Traub, C. and Williams, T. 2017. Nomadic Ecology Shaped the Highland Geography of Asia's Silk Roads, *Nature* 543, 193–198.
Goh, B. and Gordeyeva, M. 2019. Shuttered at Home, Cement Plants Bloom along China's New Silk Road, *Reuters Business News*, January 30.
Heissig, W. 1980. *The Religions of Mongolia*, London.
Honeychurch, W. 2014. From Steppe Roads to Silk Roads: Inner Asian Nomads and Early Interregional Exchange, in: Amitai, R. and Biran, M. (eds), *Nomads as Agents of Cultural Change. The Mongols and Their Eurasian Predecessors*, Honolulu, 50–87.
Ito, T.Y., Lhagvasuren, B., Tsunekawa, A. and Shinoda, M. 2017. Habitat Fragmentation by Railways as a Barrier to Great Migrations of Ungulates in Mongolia, in: Borda-de-Agua, L., Beja, P. and Pereira, M. (eds), *Railway Ecology*, Cham, 229–246.
Ito, T.Y., Lhagvasuren, B., Tsunekawa, A., Shinoda, M., Takatsuki, S., Buuveibaatar, B. and Chimeddorj, B. 2013. Fragmentation of the Habitat of Wild Ungulates by Anthropogenic Barriers in Mongolia, *PLoS ONE* 8, 1–10.
Ito, T.Y., Miura, N., Badamjav, L. and Enkhbileg, D. 2005. Preliminary Evidence of a Barrier Effect of a Railroad on the Migration of Mongolian Gazelles, *Conservation Biology* 19, 945–948.
Jackson, S.L. 2015a. Imagining the Mineral Nation: Contested Nation-building in Mongolia, *Nationalities Papers* 43, 437–456.
Jackson, S. 2015b. Dusty Roads and Disconnections: Perceptions of Dust from Unpaved Mining Roads in Mongolia's South Gobi Province, *Geoforum* 66, 94–105.
Janes, C.R. 2010. Failed Development and Vulnerability to Climate Change in Central Asia: Implications for Food Security and Health, *Asia-Pacific Journal of Public Health* 22, 236S–245S.
Karrar, H.H. 2019. From Central Asia to the World: China's Regional Diplomacy as a Precursor for Global Connectivity, in: Syed, J. and Ying, Y.-H. (eds), *China's Belt and Road Initiative in a Global Context*, Cham, 61–86.
Kenderdine, T. 2017. Mongolia Gets on Board with China's Belt and Road Initiative, *Eurasianet* July 7. www.eurasianet.org/node/84266 (accessed 8 May 2020).
Khazanov, A.M. 1994. *Nomads and the Outside World*. 2nd ed, Madison.
Krätli, S., Hülsebusch, C., Brooks, S. and Kaufmann, B. 2013. Pastoralism: A Critical Asset for Food Security under Global Climate Change, *Animal Frontiers* 3, 42–50.
Levin, T. and Süzükei, V. 2007. *Where Rivers and Mountains Sing*, Bloomington.
Levin, T. and Süzükei, V. 2017. Timbre-Centered Listening in the Soundscape of Tuva, in: Dolan, E. and Rehding, A. (eds), *The Oxford Handbook of Timbre*, Oxford. DOI: 10.1093/oxfordhb/9780190637224.013.15.
Li Peng. 1994. Li Peng addresses Uzbek parliament on China's relations with Central Asia, *BBC Summary of World Broadcasts*, April 22, 1994.
Lindskog, B.V. 2016. Ritual Offerings to Ovoos among Nomadic Halh Herders of West-Central Mongolia, *Études Mongoles et Sibériennes, Centrasiatiques et Tibétaines* 47, 1–20.
Marin, A. 2008. Between Cash Cows and Golden Calves: Adaptations of Mongolian Pastoralism in the 'Age of the Market', *Nomadic Peoples* 12, 75–101.
Marin, A. 2010. Riders under Storms: Contributions of Nomadic Herders' Observations to Analysing Climate Change in Mongolia, *Global Environmental Change* 20, 162–176.

Millward, J.A. 2005. Uyghur Art Music and the Ambiguities of Chinese Silk Roadism in Xinjiang, *Silk Road* 3, 12–14.

Millward, J.A. 2009. Positioning Xinjiang in Eurasian and Chinese history: Differing visions of the 'Silk Road', in: Mackerras, C. and Clarke, M. (eds), *China, Xinjiang and Central Asia: History, Transition and Crossborder Interaction into the 21st Century*, New York, 55–74.

Mineral Resources and Petroleum Authority of Mongolia. 2017. The Annual Report, Ulaanbaatar, 19.

Ministry of Road, Transport, Construction and Urban Development. 2009. *Road, Transport Sector of Mongolia*. www.unece.org/fileadmin/DAM/trans/doc/2009/wp5/GE2-wkshp1-Mongolia.pdf (accessed 8 May 2020).

Moscoso, P., Peck, M. and Eldridge, A. 2018. Emotional Associations with Soundscape Reflect Human-environment Relationships, *Journal of Ecoacoustics* 1, 1–19.

Northwest to Revive the Silk Road. 1992. *Beijing Review* 35.37, September 14–20, 7.

Ogden, L.A., Hall, B. and Tanita, K.H. 2013. Animals, Plants, People, and Things: A Review of Multispecies Ethnography, *Environment and Society: Advances in Research* 4, 5–24.

Olson, K.A. and van der Ree, R. 2015. Railways, Roads and Fences across Kazakhstan and Mongolia Threaten the Survival of Wide-ranging Wildlife, in: van der Ree, R., Smith, D.J. and Grilo, C. (eds), *Handbook of Road Ecology*, Oxford, 472–478.

Otgonbayar, M., Atzberger, C., Chambers, J., Amarsaikhan, D., Böck, S. and Tsogtbayar, J. 2017. Land Suitability Evaluation for Agricultural Cropland in Mongolia Using the Spatial MCDM Method and AHP Based GIS, *Journal of Geoscience and Environment Protection* 5, 238–263.

Otgonsuren, B. 2015. Mongolia-China-Russia Economic Corridor Infrastructure Cooperation, *Erina Report* 127, 3–7.

Parris, K.M. 2015. Ecological Impacts of Road Noise and Options for Mitigation, in: van der Ree, R., Smith, D.J. and Grilo, C. (eds), *Handbook of Road Ecology*, Oxford, 151–158.

Pederson, N., Hessl, A.E., Baatarbileg, N., Anchukaitis, K.J. and Di Cosmo, N. 2014. Pluvials, Droughts, the Mongol Empire, and Modern Mongolia, *Proceedings of the National Academy of Sciences* 111, 4375–4379.

Pegg, C. 2001. *Mongolian Music, Dance, & Oral Narrative: Performing Diverse Identities*, Seattle.

Post, J.C. 2007. 'I Take my Dombra and Sing to Remember my Homeland': Identity, Landscape and Music in Kazakh Communities of Western Mongolia, *Ethnomusicology Forum* 16, 45–69.

Post, J.C. 2017. Ecological Knowledge, Collaborative Management and Musical Production in Western Mongolia, in: Post, J.C. (ed.), *Ethnomusicology: A Contemporary Reader Volume II*, New York, 161–179.

Post, J.C. and Pijanowski, B.C. 2018. Coupling Scientific and Humanistic Approaches to Address Wicked Environmental Problems of the Twenty-first Century: Collaborating in an Acoustic Community Nexus, *MUSICultures* 45, 71–91.

Post, J.C. 2019a. Climate Change, Mobile Pastoralism, and Cultural Heritage in Western Mongolia, in: Cooley, T.J. (ed.), *Cultural Sustainabilities: Music, Media, Language, Advocacy*, Urbana, 75–86.

Post, J.C. 2019b. Songs, Settings, Sociality: Human and Ecological Well-being in Western Mongolia, *Journal of Ethnobiology* 39, 1–21.

Rao, M., Davi, N.K., D'Arrigo, R.D., Skees, J., Nachin, B., Leland, C., Lyon, B., Wang, S.-Y. and Byambasuren, O. 2015. Dzuds, droughts, and livestock mortality in Mongolia, *Environmental Research Letters* 10, 1–12.

Rath, R.C. 2003. *How Early America Sounded*, Ithaca.

Reading R.P., Wingard, G., Tuvdendorj, S. and Amgalanbaatar, S. 2015. The Crucial Importance of Protected Areas to Conserving Mongolia's Natural Heritage, in: Wuerthner, G., Crist, E., and Butler, T. (eds), *Protecting the Wild: Parks and Wilderness, the Foundation for Conservation*, Washington, DC, 257–265.

Revill, G. 2014. El Tren Fantasma: Arcs of Sound and the Acoustic Spaces of Landscape. *Transactions of the Institute of British Geographers* 39, 333–344.
Revill, G. 2016. How is Space Made in Sound? Spatial Mediation, Critical Phenomenology and the Political Agency of Sound. *Progress in Human Geography* 40, 240–256.
Rovnick, N. 2010. Mongolia Revokes Mining Licences, *South China Morning Post* November 20.
Rice, T. 2003. Soundselves: An Acoustemology of Sound and Self in the Edinburgh Royal Infirmary, *Anthropology Today* 19, 4–9.
Silk Road Cultural Journey Launched in NW China, *China Daily* Sept 21, 2014. http://www.chinadaily.com.cn/culture/2014-09/21/content_18635089_2.htm.
Sneath, D. 2006. The Rural and the Urban in Pastoral Mongolia, in: Bruun, O. and Narangoa, L. (eds), *Precious Steppe: Mongolian Nomadic Pastoralists in Pursuit of the Market*, Copenhagen, 140–161.
Soma, T. and Schlecht, E. 2018. The Relevance of Herders' Local Ecological Knowledge on Coping with Livestock Losses during Harsh Winters in Western Mongolia, *Pastoralism* 8.1, 1–14.
Soni, S.K. 2018. China–Mongolia–Russia Economic Corridor: Opportunities and Challenges, in: Deepak, B.R. (ed.), *In China's Global Rebalancing and the New Silk Road*, Singapore, 101–117.
SouthGobi Resources. 2019. SouthGobi Announces Termination of Mining Licenses for the Soumber Deposit, *GlobeNewswire*, August 28. www.globenewswire.com/news-release/2019/08/28/1907753/0/en/SouthGobi-Announces-Termination-of-Mining-Licenses-for-the-Soumber-Deposit.html (accessed 7 May 2020).
Squires, V.R. (ed.) 2012. *Rangeland Stewardship in Central Asia Balancing Improved Livelihoods, Biodiversity Conservation and Land Protection*. Dordrecht, New York.
Stanway, D. 2018. Mongolia Plans Coal Rail Link to China by 2021. *Reuters*, November 7.
Suzuki, Y. 2013. Conflict between Mining Development and Nomadism in Mongolia, in: Yamamura, N. and Fujita, N. (eds), *The Mongolian Ecosystem Network: Environmental Issues Under Climate and Social Changes*, Tokyo, 269–294.
Tracy, E., Shvarts, E., Simonov, E. and Babenko, M. 2017. China's New Eurasian Ambitions: The Environmental Risks of the Silk Road Economic Belt, *Eurasian Geography and Economics* 58, 56–88.
Trankmann, B. 2018. Nomads No More, *United Nations Development Programme*. www.undp.org/content/undp/en/home/blog/2018/Nomads-no-more.html (accessed 8 May 2020).
van der Kley, D. 2016. China shifts polluting cement to Tajikistan, *China Dialogue*, August 8. www.chinadialogue.net/article/show/single/en/9174-China-shifts-polluting-cement-to-Tajikistan (accessed 8 May 2020).
Walsh, J. 1993. China and the New Geopolitics of Central Asia, *Asian Survey* 33, 272–284.
World Bank. 2018. Rural Population Mongolia, *The World Bank DataBank*. data.worldbank.org/indicator/SP.RUR.TOTL.ZS?locations=MN (accessed 8 May 2020).
Yoon, S. 2019. Mobilities, Experienced and Performed in Mongolia's *Urtyn Duu* Tradition, *Asian Music* 50, 47–77.

Chapter 12

Erasing the Local, Celebrating the Local: Tracing the Contradictions of the Silk Road in Pakistan

Chad Haines

The idea of the Silk Road invokes provocative images of exotic goods, trade caravans, and bustling markets; it is a powerful symbol of social interconnectivity, cultural exchange, and ethical cosmopolitanism. Historically, the Silk Road is in fact the Silk Roads, providing a historical frame for interpreting and representing a Braudelian *longue durée* of movement and interaction, linking disparate corners of the old world, across a rich diversity of interconnected routes. The Silk Roads, as documented, analyzed, and celebrated in this volume, is a plurality of interconnections across time and space that are at the very heart of human history. Silk Roads provide us an avenue of moving outside the box of the nation-state frame of history and historiography. Yet, as an idea, the Silk Road is purely modern, first imagined in the late 19th century by a German geographer. As a modern frame for interpreting the past, the Silk Road idea is in fact laden with contemporary political interests employed to advance nation-state formation processes, historiographies of difference and exclusion, and particularistic global interests and aspirations. The modern construction of the Silk Road is completely contradictory to the histories and values it invokes.

In unravelling these contradictions, this chapter focuses on the multiple employments of the Silk Road symbol in Pakistan. The Silk Road, as an idea and symbol, has multiple lives and uses in Pakistan, advancing national and global interests. First, it was employed in the 1960s to reorient Pakistan's national horizon, deepening its connections to Iran and Turkey, Afghanistan and Central Asia, and China and Inner Asia. Second, the image of the Silk Road became intricately merged with the Karakoram Highway (constructed in the 1970s) that links Pakistan and Xinjiang province of western China. The Silk Road imagery masked the internal colonization of the Gilgit-Baltistan region of northern Pakistan, becoming a process of erasing the local. Finally, we observe the Silk Road being employed to advance particular global interests, from that of international tourism to that of Chinese imperialism, with a tinge of neoliberal globalization supported by the United Nations Education, Scientific, and Cultural Organization (UNESCO). The Silk Road is a hot global commodity today.

Here I differentiate between the Silk Road in the singular and Silk Roads in the plural. In the singular, the Silk Road is a symbol, an idea, a popular image that has little to do with history. In the plural, the Silk Roads are lived histories of interconnections and exchanges, of the unfolding spread of ideas, goods, and people. It is plural, as there were multiple Silk Roads waxing and waning across human histories. As a symbol, the Silk Road also has multiple lives, but as a tool of representation, of inter-state politics, and of the advancement of narrowly defined national interest, not to mention of globalization of particular ideologies, such as neoliberalism. The Silk Road symbol, though claiming historical connections, ultimately has very little to do with actual histories and lived experiences. It is a powerful symbol for state and commercial interests, but provides an avenue for resisting one-sided, exploitative integration. Given the Silk Road, as symbol and as histories, is a bundle of contradictions, I question the utility of the label for the study of history – does it confuse or clarify historical processes of interconnections and movements, exchanges and borrowings?

The Birth of the Silk Road: Colonialism, Antisemitism, and German Romanticism

As an idea, the Silk Road commenced in 1877, brought into the world by Ferdinand von Richthofen (1833–1905). Richthofen is known as a geographer, providing colorful descriptions of different regions of Asia, particularly China, detailing rich geological resources. The idea of the Silk Road was born at a time of European colonial expansion into China with the weakening Qing Dynasty (1644–1912). The idea was also born when European historiography was being reformed. History was molded into serving emergent nationalist ideas and values, tracing romanticized visions of self, bounded by discrete territories. The idea of the Silk Road provided an avenue for the advancement of both colonial interests and national imaginations.

Richthofen's Silk Road was not a historical imagining of exchange but rather a blueprint for German colonial expansion into China. 'Richthofen's writings on the Silk Road belonged to a multivolume geological survey of China that guided the German colonial seizure of Qingdao[1] in northern China in 1897.'[2] Richthofen's expeditions across Central Asia and geographical studies of China were funded by American and German commercial interests. His reports and publications detail locations of natural resources, trade routes, and possible routes for railway lines.

More than a road for material advancement, the Silk Road idea and its mappings contributed significantly to the reimagining of Europe's ancient past. Of particular concern for German historians was moving away from the Mediterranean-centric, Greco-Roman, and Semitic pasts of Europe and forefronting an ancient German place in world history. The Silk Road, combined with myths of ancient Aryan people, provided Germany that past, an avenue of de-Semiticizing Germany.[3] While Richthofen expressed very derogatory ideas about China and Chinese people, it was his student, the famed Silk Road explorer Sven Hedin, who interwove the Silk Road idea with his personal anti-Semitism and political support of the National Socialist regime.

The Silk Road provided an avenue for rewriting an ancient Germanic past along two tracks. The first was the decentering of the Greco-Roman past. The Silk Road idea did this by highlighting a rich and dynamic history that bypassed the Mediterranean. Overland trade routes meant that there were histories other than that of Greeks, Romans, and Hebrews.

The Silk Road also provides an epistemological decentering that opened up intellectual knowledge to a diversity of sources and not just Greco-Roman scholarship and philosophy. While this diversity of intellectual sources was conducted through the colonial project of 'discovery' and the colonial lens of inherent western superiority, it nonetheless provided a means of exploring and incorporating new kinds of knowledge for the service of western conquest and exploitation of the world.

Thus, infused with the very idea of the Silk Road are three contradictory ideas: nationalist historiography, colonial expansion, and alternative methodologies and historical resources. This later contribution is significant in moving away from the nationalist and colonialist idea of the Silk Road. The Silk Road idea opens up possibilities for intensive and extensive research and comparisons across time and place, forging new methodological possibilities. Yes, the Silk Road continues to live on as a nationalist and colonialist project, today infused with neoliberal values of globalization that privilege particular economic interests. In so doing, these expressions of the Silk Road erase local histories and cultures rather than incorporate them into our knowledge base.

The Silk Road is two radically different roads, one to be celebrated and explored, one to be critiqued and diminished. Yet, the terminology is the same, the basic concept is the same. As such, perhaps it is time to rethink the appellation Silk Road and reimagine global histories through a different frame of interconnectivity, flows, and exchanges. To fully understand these two diverging Silk Roads, I focus on contemporary expressions of the Silk Road in Pakistan.

The Silk Road and the Reimagining of Pakistan's National Horizons

Emerging out of partition, divided into two incongruent and disparate territorial units separated by a hostile neighbor, missing what it considers an integral part of the national territory, one region in revolt, another neighboring country claiming a significant part of its territory, and having many of its international borders as undefined, Pakistan was born in a state of 'territorial liminality'.[4] The country's location at the crossroads of South, Inner, Central, and West Asia with close maritime connections to the Arabian peninsula and the Middle East, not to mention East Pakistan's linkages to Southeast Asia, contributed to creating an anxiety in forging its place in the world.

Pakistan's territorial liminality and spatial anxiety was compounded by the eruption of violence and outright war with neighboring India. India loomed, and continues to loom, large in the Pakistani imagination, informing a number of social developments including the domineering place of the military in Pakistan's political and economic life. Strategically, these aspects of Pakistan's postcolonial condition informed a series of geopolitical moves, realigning Pakistan's relationships with its neighbors. In the 1960s the military dictator Ayub Khan (r. 1958–1969) put into motion a series of geopolitical shifts. The movement occurred in two spheres, the first of which was a reimagining of Pakistan's national horizon away from South Asia and India and towards different parts of the Muslim world, asserting a modern Muslim identity. The second sphere was grounding these shifts through advancing economic relationships with neighboring countries. This reorientation was to be realized along three infrastructural linkages, three new and modern Silk Roads. Ultimately, only one of the Silk Roads imagined in the 1960s was actually constructed and completed – the Karakorum Highway, essentially linking Pakistan's new capital, Islamabad, with the western Chinese city of Kashgar, in the Xinjian province.

These modern Silk Roads envisioned by Pakistan are born out of Pakistan's postcolonial condition – they are modernist projects, solving modern problems, for modern usages. Yet, the Silk Road appellation provides a significant symbolic value to the projects and to Pakistan. The Silk Road allusion provides Pakistan historical continuity, particularly significant for a country carved out of partition and in a state of territorial liminality. The Silk Road imagery simultaneously provides Pakistan a strategic present in its efforts to forge territorial coherency and define its place in the world. The Silk Road idea also provides Pakistan a future as avenues of economic development.

For its national horizon, the Silk Road idea is an avenue for Pakistan to reorient its identity away from South Asia and Hindu India. In less than 20 years after independence from British colonial rule, Pakistan had fought, and lost, two wars with India over the disputed region of Kashmir. India's mass population and economic productivity allowed it to loom large within South Asia and the world. As a leader of the Non-Alignment Movement, India forged a strategic place for itself in the world. These factors deepened Pakistan's sense of national anxiety. Under Ayub Khan, Pakistan crafted a modernist Islamic vision.[5] Looking south and west, Pakistan forged new relationships and an Islamic identity, all made real by the Silk Road.

Rather than looking east to India, Pakistan turned west to the Muslim countries of West Asia, the Middle East, and the Near East. In 1964 Pakistan, with Iran and Turkey, created the Regional Cooperation for Development (RCD). As part of the realization of this modernist Islamic trading bloc, a diversity of infrastructural linkages were planned. One of them was the 803-mile long RCD Highway. Ayub Khan's foreign minister, Zulfikar Ali Bhutto (later to become Prime Minister from 1973–1977) hailed the highway as reestablishing the historical linkages with West Asia along the 'Silk Route'.[6]

Strategically this Silk Road connected Pakistan and Iran through the province of Baluchistan. Almost from Independence, the region of Baluchistan had been in armed resistance against the government of Pakistan. Rich with natural resources, but one of the poorest populations, the Baluch felt marginalized and exploited. The new Silk Road through the region would advance the interests of the Pakistani state not only to connect the country with Iran and Turkey, but to provide a road to transport extracted resources out of Baluchistan and to transport necessary military personnel in to help suppress the militant movement. In the end, very little progress on the RCD Highway was made as maritime routes were much more economical in connecting Pakistan, Iran, and Turkey.

Just as Pakistan was looking for new Silk Roads across the southwestern part of the country, they also looked to the northwest. The Khyber Pass linking Pakistan and Afghanistan has a long history as a major trade route, going back millennia. Buddhism flowed across this region into Central and Inner Asia and beyond into East Asia. It is said that Alexander the Great made it to the region as well before falling short of his aspiration to conquer various Indic dynasties and kingdoms. The trade routes across the Hindu Kush mountains, of which the Khyber Pass is just one of many, has a legitimate place in the history books as being an integral component of the Silk Roads.

Though firmly in the sphere of western influence during the Cold War, Ayub Khan made overtures to the Soviet Union and visited in 1967. The talks he conducted included agreements to revive 'Silk Road trade' between the two countries through Afghanistan.[7] The Soviet Union saw a possible avenue for the shortest route to the ocean for natural gas drilled from the Central Asian region. For Pakistan, this Silk Road provided an opportunity to assert its rights over the Northwest Frontier Province (NWFP)[8] that Afghanistan laid claim to. The border between the two regions, known as the Durand Line, was mapped by the British in 1893 following their conquests and integration of the Peshawar region. The Durand Line essentially divided the various tribal lineages of Pakhtuns between Afghanistan and British India. When Pakistan gained independence from British colonial rule in 1947, Afghanistan claimed the Northwest Frontier Province as theirs based on ethnic affinity. When Pakistan sought membership to the United Nations, Afghanistan was the only country to vote against Pakistan.

The two countries continued to have conflictual relationships and the Silk Road planned linking Pakistan, Afghanistan, and Soviet Central Asia was an avenue for diminishing those tensions. While the Pakhtuns of the Northwest Frontier Province never openly revolted as the Baluchis, there was always a tense relationship between the region and the central government and often open hostile ethnic relationships between Pakhtuns and other communities in Pakistan. Thus, like the southern, RCD Silk Road, the northwestern Silk Road had an internal component of pacifying and integrating a region and people more fully into the emerging nation-state of Pakistan.

By the late 1970s though neither the RCD Silk Road nor the NWFP Silk Road were realized in their full capacity. The Soviet invasion of Afghanistan in 1978 and the Iranian revolution the following year completely disrupted those plans. Two modern Silk Roads were left incomplete due to geopolitical realities. The very idea of such roads was rooted in the interests and concerns of postcolonial Pakistan, they are the product of particular histories. The Silk Road is a tool of national ideology and state interest, advancing identity politics, economic development agendas, and foreign relations.

But the RCD and NWFP Silk Roads were not the only ones envisioned by Ayub Khan and Zulfikar Bhutto. A third Silk Road was planned, linking Pakistan with China. This Silk Road became a strategic advantage following the Sino-Indian war of 1962, providing Pakistan an important ally against hostile neighboring India. Unlike the southern and northwestern Silk Roads, the northern Silk Road across the remote, mountainous region of Gilgit was actually built and continues as a thriving economic corridor linking western China with the sea port in southern Pakistan.

Landscaping the Silk Road: The Karakoram Highway and Routes of Integration

At the time of Partition, one of the greatest existential challenges for Pakistan, and continues today, is over the disputed region of Kashmir.[9] At independence the many rulers of the over 400 Princely States the British created in their Indian colony were given the choice of either joining Pakistan or India. For most it was a clear choice, but for several, including for the Maharaja of Jammu and Kashmir, the decision was not easy. Though ruled by a Hindu, the region was majority Muslim and thus, according to the policies of the division of territories, should have been mapped within Pakistan. Rather than deciding, Maharaja Hari Singh (r. 1925–1949) played a waiting game, many thinking in the hopes of carving out a sovereign state of his own. War broke out and under pressure from the new prime minister of India, Jawaharlal Nehru, himself of Kashmiri Hindu descent, Maharaja Hari Singh hastily signed articles of accession to India. The United Nations stepped in to broker peace between Pakistan and India over the region of Kashmir and divided the region in two – Indian Administered Kashmir comprised of three subregions (Jammu, Ladakh, and the vale of Kashmir) and Pakistan Administered Kashmir comprising three subregions (Azad Kashmir, the Gilgit Agency, and Baltistan, the later two regions today are merged into one region known as Gilgit-Baltistan).

In the story of the politics of discovering, (re)opening, and constructing the Silk Road, the region of Gilgit-Baltistan takes center stage. Gilgit-Baltistan is a remote mountainous region defined by narrow riverine valleys winding their way around some of the highest mountains in the world. It is also the most glaciated region in the world outside the polar caps. Ethnically, the region is highly diverse, with speakers of Shina (related to Kashmiri), Balti (a dialect of Tibetan), Burushaski (seemingly unrelated to any other language), Wakhi (a Turkic language), and several others populating the region. Religiously, they are all Muslim, though divided between Twelver Shi'as, Ismailis (whose leader, the Aga Khan, plays a strategic role in the politics and development of the region), and Sunnis of various orientations. Prior to the final British conquest of the region in 1891, when they established the Gilgit Agency, different valleys of the region maintained a diversity of political and economic relationships including with Kashgar (China), Badakhshan (now in Afghanistan), Ladakh (now in India), and Peshawar and Rawalpindi (now in Pakistan).

When Pakistan was granted administrative control over the Gilgit Agency, the main road into the region was via Srinagar, which was now in India. Pakistan constructed the Indus Valley Road, the only infrastructural linkage into and out of the region, connecting Gilgit with the garrison town of Abbottabad. As part of its forging a new national horizon and overcoming the anxieties of its inherited territorial liminality, besides Iran, Turkey, and the USSR, Pakistan also forged close ties with China in the 1960s. In 1963 they jointly signed an agreement delineating the border between the two countries that the British left undefined during their colonial rule. Further cementing their strategic relationship, in 1967 Pakistan and China signed a trade agreement, part of which was to open the border between Xinjiang province and the Gilgit Agency for cross-border trade. The signing was celebrated in Pakistan, as 'the centuries-old overland trade route more popularly known as 'the Silk Route' between Gilgit Agency and Sinkiang will be re-opened'.[10]

The cross-border trade idea was advanced with the signing of another bilateral agreement in 1969. The agreement was to jointly construct a motorable highway that would link Rawalpindi (the military cantonment outside the new capital of Pakistan, Islamabad) with Kashgar. The highway, known popularly as the Karakoram Highway or the KKH, runs along the Indus River, then to the town of Gilgit and beyond along the Hunza River through the valleys of Nagar, Hunza, and Gojal to the border village of Sost and then across the Khunjerab Pass on the Pakistan-China border. The agreement to advance cross-border trade and to construct the KKH was signed in 1969 in the town of Gilgit. To mark the historic agreement the *Pakistan Times* ran a number of stories about the remote region, its history, and the significance of the linkage with China. With the signing of the 1969 agreement, their major story was explaining how:

> The Silk route has been opened again after 20 years. The scene as the first Chinese caravan – with its camels and horses laden with silk and tea – arrived in Gilgit on Sunday must have been evocative of the more romantic days of yore when states had not become huge economic apparatuses and local people could still trade with one another. The opening of the route is thus at once an affirmation of the old values that bound the people of the two regions – values that were shattered by alien rule – and proof of the new ties of friendship and good will that once again unite the people of Pakistan and China. Nor is the value of the route merely symbolic.... Cottage industry in the Gilgit-Hunza area should get a fillip, and there will moreover be a new incentive for Pakistani exports to enter the Chinese market.[11]

The news item plays upon prevailing ideas about the Silk Road, an idea that invokes a history of the free flow of goods and people, 'evocative of more romantic days of yore when states had not become huge economic apparatuses', and cosmopolitan values that connected people across cultural worlds. Yet, the idea of a contemporary Silk Road is intricately interwoven into the interests of the modern nation-state, providing an avenue (literally and figuratively) for economic development and, unstated in the article, the movement of military personnel and equipment. The symbolic value of the Silk Road is as both a unifying force for national integration and as a veil for advancing the particular interests of the state.

It is necessary more fully to interrogate the sentiment expressed in the romanticized vision of the Silk Road portrayed in *The Pakistan Times* article. These sentiments express popular attitudes and knowledge about the Silk Road, circulated in media, by popular nationalist historians, and tourist information that continue to be held today. First, the story of the Chinese caravan is not even accurate. According to a story in the same edition of *The Pakistan Times* several pages later, due to bad weather, the Chinese delegation flew into the town of Gilgit by military helicopter![12] The delegates were Han Chinese, representatives of the People's Republic of China, with no members of the local Uyghur population. Likewise, on the Pakistan side, the Chinese delegation met with representatives of the Pakistani state, bureaucrats from the capital. The local populations had nothing to do with the reopening of the Silk Road, it was a state project. Locals were merely cultural performers, displaying traditional clothes and dances. Though the local is the context of the opening of trade between Pakistan and China, the actual local is reconstructed as a colorful backdrop to larger national and global interests.

Second, curiously, the article refers to how the Silk Road will *again* 'unite the people of Pakistan and China'. Only, prior to 1947 there were no people of Pakistan, as Pakistan did not exist. More importantly, the people who did participate in such movements had no linkage whatsoever with the emerging dominant populations of Pakistan. The linkages of the various people of the region were north, west, and east, with minimal connection to the south, what is today Pakistan. Again, as with the erasure of who the first trade delegates actually were, the local is usurped for the national, a theme more fully discussed in the next section. Third, other historical confusions include that somehow the trickle of trade and interaction that did occur across the western Karakoram mountains was 'cross-border'. Only, as with most of the pre-colonial world, there were no borders delineating discrete territories. It should be noted that there is very little indication of any extensive trade actually occurring across the western Karakoram mountains at any time in history. The British attempted to open the passes of the western Karakorams in order to shift the trickle of trade away from Kashmiri-controlled Leh, the capital of Ladakh, to British-controlled Gilgit. However, they failed miserably. The Khunjerab pass was extremely difficult to cross as compared to the Karakoram Pass that linked Leh with Kashgar. Further, it was only possible to pass during the summer months. However, moving from the northern valley of Gojal down to Gilgit was only possible during the winter months when the Hunza River froze, otherwise it was unfordable. The region also did not have the resources to host trade caravans. Land is minimal in the narrow riverine valleys, the pastures suitable only for local livestock, making the availability of fodder for additional animals unavailable. Despite the geographical realities, the idea of an ancient Silk Road still circulates as a historical reality.

These multiple historical erasures of the local, anachronistic projections of the nation back in time, and romanticized visions of the past are the frame through which the Silk Road is projected. What this nationalist frame leaves out of the picture are the interests of the state. More than a fillip for the local people, the KKH is an avenue for deeper political and social integration of the remote region. The opening of the Silk Road allowed the Pakistani military to more easily move troops into and out of the region and provide a secure and continuous supply route. As part of greater Kashmir, the region was under dispute and Pakistan's actual administrative control was minimal until the new Silk Road was built.

Prior to the agreement to construct a new highway, Pakistan ruled the Gilgit Agency through the traditional rulers of the various valley kingdoms, just as the British had. In the early 1970s, the Pakistani state did away with the local rulers and established direct, bureaucratic administration over the region and people. In addition, the colonial nomenclature of the region was changed from the Gilgit Agency to the Northern Areas (which has since been renamed as Gilgit-Baltistan) firmly mapping it within Pakistan.

The trade and infrastructure agreement between China and Pakistan was more than a reopening of the Silk Road – it was an avenue along which the Pakistani state could travel, integrating the disputed and remote region, giving a more fixed definition to its postcolonial territorial liminality. The Silk Road provides a veil to state interests. In so doing, it erases the local as it tries to nationalize their histories, singularizing them through the Silk Road frame.

Erasing the Local, Discovering the Nation

While geography in the form of bounded territory and infrastructural linkages are defining features of the modern state, features that the Silk Road idea provided the landscape for, the modern state is also interwoven into the idea of a nation, a community of people with a shared identity. Only, across the world, be it in former colonies or colonizers, there is no true singular nation bounded within a given territory. The nation, like the state, is a project, an unfolding series of attempts to craft a common identity for all citizens of the nation. For a country like Pakistan, born out of separation from a larger political entity (itself an artificial cobbling together of diverse regions and peoples by the British colonial forces) imagining a nation is that much more difficult. Though Islam looms large as a marker of Pakistani identity, the Silk Road also plays its part, providing Pakistan a longer and richer history than Islam.

By asserting a Silk Road past, Pakistan forges a particular kind of national belonging. In so doing though, the local communities, cultures, and histories through which the imagined Silk Road traverses are erased, supplanted by a national community, culture, and history.

The Karakoram Highway as Silk Road is a dominant motif in the Pakistani imagination, forging a linkage to a rich ancient past where Buddhism flourished, Greco-Indic art took form, and interweaving the region into the global fabric of diverse civilizations. For some Pakistani scholars, this diverse, multicultural path is an avenue for imagining Pakistan as a plural, inclusive nation critiquing those who want to see Pakistan as singularly Islamic as well as providing another differentiation from the closed, caste-centric culture of Gangetic India.[13]

The Silk Road looms large in these national imaginings of Pakistan's ancient past. The Silk Road historical frame provides Pakistan a long narrative and material artifacts to populate the imagination with. In looking for the Silk Road, historians and archaeologists uncover Pakistan's past, fill museums with evidence of that past, and offer physical locations for citizens to observe and experience 'their' history. Through the Silk Road, the past becomes a living memory of Pakistan, situating Buddhist history of Gandhara within an understandable frame, though one often left to the dustbins of time, as with the Buddhist art collection in the Peshawar Museum. However, the town of Taxila, an ancient center of Gandharan social and cultural life, remains a popular picnic spot amidst the Buddhist ruins on weekends for families from Islamabad.

Museums are key to advancing and naturalizing the myth of the Silk Road as Pakistan's ancient history, though they remain rarely visited. More popular are a number of historical studies by the nationalist historian A.H. Dani. With titles such as *Human Records on Karakorum Highway*, *The Historic City of Taxila*, *History of Northern Areas of Pakistan*, *Romance of the Khyber Pass*, and *New Light on Central Asia*, Dani populates the national imagination with the Silk Road idea. His works are standard required readings in most university history and Pakistan studies courses and are popularized in newspaper and magazine reviews. Few educated Pakistanis are unaware of his work and certainly most people in the Gilgit-Baltistan region (what used to be called the Northern Areas) are more than aware of the details of their history, as told by Dani.

When conducting my own research on the spatial history of Gilgit, I was impressed with the specifics of the region's ancient past of which so many of my informants were aware. The more interviews I conducted though I began to discern similar stories, almost told verbatim by diverse people. When I finally purchased a copy of Dani's *History of Northern Areas of Pakistan*, I discovered the history I was being told was the history written by Dani. No other historical memory existed except that which framed the region within its Buddhist past as part of Pakistan as retold by Dani.

The singularized frame and narrative though are not the full story of the northern region's past. There are alternative narratives, only they are being erased by nationalist history concerned with the anachronistic imagining of an ancient Pakistan. There are diverse histories laying under the surface in northern Pakistan. One alternative history is actually a usurpation of the Silk Road past, not as national history but rather as local history. A second source of alternative history is in the rediscovery of local myths and tales as integral components of the social and cultural world. While nationalist historians map local myths and tales as folklore, placing it on a lower rung of factuality as compared to history, there are movements to re-employ folk stories, ideas, and values as a means of grappling with contemporary issues.

One example of this is in the northern valley of Shimshal that runs along the China border east of the Karakoram Highway and Khunjerab Pass. As a pastoralist community they move their livestock between the lowland winter pastures around their villages and the high alpine summer pastures several days hike away. As their communities have grown so have their demands on the fragile pasturelands, leading to areas of erosion and destruction. Starting in the 1990s with several local activists the communities started to re-engage their past and their oft forgotten traditional values and practices of environmental stewardship. School children were tasked with collecting folk tales and knowledge from their elders and the tales were re-employed as lessons in fostering ecological balance. The context of this reassertion of local values was the usurpation of their high alpine pastures by the state to create a new bi-national ecological park. Big game hunters and trekkers were granted access to the park, but not the local population that was dependent on the land. Local cultures and traditions became a mode of resisting state and global forces and forging more environmentally friendly practices.[14]

Running parallel to the Silk Road history are alternative, local histories and traditions that are often erased or frozen as 'folk' ideas. Along the Karakoram Highway there are a number of significant Gandharan Buddhist artifacts, and others, indicating a long and rich history. There are the ruins of a Buddhist monastery outside the village of Chilas, south of Gilgit; thousands of inscriptions, including the documenting of a Tibetan invasion in the 8th century; and a number of carvings of the Buddha from different eras. Though not a single one indicates anything about trade and trade routes, from them the Silk Road history is imagined.

But there exist other histories, other stories. One such story reflects the contested meanings over a rock engraving in a small nullah just outside the town of Gilgit. A weatherworn figure appears high up on a cliff face. The local version of the history of the figure is related through a story. An abbreviated version of the story goes as follows:

> In a time long ago there was a small village located in a narrow valley high up in the mountains. Life was difficult but peaceful.
>
> One year catastrophe came. A demon appeared and established himself just outside the village. He demanded from the people that every spring a young damsel be brought to him. The people had no choice for fear of the wrath of the demon and sure destruction. To keep their peaceful life, every year a young girl was selected and given to the demon. But soon there were no young girls left in the village – nothing but doom lay ahead. The demon became impatient and demanded a young girl be sent to him immediately, or else. As the villagers

> gathered to meet and discuss their dire situation, a young prince rode into the village. Upon hearing the fate of the villagers, he hatched a plan.
>
> The very next day two villagers presented a young damsel to the demon and quickly fled. The demon emerged from his cave to grab hold of the girl. The girl removed her outer cloak and low and behold, it was the prince. Taking advantage of the element of surprise, the prince pounced upon the demon and there they tumbled, wrestling about. Back and forth they fought, for hours and days the villagers could hear the thrashing and crashing about. At one point the demon had pinned the exhausted prince to the ground and raised a rock to smash him. The prince, garnering all the strength that remained in him, kicked the demon with all his might. The demon flew back and smashed into the hillside. The prince quickly leaped up and pinned him to the cliff wall.
>
> And there the demon remains to this day, a reminder to all bad spirits that this village is a protected place.[15]

This story is quite different than the story of the figure on the cliff provided by A.H. Dani. Dani, steeped in the frame of the Silk Road and Buddhist path, does not see a demon, but rather 'On the eastern face of this hill is seen the standing figure of Buddha in *abhaya* (fearlessness) pose ... The Buddha is sculptured and is in Tibetan style' from approximately the 8th century.[16] The story of the demon hardly exists today, erased by a more 'scientific' historical explanation, one recounted by every local historian, tour guide, and school child. I suggest that the demon story is part of the Silk Roads while Dani's Buddha figure is part of the Silk Road. That is to say, the demon reflects a plurality of pasts, of ways of studying the past, of rethinking historiography, while Dani's approach is an anachronistic approach that projects the modern nation-state back in time, documenting bounded histories.

This singular, bounded Silk Road of Dani, Bhutto, and the media integrates the Gilgit-Baltistan region into the ancient history of Pakistan. Indeed, the region provides the very avenue for that history along the Karakoram Highway. The highway opens up new vistas of identity by providing the opportunity for one literally to walk along the ancient Silk Road, at least according to much of the tourist literature on the region. The highway and the Silk Road past at once integrate the region into Pakistan, but also provides a pathway of contemporary global interconnectivity.

Commodifying the Silk Road: Development, Tourism, and the Neoliberal Dream

The KKH was officially opened in 1979 though construction continued, as the highway is more of an ongoing project than a fixed road. Even today it is being reconstructed as part of the China-Pakistan Economic Corridor (CPEC). In 1986 the highway was opened to foreigners all the way to the China border. This was a time of the first phase of the Afghanistan war following the USSR invasion in 1978 and the CIA's proxy war administered by Pakistan. It was the time of Ronald Reagan in the United States and Margaret Thatcher in the United Kingdom, together implementing dramatic economic policies around the world, freeing capital and labor to flow across borders and restructuring national economies away from the public sector to private corporations. These processes are generally known as neoliberalism. Following the collapse of the Soviet Union neoliberal policies were naturalized as global ideology, imposed by an array of international organizations, including the United Nations.

It was at this same time that the United Nations Educational, Scientific, and Cultural Organization (UNESCO) developed an interest in the Silk Roads. UNESCO funded a number of projects documenting the historic Silk Roads, refurbishing historical monuments in the newly independent republics of Central Asia, and funding cultural groups to recreate the arts of the Silk Road region.[17] UNESCO's Silk Roads project is subtitled 'Dialogue, Diversity, and Development' clearly mapping it within the neoliberal turn of the UN. While UNESCO employs the plural Silk Roads and supports a rich diversity of studies and celebrations of the historical Silk Roads, their overall project is infused with neoliberal ideology of globalization. As a modern ideological project, UNESCO's Silk Road is clearly singular, a projection back in time of a contemporary ideology.

All of these global forces gave shape to the Karakoram Highway. Born out of a state interest to give fixity to its fluid territorial boundedness and national concerns for crafting a Pakistani identity, the highway flourishes as an avenue of neoliberalism, providing a pathway for globalization. The KKH provides a route for various non-state forces to penetrate into the Gilgit-Baltistan region including agents for economic and social development as well as domestic and international tourists. Through these processes the Silk Road is transformed into a commodity.

One of the first travelers up the KKH besides the Pakistani military was the Aga Khan Rural Support Program (AKRSP) and eventually its many affiliated organizations including the Aga Khan Trust for Culture. Though focused primarily in the Ismaili communities of Hunza and Gojal, the northern-most valleys along the KKH, the AKRSP works across religious communities, fostering significant development of the region. Though its work is widespread, one of the main programs is to advance economic development by training locals to integrate into the global economy. Through micro-finance programs and others, AKRSP provides a platform for allowing locals to capitalize off the flow of traffic along the KKH.

The development vision of the region is intricately interwoven into the life of the Karakoram Highway which has two economic forces: trade and tourism. Trade is defined by connections with China, bringing across cheap Chinese consumer goods and construction materials, all replacing local crafts and goods. For Pakistani domestic tourists the 'Cross-Border China Trade' is marketed as the Silk Road. Families and school groups from the Punjab and beyond travel up to Gilgit to wander the bustling markets full of Chinese goods. For them, the Silk Road is a living reality, to be experienced through shopping.

For international tourists, the Silk Road provides an opportunity to consume history and culture. Various Buddhist sites along the KKH are documented in every guidebook, where they can go on special jeep trips marketed as Silk Road tours and stay in hotels such as the Silk Road Lodge and the Marco Polo Inn, and, if they desire, even buy a cultural performance of a local wedding or even a polo match (a traditional sport of the region from where the British learned the game).

The Karakoram Highway as Silk Road does not merely provide an avenue to consume the past, linking disparate places, but is itself a place, a destination, a commodified tourist spot. As a commodity produced for and by foreign tourists, the local is further erased. Local people are predominately left out of touristic discourses, at best mere coloring to the otherwise dramatic landscape of the region. They are given little voice in the representation of the region, dictated by global guidebook publishers and online forums for tourists themselves. The local are reduced to quaint cultural artifacts, waiters and drivers to serve, or nuisances that spoil an otherwise rich travel experience.

Conclusion

The Silk Road is a contradiction. The idea was born out of colonialism and ideologies of racial and civilizational superiority and imagined within the interests of nation-state formation and neoliberal globalization. Yet, in popular and scholarly imaginations, despite the history of the idea, the Silk Roads invoke something radically different. The Silk Roads are an avenue of interconnection, exchange, and wandering across social and cultural boundaries. They are places of cosmopolitan values where difference is celebrated rather than perceived as a threat. This contradiction allows the Silk Road to be employed by multiple actors all with very different agendas. Though ultimately, it is only through state/government support and UN backing that the Silk Road continues to exist.

The Silk Roads though are more than a mere symbol, they are history. There are very real histories of long-distance exchange throughout the world. Trade and the flow of goods, people, and ideas is what history is made of. Which begs the question, why give such undue attention to one particular expression of such flows? Why do we see intercivilizational flows as even unique? Since when do scholars so uncritically employ the same conceptual frame to document historical phenomena across millennium? Is the Silk Road a necessary frame for studying the history of trade?

Given the concept's troubled history, given the ways in which the idea is employed by states and particularistic global interests to advance political agendas, given the ways in which the idea is commodified, given the way local histories and cultures are erased, we must ask – is the Silk Roads frame even a viable scholarly notion? The Silk Roads frame of history, of situating ancient pasts into larger global processes, erases local histories and meanings. Yes, many historians and others are now documenting local processes along the great avenues of inter-civilizational encounters and exchanges, as most of the contributions to this volume attest. But the frame itself remains that allow us to focus on certain phenomena while ignoring others. We search for exchanges and borrowings, we seek for expressions of cosmopolitanism and interconnectivity, but we rarely see alternative histories. We gladly accept the story of the Buddha and relegate the story of the demon to lesser than history, to myth and folklore. Is it possible to rethink our demon past and employ the Silk Roads not as history but as historiography, of creating a series of questions, issues, and approaches to studying pluralistic pasts rather than singular pasts?

Notes

1 Interestingly, today, Qingdao figures strategically into China's One Belt One Road global initiative, also known as the Silk Road Economic Belt and 21st Century Maritime Silk Road.
2 Chin 2013, 196.
3 For a full discussion on the intellectual history of German anti-Semitism, see: Poliakov 1974.
4 See Haines 2012, particularly chapter 3.
5 Zaman 2018, 62–65.
6 Bhutto 1976, 3–12.
7 *The Pakistan Times*, October 27, 1967, 8 in Haines 2012, 65.
8 Since 2010, the region is known as Khyber Pakhtunkhwa.
9 There are countless books and articles on Kashmir, most of them partisan, leaning towards justifying either India's or Pakistan's claim to the entire region. For a detailed history of Kashmir see: Rai 2004 and for a long history see Zutshi 2014.
10 *The Pakistan Times*, October 27, 1967, 8 in Haines 2012, 65.
11 *The Pakistan Times*, August 26, 1969, 4 in Haines 2012, 73.
12 *The Pakistan Times* August 26, 1969, 8 in Haines 2012, 73.
13 For a rich and highly creative reimagining of Pakistan's pluralistic past see: Ahsan 1996.
14 Butz 1996.
15 As related to me and translated by Shakil Ahmed Shakil of Gilgit.
16 Dani 1995.
17 The UNESCO Silk Roads Project.

Bibliography

Ahsan, A. 1996. *The Indus Saga and the Making of Pakistan*, Karachi.
Bhutto, Z.A. 1976. RCD: Challenge and Response, *Pakistan Horizon* 29.2, 3–12.
Butz, D. 1996. Sustaining Indigenous Communities: Symbolic and Instrumental Dimensions of Pastoral Resource Use in Shimshal, Northern Pakistan, *Canadian Geographer* 40.1, 36–53.
Chin, T. 2013. The Invention of the Silk Road, 1877, *Critical Inquiry* 40.1, 194–219.
Dani, A.H. 1995. *Human Records on Karakorum Highway*, Lahore.
Haines, C. 2012. *Nation, Territory, and Globalization in Pakistan: Traversing the Margins*, London.
Poliakov, L. 1974. *The Aryan Myth: A History of Racist and Nationalist Ideas in Europe*, New York.
Rai, M. 2004. *Hindu Rulers, Muslim Subjects: Islam, Community, and the History of Kashmir*, London.
The UNESCO Silk Roads Project [n.d.]. https://en.unesco.org/silkroad/unesco-silk-roads-project.
Zaman, M.Q. 2018. *Islam in Pakistan: A History*, Princeton.
Zutshi, C. 2014. *Kashmir's Contested Past: Narratives, Sacred Geographies, and the Historical Imagination*, New Delhi.

Section Four

Cultural Transactions

Chapter 13

Arsacid Economic Activity on the Silk Road

Touraj Daryaee

Arsaces I, leader of the Aparni established his power in eastern Iran and crowned himself in Ghouchān/Hecatompylos as king in 247 BCE. It was, however, during the rule of Mithradates I with his conquests of Bactria and the Western Iranian Plateau against the Seleucids that he took the title of King of Kings, where an Arsacid Empire was established.[1] The Arsacids ruled over a vast area from its early dynastic center in Nisa in Central Asia to Ctesiphon as first the winter residence of the king of kings, and then the empire's capital in Mesopotamia. In the Caucasus, through dynastic marriage and political allegiance, the Arsacids became dominant in Colchis/Georgia, Armenia and Balāsagan/Republic of Azarbījān. In the southern direction, the Arsacids allied themselves with Arabs to control the Persian Gulf region. An *Arabarch* was set to regulate the Persian Gulf and the various tribes who roamed the inland and those Arabs who lived close and along the caravan cities.

As the Arsacid Empire came into being, it came into contact with China, but the process was gradual until the reign of Mithradates I and the embassy of Zhang Qian. Some of these embassies were large, carrying with them a large amount of goods, including, of course, silk. At the end of the 2nd century BCE, contact between Parthia and China became intense enough that between 5 to 10 parties at any given time were sent to the Arsacid Empire.[2] Of course Chinese interest did not only concern commodities and trade with the Arsacids, but also the Roman world. There are many studies on the relations between China and Rome, but the fact that this trade mainly went through the Arsacid Empire appears to be altogether missed with just a few studies pointing to this fact.[3] One only needs to read Isidore of Charax's *Parthian Stations*, penned in the 1st century CE, to understand the importance of this region for the Silk Roads trade.[4] But Chinese sources also attest to the fact that the Arsacid Empire was a powerful empire, where the king of kings minted coins and ruled over a vast area with many cities and a large population.[5]

The Arsacids mainly benefited from the Silk Roads trade on sea and land by creating a loose, and sometimes a not so loose, collaboration, which grafted the Caucasus with the Arsacid Empire. This connection was made through dynastic marriages, or the Arsacids appointed officials in such cities as Hatra and Dura Europos and even in Palmyra to keep their economic interests viable. This system, which has come to be known as the 'Parthian Commonwealth',[6] was effective wherever the Arsacids controlled a large mass of Eurasia. In the Persian Gulf region the Arsacids appointed an *Arabarch* to deal with the nomadic tribes and regulate the trade in the region.[7] To grasp the importance of such trade routes throughout the Arsacid Empire, it is beneficial to look at the cities that connected Arsacid Mesopotamia to the well-known 'Caravan Cities',[8] and others on the eastern front of the empire.

Isidore of Charax demonstrates that the route, which was most important for trade, ran from China through Central Asia to the Arsacid Empire. From Hyrcania/Gorgān we encounter many cities and stations which in the Iranian world may be translated as (Middle Persian) *kārwān*, *i.e.*, caravan stops. Some of these caravansaries could be within cities, for example those in Adrapana/Ecbatana, where the author states that a custom's house is located there; that is, the capital city of the Arsacids called Bazigraban.[9] I would like to tentatively suggest that it is possible to read this as *Wāzarg-bān*, 'Merchant Guard', or simply the term for a merchant, Wāzargān/Bāzargān, from Pahlavi *Wāzār*/Persian *Bāzār*/English Bazaar or market.[10] Also, according to Isidore of Charax, all along the Silk Roads beginning at Arsacid Nisa through Astauena, Hyrcania, Ragha, and Ecbatana to Mesopotamia a strand of cities, villages and stations line the way.

Although more than 400 caravansaries have been identified on the Iranian Plateau, because of the unchanging nature of the roads, many were rebuilt in later periods. Most of these caravansaries date from the early modern and modern era, namely the Safavid and Qajar periods (16th to 19th centuries CE).[11] So far the oldest caravansary dates from the Sasanian period which according to various sources was either established by the vanquisher of the Arsacid Empire, Ardašīr I (224–240 CE), or to the 6th century CE. The Deyr-e Gačīn caravansary is located on the great Khorasan Silk Road, close to Isidore of Charax's Rhaga/Reyy. Hence, it is not an in-city caravansary, but rather the more common caravansary situated along the east–west route. While the original plan of the caravansary can be gleaned, it was reconstructed during the Seljuk and the Safavid period (12th and 17th centuries CE). But if the size of this caravansary is any indication of the measure of trade that passed through it, then the enormity of this site attests to the vibrant nature of the Silk Roads trade in antiquity.[12]

The importance of the Silk Roads on the Great Khorasan Road never suffered as a result of the coming of the Sasanians in the 3rd century CE. This was also the period in which the sea borne trade of the Silk Roads was flourishing with Persians and Romans taking to the seas to compete for exotica in foreign markets.[13] Cosmas Indicopleustes states that the quickest way in which Silk came from China to Persia was by land, as the sea route took much longer to travel.[14] Hence, we can surmise that the Silk Roads land route was of utmost importance, where merchandise from China and India, including frankincense, cassia, calamus and many articles from Africa ended up on the Iranian Plateau.[15]

In a recent work, Fabrizio Sinisi has shown that the main mints of the Arsacids reflect an east–west route, which is in line with the economic activity of the empire.[16] If we think of the Silk Roads trade that was taking place at the time of the Arsacid Mithradates I, it makes sense that the worlds of Indo-China and Rome would have had the Arsacid empire at the nexus of this commercial exchange both on land and on sea (Fig. 13.1).[17]

In regards to the caravan city trade of the Arsacids, the existence of Palmyrene traders as indicated by their inscriptions found at Charax on the mouth of the Persian Gulf, suggests the importance of this commercial highway forming as it were a 'network empire' in the Arsacid kingdom.[18] The kingdom of Characene, usually under Arsacid control, was key for the empire to maintain its eastern trade routes leading to India. This network also included a series of harbors and terminals that connected Charax to ports in the Persian Gulf, like Bahrain, and others, such as Oman, in the Indian Ocean. The *Periplus* states that beyond the Strait of Hormoz in the west at the mouth of the Euphrates there is a market town 'designated' by law as a port of entry, called Apologus, near Charax Spasinou, connecting it to Ommana, where vessels

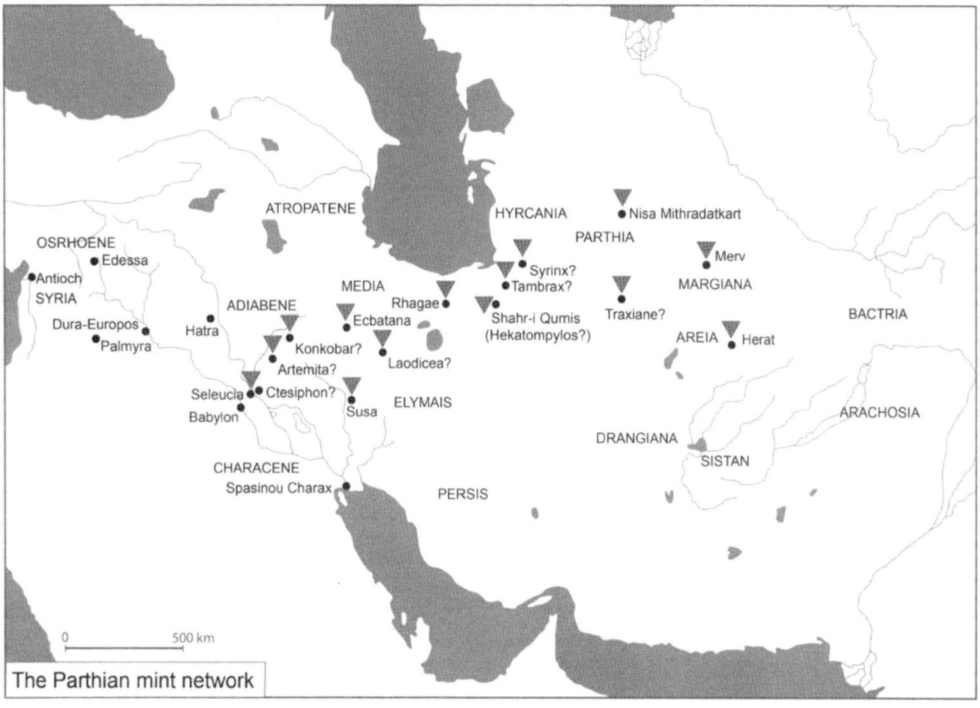

Fig. 13.1. The Parthian mint network (after Sinisi 2018).

made their way from Barygaza in India.[19] According to Leonardo Gregoratti, by the mid-2nd century Palmyrene merchants would have regularly sailed to and from the ports of Barbarikon and Barygaza on the northeast coast of the Indian sub-continent.[20] The Arsacids allowed the thalassocracy of the Characenians to continue, because as beneficiaries of the trade and commerce it served their interest. This situation changed, however, as a result of Trajan's conquest of southern Mesopotamia when Attambelos VII (ca. 113/114–116/117 CE) switched his allegiance to Rome. Following this short-lived interlude of the region, the Arsacids replaced the Charecene political elites with the Arsacid Meredat (ca. 131–151 CE), son of the Parthian king Pakorus, over the kingdom.[21]

For their part, the Palmyrenes were extended the privilege of conducting international trade with their mercantile organizational system,[22] unlike the Charecenes who were now ruled directly by the Arsacid family.[23] The Arsacids through their Charecene representatives in Palmyra made sure that this route remained open and under their control.[24] Although ostensibly in the hands of the Arsacids in the 2nd century CE, in reality the Palmyrene merchants controlled the Persian Gulf trade all the way to Oman (ed-Dur) and the United Arab Emirates (Dibba and Mleiha).[25] There are inscriptions attesting to the fact that the Palmyrenes were quite involved in Charax due to its importance as an entrepôt and as a hub where caravan chiefs protected caravans and lent money to boost commerce.[26] As mentioned, the Arsacid Arabarch was responsible for the safety of the caravan roads between the caravan cities and the Persian Gulf (Fig. 13.2).

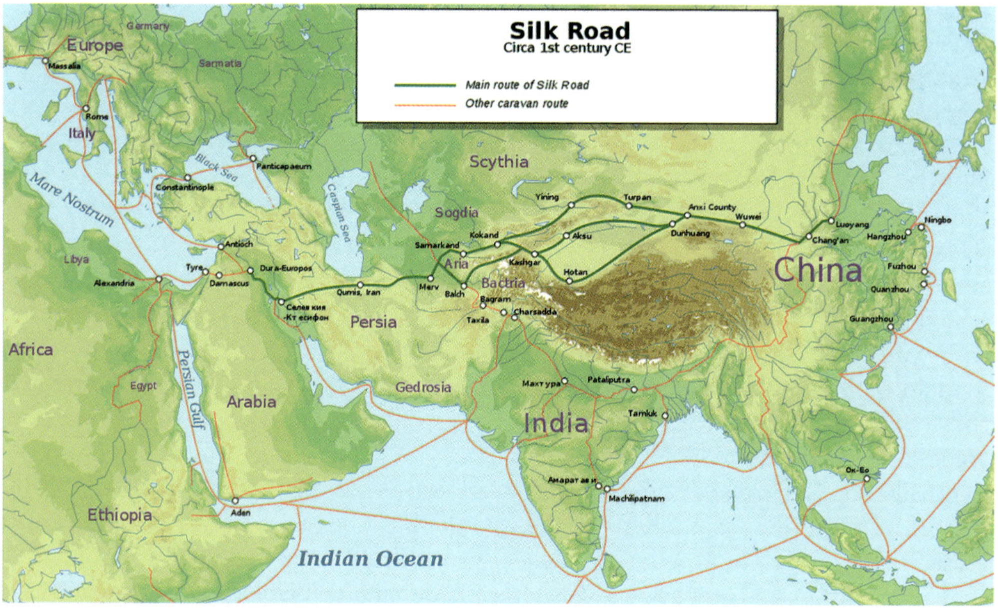

Fig. 13.2. Silk Road trade in the 1st century CE.

I would contend that in the 2nd century CE a number of events contributed to the decline and downfall of the Arsacid Empire. These causes include the empowerment of the Kushan Empire in the east, under Kanishka (127–153 CE), coupled with the change in Roman policy toward the Arsacids at the end of the 2nd century CE. In the east, it was Kanishka's enthronement, which brought changes to the power and the economic reach of the Arsacid Empire. Kanishka expanded his realm at the expense of Indian rulers to the south, and the Arsacids to the west. In the ensuing wars that took place, Chinese sources exaggerate the numbers of the Arsacids killed by the Kushans.[27] That the Kushans were the beneficiaries of these conflicts is seen in their minting of gold coins, not only during the reign of Kanishka,[28] but also during the reigns of his successors – Huviška and Vasudeva – at the beginning of the 3rd century CE.[29]

With Alexander Severus (193–211) and Caracalla (211–217) at the end of the 2nd century CE, we begin to see a change in Roman foreign policy against the Arsacid Empire.[30] It is quite possible that the Romans had decided to once and for all put an end to the Arsacid Empire, as it appeared internally fractured and weak.[31] The continuous defeats and the sacking of the Arsacid capital by the Romans had signaled the weakening and loss of Arsacid military aggression in the 2nd century CE.[32] With the victories of Severus there came the annexation of the territory of northern Mesopotamia and the building of fortresses across the Euphrates.[33] While the strategic implication of such a move by the Romans is clear, the economic impact was far more devastating for the Arsacids. Mesopotamia was one of the richest sources of agriculture for the Arsacid Empire, where combined with incessant political disruptions, led to the deaths of up to one fourth of the population along with plague,[34] that led to the sharp decline of the Arsacid economy. It may well be that such economic decline on both the eastern and western limits of the kingdom no longer allowed the state to effectively maintain its empire,[35] resulting in the eventual collapse at the hand of a local potentate, Ardashir Ī Pābagān, from the province of Persis, at the edge of the Iranian Plateau.[36]

The once lucrative Silk Roads trade for the Arsacids was blocked at both ends of their empire, having fallen into the hands of competing empires, which restricted the movements of Arsacid merchants and led to an economic decline. Once upon a time, Seleucia, Babylon and Vologesis were the beneficiaries of the caravan trade, which ended at Spasinou Charax. As already indicated, in the early 2nd century CE, there were attempts at independence and the wresting away of any control from the Arsacid realm.[37] The Arsacids had placed members of their own family to rule over these kingdoms, or through their agents to exercise a modicum of control over the trade and commerce of localities. This scheme, however, was not able to sustain the main Silk Roads trade and consequently to allow the Arsacids to survive economically. It is no wonder then that when the Sasanian dynasty came to power on the Iranian Plateau, after defeating the last Arsacid King of Kings, Artabānus V, Ardašīr I in 225 took it upon himself to attack the Kushans and defeat and displace Vasudeva with the Kushano-Sasanian rulers.[38] Ardašīr I then moved to Mesopotamia and made war on the

Romans to check their encroachment to wrest control of the Silk Roads trade away from them. The Roman authors may have thought that the Sasanians were making war to claim Achaemenid territorial heritage, but they were simply attempting to take back what had been taken by the Romans during Arsacid rule,[39] which had suffocated their economy and ultimately led to their demise.

Notes

1 Lerner 2015, 2.
2 De la Vaissière 2005, 29.
3 Benjamin 2018, 148–176.
4 Isidore of Charax and Schoff 1914.
5 Isidore of Charax and Schoff 1914, 40–41.
6 de Jong 2009, 143–160.
7 College 1967, 81.
8 Rostovtzeff 1932.
9 Isidore of Charax and Schoff 1914, 6.
10 MacKenzie 1986, 89.
11 Kiani/Kleiss 1990; for a list of existing caravansaries on the Khorasan plain, see http://www.caroun.com/Architecture/Caravansary/00-Khorasan.html. For a complete survey see, Kiani 1978.
12 Shokoohy 1983, 445–461.
13 Daryaee 2003, 1–16.
14 Cosmas Indicopleustes in McCrindle 2010, 48–49.
15 Cosmas Indicopleustes in McCrindle 2010, 51.
16 Sinisi 2018, 475.
17 Traina 2018, 123.
18 Gregoratti 2019, 53.
19 Colledge 1967, 78; Gregoratti 2019, 55 (*Periplus* 35–36).
20 Gregoratti 2019, 59.
21 Potts 1988, 143.
22 Gregoratti 2019, 63.
23 Colledge 1967, 88.
24 For example the inscription of Yarhai from April 131 CE at Palmyra, Potts 1988, 144.
25 Gregoratti 2019, 64.
26 Colledge 1967, 81.
27 Benjamin 2018, 192.
28 Jongeward, Cribb and Donovan 2015, 65–88.
29 Benjamin 2018, 193.
30 Maksymiuk 2004, 132.
31 Birley 1988, 129–130.
32 Gardoni 2013, 16.
33 Millar 1993, 118–119.
34 Gardoni 2013, 5–6.
35 Maksymiuk 2004, 133; Gradoni 2013, 19.
36 Daryaee 2010, 236.
37 Maksymiuk 2004, 133.
38 Bivar 1983, 209.
39 Potter 1987, 156–157.

Bibliography

Benjamin, C. 2018. *Empires of Ancient Eurasia: The First Silk Roads Eras, 100 BCE–250 CE*, Cambridge.

Birely, A.R. 1988. *Septimius Severus: The African Emperor*, London.

Bivar, A.D.H. 1983. The History of Eastern Iran, in: Yarshater, E. (ed.), *The Cambridge History Iran*, vol. 3.1, Cambridge, 181–231.

College, M.A.R. 1967. *The Parthians*, New York/Washington.

Daryaee, T. 2003. The Persian Gulf Trade in Late Antiquity, *Journal of World History* 14.1, 1–16.

Daryaee, T. 2010. Ardaxšīr and the Sasanians' Rise to Power, *Anabasis: Studia Classica et Orientalia* 1, 236–255.

de Jong, A. 2013. Hatra and the Parthian Commonwealth, in: Dirven, L. (ed.), *Hatra: Politics Culture and Religion between Parthia and Rome*, Oriens et Occidens 21, Stuttgart, 143–160.

De la Vassière, É. 2005. *Sogdian Traders: A History*, Leiden/Boston.

Gradoni, M.K. 2013. The Parthian Campaigns of Septimius Severus: Causes, and Roles in Dynastic Legitimation, in: De Sena, C. (ed.), *The Roman Empire During the Severan dynasty: Case Studies in History, Art, Architecture, Economy and Literature*, American Journal of Ancient History 6–8, 3–24.

Gregoratti, L. 2019. Indian Ocean Trade: The Role of Parthia, in: Cobb, M. (ed.), *Indian Ocean Trade in Antiquity. Political, Cultural and Economic Impact*, London/New York, 52–72.

Isidore of Charax and Schoff, W.H. 1914. *Parthian Stations by Isidore of Charax. The Greek text, with a translation and commentary by W.H. Schoff*, Philadelphia.

Jongeward, D., Cribb, J. and Donovan, P. 2015. *Kushan, Kushano-Sasanian and Kidarite Coins. A Catalogue of Coins from the American Numismatic Society*, New York.

Kiani, M.-Y. 1978. *Iranian Caravansarais with Particular Reference to the Safavid Period*, Tokyo.

Kiani, M.-Y. and Kleiss, W. 1990. Carvansaray, *Encyclopaedia Iranica*. www.iranicaonline.org/articles/caravansary.

Lerner, J.D. 2015. Mithradates I's Conquest of Western Greek-Baktria, *Journal of Historical, Philological and Cultural Studies* 47.1, 45–54.

Maksymiuk, K. 2004. Some Remarks on Political Relations Between Sasanian Persia and Rome during the Principate, *Eos* 91, 131–139.

McCrindle, J.W. 2010 [1897]. *The Christian Topography of Cosmas, an Egyptian Monk*, Cambridge/New York.

MacKenzie, D.N. 1986. *A Concise Pahlavi Dictionary*, Oxford.

Millar, F. 1993. *The Roman Near East 31 BC–AD 337*, Cambridge, MA.

Potter, D. 1987. Alexander Severus and Ardashir, *Mesopotamia* 22, 147–157.

Potts, D.T. 1988. Arabia and the Kingdom of Characene, in: Potts, D.T. (ed.), *Araby the Blest: Studies in Arabian Archaeology*, Carsten Niebuhr Institute of Ancient Near Eastern Studies 7, Copenhagen, 137–167.

Rostovtzeff, M. 1932. *Caravan Cities*, Oxford.

Shokoohy, M. 1983. The Sasanian Caravanserai of Dayr-i Gachīn, South of Ray, Iran, *Bulletin of the School of Oriental and African Studies* 46.3, 445–461.

Sinisi, F. 2018. Some Remarks on the Patterns of Coin Production in the Parthian Empire, in: Woytek, B. (ed.), *Infrastructure and Distribution in Ancient Economies: Proceedings of a Conference Held at the Austrian Academy of Sciences, 28–31 October 2014*, Wien, 473–496.

Traina, G. 2018. Central Asia in the Late Roman Mental Map, Second to Sixth Centuries, in: di Cosmo, N. and Maas, M. (eds), *Empires and Exchanges in Eurasian Late Antiquity. Rome, China, Iran, and the Steppe, ca. 250–750*, Cambridge, 123–134.

Chapter 14

Pearls and Power: Chōla's Tribute Mission to the Northern Song Court within the Maritime Silk Road Trade Network

James A. Anderson

In the Xining reign period (1068–1078) of the Song emperor Shenzong, the kingdom of Chōla (zhunian 注輦)[1] *sent an emissary to China bearing tribute. The emissary requested that, in accordance with his own custom (bensu 本俗), he might perform 'the casting (of pearls) into the court (sadian 撒殿)'. A Chinese court official announced that this custom could be performed.*

The emissary carried a golden plate filled with pearls, and he knelt before the doorway's front step (kan), raising the plate over his head. Using a golden lotus, he scooped up the pearls and spread them across the court in the direction of the throne. This custom 'sadian' is the country of Chōla's most venerable ritual (li 禮).

After the court session had ended, those officials in charge of cleaning up the court gathered more than ten liang of pearls. They distributed these pearls as gifts (fenci 分賜) to the Audience and Palace Attendants[2] *who had served that day in court.*

Peng Cheng, 'The Casting of Pearls into the Court'[3]

This paper begins with a close reading of an anecdote written in the 1070s by the Chinese scholar, Peng Cheng, a minor official of the Northern Song dynasty (960–1126).[4] Peng's account describes the arrival of a tribute mission from the distant South Indian kingdom of Chōla (*zhunianguo* 注輦國), which flourished from ca. 850–1279, at the Chinese court.[5] According to Peng, an emissary from this foreign mission requested that the Chinese court allow him to perform a native custom, 'casting into the court (*sadian*)'. Following this Chōla practice, the emissary knelt before the chamber in which the emperor was seated. Using a golden plate and a lotus-shaped ladle, he cast several tens of *liang*[6] of pearls across the court's floor toward the emperor's throne. Reflecting in private on this event, Peng Cheng found the initial request, and the fact that the Chinese court permitted a foreign custom to be included in Song court ceremony, curious enough to recount. Peng's tale is a chapter in his one known work, *Scattered Illuminations from the Literatus* (*Moke huixi* 墨客揮犀),[7] along with other anecdotes the Chinese scholar likely circulated among his friends.

At first glance, one sees reflected in this 11th-century anecdote the nature of Song China's relations with the kingdoms beyond its borders. Descriptions of foreign cultures and their interaction with the imperial court often appeared in detail in the *History of the Song Dynasty* (*Songshi* 宋史) historical record. From these records came an official understanding of 'barbarian decorum', against which the Song state continued to measure its own cultural norms. This approach to the study of Song foreign relations follows the conventional assumption that the Confucian ideals behind the Song state's worldview can be applied unaltered to both public and private realms of discourse. Does the official account of a foreign mission necessarily tell the same tale as does an account recorded privately by a Song official, even if both accounts are decidedly Confucian? If the terms are the same, is the meaning necessarily so? A study of Chinese tributary relations with its neighbor from across the Indian Ocean network reveals the many levels of images employed by the Song state as expressions of the power and prestige in a Chinese cultural empire.

Regarding the protocol of tributary relationships, the Song court's reception of foreign visitors was clearly limited by imperial ritual, particularly the tenets of Guest Ritual (*binli* 賓禮). The *Songshi* gives the following description of *binli*:

> In early times the kingdom of Zhou defeated and eliminated the kingdom of Yin.[8] The Zhou rulers honored Wei Zi 微子 as a descendant of the Yin rulers. Reforming the practice of their *li*, the Zhou rulers made room for Wei Zi as a guest in the king's home, and thereby ensured peace for the kingdom. The Song rulers, desiring to employ the ceremonies of Zhou and to include themselves as descendants of Confucius, took as their guiding principle the practices and spirit of these former kings and worthies. These practices came to be known as *binli*. These practices included arranging feasts for emissaries to the court, as well as providing seasons during which to provide gifts and goods to foreign embassies and to receive tribute from faraway regions. As for the proper method of caring for and banqueting (guests) and (for the guests themselves) the propriety of ascending and descending from the throne, bowing in appearance at and withdrawing from court, *binli* orders these practices.[9]

In the Song government, the ministries of War, Household (finance), and Rites all handled tribute missions to the Chinese court. However, while the first two ministries resembled modern-day institutions of diplomacy, the Ministry of Rites also served a worldly role by arranging and regulating the cosmic and moral orders in which all kingdoms functioned. Diplomatic protocol was intended to reflect these orders, and cosmological changes could require diplomatic responses. The example of the Zhenzong 真宗 Emperor's (r. 997–1022) insistence, with the controversial 1008 appearance of Taoist text titled the 'heavenly book' (*tianshu* 天書) on visits from a large contingent of vassal states, is a clear example of a diplomatic response to a shifting cosmic order.[10]

Although many aspects of Imperial China's foreign relations, including the existence of an unchanging 'tribute system', have been debated vigorously by scholars in recent years, James Hevia's discussion of *binli* in his study of the Qing court's reception of the MacCartney Embassy brings us greater understanding of this ritual's use in earlier periods.[11] Hevia contends that relations between China and its neighbors were a flexible matter. As he writes, 'rather than reflecting or representing an unchanging tradition of foreign relations replicated through a succession of Chinese dynasties, supreme lord/lesser lord relationships were unique achievements produced and reproduced, made and altered, in response to changing circumstances.'[12] Peter van der Veer describes Hevia's understanding of *binli* with Clifford Geertz's observation that '(pomp) serves power, not power pomp'.[13] Although such flexibility also applied to the spirit in which Song policy was formulated, the notions of historical precedent and the relative positions of 'lord and vassal' tempered specific shifts in the court's external policy. Moreover, individual officials who themselves identified with ideal Confucian morality would likely see the faithful observance of *binli* as the only trusted path toward universal harmony.

Perhaps Peng Cheng sensed that the Song court had faltered in its efforts to uphold the Confucian values to which Peng himself was dedicated. Peng as an historical figure is a mystery; little is known about his personal biography. However, the period of Chinese history during which he wrote is well known as a time of radical political experimentation at court, led by the reformer official Wang Anshi 王安石 (1021–1086) with the support of the Shenzong 神宗 Emperor (r. 1067–1085). Open opposition to Wang's reform policies could be costly politically, and many officials were sent into exile for expressing their views. There is no record of Peng Cheng being censored for his politics. However, he professes in his writing a great admiration for the leading official and literary critic Su Shi 蘇軾 (1037–1101), and Su stood squarely in the anti-reformers' camp. There are indirect suggestions in Peng's work that he may have held similar political convictions.

The reader will note in examining the terms Peng chose to describe the events of *sadian* and the context in which he placed the story in the *Moke Huixi* collection, that the author imparts a morally charged message on his audience. As Michael Freeman describes the anti-reformist writings of the 1068–1086 reform period, 'temporal events, the manifestations of the workings of basic principles, were both held together by a basic order and charged with moral import. Moral meaning was not imposed by the historian; it inhered in events themselves and would stand revealed when the true story was recounted.'[14] Peng's own protest was lodged in just such a manner.

There are many questions to ask of this anecdote, *sadian*. First, what did Peng mean by concluding that the 'casting of pearls into the court (*sadian*)' constituted 'their own custom (*bensu*本俗)'? Here, there are interesting issues of translation and untranslatability, for which we can refer to an early work of Vincente Rafael. Regarding the Philippine nationalist hero Jose Rizal's fictional account of 'Father Damaso's sermon', Rafael describes the conditions under which a Tagalog-speaking audience is drawn to unrecognized fragments of the sermon 'that slide away from their context and so are susceptible to recontextualization'.[15] Translation here involved the transmission of the 'untranslatable' from one cultural or political context to another. The 'pearl-casting' of Peng Cheng's tale appears as just such a fragment, taken from its East Indian context by a Chinese scholar, categorized as 'their own'/non-Chinese, and then incorporated as part of Chinese tributary protocol. However, one might ask whether this act is a re-affirmation of a trans-cultural link between the ruler and the ruled, or if something else is happening here. Does Peng notice here that 'pearl-casting', as was the case for Father Damaso's Catholic Christian imagery in Rizal, has found meaning for the Chinese court that may be quite different from the meaning intended by its East Indian performers?

Multiple readings of the Chinese characters in Peng's account result in a variety of interpretations that depend on the reader's approach to the text. *Su* 俗 is a slippery term to translate from Classical Chinese. Defined primarily as 'custom', this character carries with it a secondary meaning of 'vulgar' or 'worldly'. Such a translation might betray Peng Cheng's chauvinistic disdain for pearl casting as a foreign practice. The poet and statesman much admired by Peng, Su Shi, once commented that, 'a man who is thin may gain weight, while an official who is unrefined (*su*) can never be cured.'[16] This characterization of barbarian practices would not be uncommon among early Song officials who saw themselves as the vanguard of a renewed Confucian idealism.

However, Peng later refers to *sadian* as 'ritual (*li* 禮)'. *Li* suggests a quality of universal order and structures and may be translated as 'the exemplification of eternal principles'.[17] Does Peng's understanding this foreign practice as *li* allow him to suppose that this East Indian kingdom has practices equal to the *li* of China? Is there an accepted universal category in Song China's political vocabulary that suggests a cross-cultural equality beyond the subjugating norm of the Chinese tribute system? Such an assertion would challenge the Sino-centric world order that places utmost importance on the performance of tributary practices that accord with Confucian beliefs.

The tribute mission Peng described also appeared in the *Songshi*. Peng does not indicate in which year the tribute mission arrived at the Song court, specifying only that the event occurred during the Shenzong Emperor's reign. The *Songshi* and *Statutes of the Song Dynasty* (*Song Hui Yao* 宋會要) both noted that a mission from Chōla arrived in Kaifeng on 27 June 1077.[18] The envoys included 'chief envoy' Qilolo, 'vice envoy' Nanbeipada, and 'staff official' Matuhualuo.[19] Three days after arriving at court, the envoys received an audience with the emperor, at which time they spread pearls before the throne. The *Songshi* account notes that the Song emperor gave 800 strings of copper cash and 52,000 *liang* of silver as gifts to the Chōla ruler in return.[20]

However, Peng Cheng was not the only Song period scholar to make note of this event. Among Peng's like-minded contemporaries were Pang Yuanying 龐元英 (fl. 1078–1082) and the well-known scholar Shen Gua 沈括 (1031–1095). Pang wrote in his account that envoys from Chōla and Sri Vijaya (Sanfoqi 三佛齊) both approached the Song court, at which time the *sadian* ritual was performed. Pang's account of the diplomatic exchange was written during his tenure as the director of the Wenchang 文昌 Ministry, the Department Head of the Reception of Foreign Guests, from 1082 to 1085.[21] Pang makes no reference to the embassy's description of *sadian* as the Chōla kingdom's 'most respectful ritual (*zhi jing zhi li* 至敬之禮)', but he specifically notes that the request to perform *sadian* had been presented in a formal memorial in advance of the audience.[22] Shen Gua's account may be found in his well-known treatise of Northern Song history and scientific achievement *Dream Pool Essays* (*Mengxi Bitan* 夢溪筆談), completed around 1088.[23] Shen Gua included all the details mentioned in Peng's account, and appended a geographical note with reference to the Chōla kingdom being located about 4,000 *li* by sea from Guangzhou and the ethnographic note that the people of Sri Vijaya were another type of 'Southern Barbarian' and neighbors of the Cham.[24]

A later generation of Song scholars, such as the prolific Ye Mengde 葉夢得 (1077–1148), the Neo-Confucian scholar Xue Jixuan 薛季宣 (1134–1173), the occasional envoy to the Jin court Lou Yue 樓鑰 (1137–1213) and the local official Jiang Shaoyu 江少虞 (*jinshi* after 1115, d. after 1145), also took a great interest in this anecdote. Ye's wide-ranging history *Dinnertime Discussions of the Stone Forest* (*Shilin Yanyu* 石林燕語), contains an account that closely follows Shen Gua's version.[25] Ye was also a young contemporary and admirer of Su Shi, a relationship he likely shared with the lesser-known Peng Cheng. However, the Chinese scholar with the account that most closely resembles Peng Cheng's version is Jiang Shaoyu, the lowly prefect whose one known literary work *Categorized Garden of Facts* (*Shishi Leiyuyuan* 事實類苑) has been described as a selection of passages chosen to illustrate aspects of 'social morality'.[26] In his account Jiang includes all of the details mentioned by Peng, elaborating only in his description of the vessel used to perform the *sadian* ritual.[27]

Certain details, included only in Peng Cheng's account, lead to multiple interpretations of the event itself. The other accounts, with the exception of Jiang's work, did not mention that the pearls were later gathered up and distributed to the Audience Attendants and minor officials at court. The Song government had put restrictions on the number of tribute missions the court would receive from certain kingdoms, and the Chinese had set duties for certain 'non-tribute' items. In 1061, there was an official proclamation that all surplus tribute must be presented first to the emperor, while goods deemed 'non-tribute' would be taxed. The notion that a valuable tribute item such as pearls was being parceled out to court officials may have suggested a minor scandal.

Moreover, pearls and a golden lotus-shaped ladle, have links to the Chinese Buddhist tradition. In North China, for example, pearls were suggestive of Buddhist remains.[28] Although Indian rulers were no longer strong patrons of Buddhism, the early Song emperors were. Whether the performance of *sadian* before the Shenzong Emperor was a conscious act of Buddhist piety is difficult to determine from official records. However, such an accusation would not be unfounded. A similar mission from the Chōla kingdom had come to China 44 years earlier, arriving in Kaifeng on

15 November 1033 and staying at court until 22 March 1034. 'Barbarian etiquette (*yili* 夷禮)' was observed when the envoys scattered pearls from a silver container before the throne.[29] Noting that the Chinese emperor of that time, Zhenzong, was a supporter of Taoism at court, the Chōla emissary *salisanwen* 娑里三文 (Soli Samudra) worded many of his comments to the throne in Taoist references.[30]

In the *Songshi* account, the practice of *sadian* by the Chōla mission is described as *yili*. Such a description distinguishes *sadian* as a form of *li*/ritual from the practices of *li* initiated by the Chinese court. A reason for making this distinction is because 'barbarian' *li* ordered the courts of particular kingdoms while Chinese *li* was seen as ordering a universal system of relations.

However, Peng clearly labeled *sadian* as the Chōla kingdom's most venerated *li*, without any 'barbarian' qualification. In this case, the practice of *sadian* at the Song court was an acknowledgment that universal moral order, which was the Chinese emperor's domain, could be expressed through a non-Confucian practice. Peng's note that the emissaries described *sadian* as their own traditional custom (*su*) complicates the picture, because such base behavior was, in Chinese terms, detrimental to the natural moral order.

Medieval Indian historical sources show that subordinate rulers at birth festivities for the Chōla ruling house performed a court ceremony similar to the one described by Peng.[31] As for the Chōla emissaries in China, performing this *li*/ritual before the Chinese throne, and labeling it as their 'most highly venerated *li*', was a gesture of diplomatic good will and an acknowledgment of the Song empire's regional importance. However, the fact that, as Peng describes it, the Song emperor would allow a barbarian vulgar custom to be performed as *li* suggests that Confucian principles were not always needed to reinforce world order.

This Chōla tribute mission and audience with the Song emperor was not an isolated incident. Beginning in summer 1015 the Chōla kingdom under Rajendra I (1014–1044) sent envoys to Kaifeng (Fig. 14.1). In his list of tribute missions to the Song, Robert Hartwell noted the *Songshi* account, where the emperor Song Zhenzong 宋真宗 (r. 997–1022) ordered the Imperial Aide (*gongfeng guan* 供奉官) and Usher at Imperial Audiences (*gemen zhihou* 閤門祇候) Shi Youzhi 史祐之 to serve as the Lesser Lord of Diplomatic Receptions (*Honglou shao qing* 鴻樓少卿) and escort the Chōla emissaries to the Song court. This first Chōla mission remained in Kaifeng, according to Hartwell, for nearly a year from 15 May 1015 to 10 May 1016, with the actual audience with the emperor dated October 16 to November 5, 1015.[32] The envoy reportedly claimed that the Chōla kingdom had received word from a visiting merchant that the Song emperors were performing the Taishan 泰山 and Fenyin 汾阴 ritual sacrifices to Heaven and Earth, and that this conduct had calmed the seas (there had been no storms) for 10 years, which prompted the Chōla ruler to send his tribute mission.[33] In the spring the Song emperor had ordered the Bureau of Protocol (*liyi yuan* 禮儀院) to compile a list of tribute missions from various kingdoms, which the bureau titled 'Chart of the Tribute Missions of the Four Barbarians'

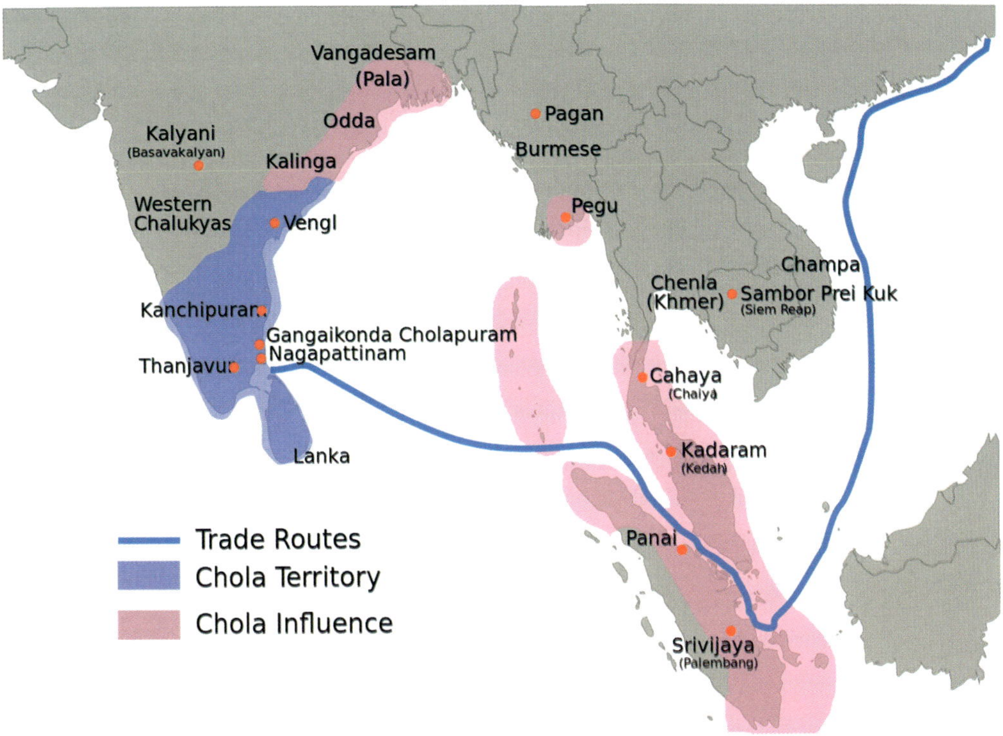

Fig. 14.1. The Chōla empire during Rajendra Chola I (ca. 1030 CE) (WikiCommons, Gregors 2011).

(*siyi shuzhi tu* 四夷述職圖). The Supervisor for the Court of Diplomatic Protocol Zhang Fu 張復 then compiled a description of the Chōla kingdom's dress and customs to present to the emperor.[34] Here, we may have the source of knowledge at the Song court of Chōla practices, from which the emperor drew notes when remarking on the practice of *sadian*.

The list of subsequent tribute missions to the Song court was not extensive.

Hartwell recorded the next mission arriving in Kaifeng on 29 February 1020, although the head of Chōla mission had become sick and died upon arriving in Guangzhou, the first port of call for southern maritime kingdoms, so the emperor ordered that a banquet be given for the remaining members of the tribute mission and that lavish gifts be sent back to the Chōla ruler. During an audience with the Song Emperor on 15 November 1033 the Chōla envoy Abu Adil (Puya Tuoli 蒲押陀離) requested that barbarian etiquette be observed, and pearls were scattered from a silver container. On 26 June 1077, as described above, pearls, and in Hartwell's account ripe camphor, were scattered from a golden lotus bowl. Following these descriptions, there are no more official accounts of Chōla missions, which align with the rise and fall of the maritime empire's influence in the western reaches of the Indian Ocean trade network.

In terms of economic benefits, the Chōla kingdom and China likely did not share similar assumptions when they entered into tributary relations. While the Song empire was a larger power at the time of their encounter, the strength and regional influence of the Chōla Kingdom had been steadily increasing from the late 10th century. The Chōla kingdom under Rajendra I had led numerous successful military campaigns against its South Indian neighbors, stripping conquered states, such as Sri Lanka and Bengalese Pala kingdom, of their treasures as war booty.[35] At the height of Rajendra I's power his navies had advanced on and captured parts of modern-day Myanmar and the Malay Peninsula, including Sumatra. By the time that their 72-man mission approached the Song for the first time in 1077, the Chōla kingdom had lost control of Sri Lanka and their regional influence had waned. However, Chōla continued to remain influential in the Tamil region of southernmost India until the 1250s when the neighboring Pandyas under Sundara Pandya deposed the last Chōla ruler.

Scholars of medieval India also contend that the Chōla rulers pursued a form of 'commercial diplomacy' during this period. This term suggests flexible diplomatic relations, aimed at fostering ties of trade and economic activity. Moreover, Chōla's confrontation and eventual military defeat of the leading contemporary Southeast Asian maritime power (and Chinese tributary vassal) Sri Vijaya could have been a bid by Chōla to wrestle away from Sri Vijaya control of the commercially strategic Malacca Straits and to replace that Southeast Asian kingdom in the higher order of states that maintained close relations with the Song court. From 971 a steady stream of tribute had flowed from Sri Vijaya to the Song, and next to the Vietnamese and Champa (*zhancheng* 占城) Sri Vijaya would be the most active Southeast Asian participant in China's tribute system throughout the early Song empire. Tansen Sen contends that the Sri Vijayan leadership actively attempted to prevent contact between the Chōla kingdom and the Song court.[36] This diplomatic rivalry may not have been well understood from Kaifeng, but it would explain the reason why the Chōla envoys wished to describe their interactions with the Song court with the greatest display of veneration and through their highest order of ritual interaction.

The participants from Chōla may have seen themselves as ambitious traders wishing to establish closer relations with a regional economic power. Therefore, the Chōla delegation entered the process without the same power of *li* in their possession. What they represented in their actions could never equal what the Chinese court officials represented in theirs. Moreover, China regarded Chōla as a 'first class' trade partner, grouping it with kingdoms in the Middle East, Sri Vijaya and Java.[37] The establishment of an official system of levies in China suggests that channels of international trade, including relations with Chōla, were quite active during the early Song. However, whether or not the Song court showed deference to the 'native customs' of its visitor on purely economic grounds is difficult to determine, and likely not the case. Tansen Sen makes the point that the Song court had revamped the tribute system when it came to interaction with maritime kingdoms, given the lucrative trade that had been flowing through southern ports since the founding of the dynasty.[38] This could well be a strong reason for the Song emperor's acceptance of 'native custom' in place of established court-based protocol when interacting with Chōla envoys. It could also be the reason why Peng Cheng, as a conservative member of the court, would have found reason to lodge his protest.

Peng himself has a possible role to play in the creation of this account. In many ways he looks like a figure resembling Stephen Greenblatt's subversive Thomas Harriot in his *Brief and True Report of the New Found Land of Virginia*, or Shakespeare, as viewed through his history plays.[39] Peng Cheng wrote nothing overtly describing an authority undermining itself, but there is a warning against moral imbalance implicit in the terms he chose to use and those he chose to omit.

This warning carries over to Peng Cheng's choice of the 'miscellaneous notes' (*biji* 筆記) literary form to describe this event. This literary form suggests that the scholar intended a certain tone of informality and intimacy in his account. The *biji* is a free and unstructured style of Chinese literature, the subjects of which ranged from short fictional narratives to brief expositions of history or philosophy. The reader audience of *biji* accounts did not extend beyond the author's immediate circle of friends, and its contents exposed more personal convictions than did a writer's official commentaries and memorials.

Regarding the anecdote *sadian*, it may be that Peng Cheng was not particularly interested in the actual ritual, its importance to the visiting delegation, and the import of its performance before the Song emperor. Most likely, it was the strangeness and non-Confucian nature of this foreign practice that attracted his attention. Moreover, the collection from which this anecdote is taken enhances the context in which one may view this tale. Peng Cheng's work is filled with descriptions of the strange and the 'unnatural', including an account of an earthen Buddha that moved, monkeys that kidnapped small children and harassed Buddhist monks, and a woman whose bloated stomach could sound like a drum. These events represented disturbances in the natural and human orders, and were themselves indicative of larger disturbances. Solutions generally came in the guise of a Confucian official or his words of counsel. Once the proper way was reestablished, the disturbance vanished.

Such a literary collection finds representational parallels with the wonder-cabinets (*wunderkammern*) of early modern Europe, in which curios were kept to mark the outer boundaries of accepted culture. As Steven Mullaney notes of these European displays, 'no system determines the organization of the objects on display or separates one variety of the marvelous from another. We are surprised upon entering the room, but our surprise is occasioned not so much by the individual items we encounter, impressive though they are, as by the immediate familiarity they show for whatever joins them.'[40] Each item shares with the others its aspects of 'strangeness'; the collection becomes a representational whole with each part sharing this overall

Table 14.1. Translation of Xue Jixuan.

晚雨在林麓，	Wǎn yǔ zài lín lù,	Evening rain on the forest hill,
朝來弄晴曦。	Zhāo lái nòng qíng xī.	The morning was clear and bright.
瀾翻漾明月，	Lán fān yàng míng yuè,	The swelling waters reflect the bright moon.
粲爛生摩尼。	Càn làn shēng móní.	Its brilliance brings forth the mani jewel.
撒殿固無事，	Sādiàn gù wú shì,	Perform the 'pearl-casting' ritual but without incident.
雨天亦何爲。	Yǔ tiān yì hé wéi.	Why conduct the ritual on a rainy day?
少須風景麗，	Shào xū fēngjǐng lì,	The young yearn for beautiful scenery,
去去將安之。	Qù qù jiāng ānzhī.	While the old desire peace.

significance. The Chōla emissary's 'casting of pearls' then fits into this scheme of strange occurrences, an event worth noting because of the disturbance it causes to the natural Confucian order.

The strength of the Chinese official Peng Cheng's protest depends on the context in which he included his anecdote. Alone, the story retold does not stray far from the official account of a vassal kingdom paying tribute to its Confucian lord. However, the strangeness of the *su*/custom performed as a supplement to the Song court's 'guest ritual' becomes more striking when the other subjects of Peng's notes are included. Nature is out of order in many of these tales, and Confucian morality is required to set things straight. The Song court under the Shenzong Emperor, and dominated by the reformist faction of Wang Anshi, suffered its share of upheaval. The incident described in *sadian*, with all its unorthodox behavior (at least according to a strict Confucian moralist) in ceremonial proceedings and its official profiteering, exemplified this upheaval. The collection of strange tales in *Moke huixi* was a mirror held up against a distorted Chinese empire. Peng Cheng, as did Harriot, lodged his protest in his handling of this mirror.

A reference to *sadian* in late Song literature provides an interesting coda to contemporary attitudes toward this ritual practice. In a short poem with a long title, 'In a 40-Character Poem Composed When Evening Rain led to Morning Clear, and the Forest Looked Glorious, and Everywhere Sparkled like Pearls (*Wan yu chao ji, lin duan yu ran, wang zhi jie bao zhu ye, xi zuo si shizi* 晚雨朝霽，林端煜然，望之皆寶珠也，戲作四十字)',[41] the aforementioned Northern Song scholar Xue Jixuan wrote (*my translation*, see Table 14.1).

A prominent thinker and critic of the Neo-Confucian Daoxue 道學 School[42] Xue Jixuan draws on the image of 'pearl-casting' that has circulated in the private writings of Song scholars since the first accounts of Chōla tributary missions. The image has been placed in a Buddhist context, and it retains its fantastic and marvelous qualities. Long after Chōla emissaries performed what they termed their most reverential ritual, and Confucian scholars have noted the practice in a variety of contexts, *sadian* remains a trope for an unorthodox act with the highest intention.

Notes

1 *Zhunian* was the Chinese name for the Chōla kingdom, located on the Coromandel Coast of East India.
2 During the Song Dynasty, the Palace Attendants were often eunuchs.
3 Peng Cheng, The Casting Pearls into the Court (*sadian*) in *Continuation of Scattered Illuminations from the Literatus* (Xu Moke huixi 續墨客揮犀), j. 8, p. 16.
4 Peng Cheng 彭乘 (fl. 1050–1080) should not be confused with Peng Cheng 彭乘 (985–1049) about whom much more has been recorded (see biography in *Songshi* 宋史, j. 298). This Peng Cheng was a native of Gao'an 高安 in Yunzhou 筠州 prefecture (a part of modern-day Jiangxi province). His official posts included *zhongshu jianzheng* 中書檢正 (Official charged with investigation and notification in the Great Imperial Secretariat [the emperor's chancellors]). Peng had a high regard for the Song literary figures, Su Shi 蘇軾 (1037–1101) and Huang Tingjian 黄庭堅 (1045–1105), both of whom were considered to be political conservatives. In his own work, many of Peng's observations were of events that took place in the Jiangxi and Guangdong region of South China. (This information was taken from Balazs and Hervouet 1978, 330.)
5 The Chōla kingdom was located in Southeastern India in modern-day Tamil Nadu. This maritime empire particularly flourished during the 11th century. The Chōla leadership established their capital at Tanjore.
6 A traditional Chinese unit of weight; approximately two ounces.
7 Balazs and Hervouet 1978, 330.
8 *Yin* is an alternate title for the latter half of the ancient Chinese Shang dynasty (ca. 1401–1154 BC).
9 Toghto *et al.* 1983, j. 119, 2795.
10 For a description of these events, see Davis 2001, 67–68.
11 Hevia 1995.
12 Hevia 1989, 77.
13 Veer and Gibson 2016, 18. Geertz's quote is found in Geertz 1980, 13.
14 Freeman 1973, 158.
15 Rafael 1993, 214.
16 The original line is 人瘦尚可肥，俗士不可醫, found in the poem Yu Qian Seng Lu Yunxuan 於潜僧绿筠轩 (For the Hermit Monk Lu Yunxuan) from Su Shi, Jizhu fenlei Dongpo shi 集注分類東坡詩 (Variorum Classified Edition of Poems of Su Dongpo).
17 Mathews 1931, 566.
18 One distinct problem with the description of the 1077 tribute mission is that this mission occurs at the same time that a mission reportedly arrived from Sri Vijaya, the kingdom then at war with Chōla. The Chinese transliteration of the name of the ruler who supposedly ordered the Chōla mission to China better fits the Sri Vijayan ruler of that time. Although Robert Hartwell concludes that the Sri Vijayan mission never existed, the question still persists.
19 Hartwell 1980, 189.
20 This detail is also noted by Karashima and Sen 2009.
21 Note by V. Velgus in Balazs and Hervouet 1978, 191–192.
22 Pang Yuanying 2006, 120.
23 Balazs and Hervouet 1978, 226–227.
24 Shen Gua 1961, 767.
25 Ye Mengde 2006, 10: 33.
26 Yoshida Tora note in Balazs and Hervouet 1978, 323.
27 Jiang Shaoyu 1993, 874: 509.
28 Liu 1988, 56.
29 Hartwell 1983, 189.

30 Luo Xianglin 1986, 382.
31 Dhar 1991, 113.
32 Hartwell 1983, 188.
33 Hartwell 1983, 188.
34 Hartwell 1983, 188.
35 Keay 2000, 220.
36 Sen 2009, 63.
37 Hall 1980, 176.
38 Sen 2009, 63.
39 Greenblatt 1988, 21–65.
40 Mullaney 1983, 42.
41 Xue Jixuan 2003, 113.
42 Goossaert *et al.* 2015, 1363.43

Bibliography

Adshead, S.A.M. 1988. *China in World History*, London.
Ames, R.T. 1983. *The Art of Rulership: A Study in Ancient Chinese Political Thought*, Honolulu.
Anderson, B. 1990. *Language and Power: Exploring Political Cultures in Indonesia*, Ithaca.
Balazs, E. and Hervouet, Y. 1978. *A Sung Bibliography (Bibliographie des Sung)*, Hong Kong.
Buxbaum, D.C. and Mote, F.W. 1972. *Transition and Permanence: Chinese History and Culture*, Hong Kong.
Davis, E.L. 2001. *Society and the Supernatural in Song China*, Honolulu.
Dhar, M.K. 1991. *Royal Life in Ancient India*, Delhi.
Ding, C. 1989. *A Compilation of Anecdotes of Sung Personalities*. Chu Djang and Jane C. Djang (trans.), Collegeville, Minn.
Diskul, M.C.S. 1980. *The Art of Srivijaya*, Paris.
Edkins, J. ca. 1879. *Chinese Buddhism: A Volume of Sketches, Historical, Descriptive, and Critical*, London.
Freeman, M.D. 1973. *Lo-yang and the Opposition to Wang An-shih: The Rise of Confucian Conservatism, 1068–1086*. PhD dissertation: Yale University.
Geertz, C. 1973. *The Interpretation of Cultures*, New York.
Greenblatt, S. 1988. *Shakespearean Negotiations: The Circulation of Social Energy in Renaissance England*, Berkeley.
Gossaert, V., Kiely, J. and Lagerwey, J. 2015. *Modern Chinese Religion* vol. 2, Leiden.
Guo, P. 1981. *Song Yuan Fojiao* 宋元佛教, Fuzhou.
Hall, K.R. 1980. *Trade and Statecraft in the Age of Colas*, New Delphi.
Hartwell, R.M. 1983. *Tribute Missions to China, 960–1126*, Philadelphia.
Hevia, J.L. 1989. A Multitude of Lords: Qing Court Ritual and the MacCartney Embassy of 1793, *Late Imperial China* 10.2, 72–105.
Hevia, J.L. 1995. *Cherishing Men from Afar: Qing Guest Ritual and the Macartney Embassy of 1793*, Durham.
Ho, Y. 1976. *The Organization and Functions of the Ministry of Rites in the Early Ming Period (1368–1398)*. PhD dissertation: University of Minnesota. *Xin Ya Xuebao (The New Asia Journal)* 7 nos. 1–2, 1965–1966.
Hunt, L. (ed.) 1989. *The New Cultural History*, Berkeley.
Jiang Shaoyu 江少虞 (d. after 1145). 1983. Categorized Garden of Facts (Shishi Leiyuan 事實類苑) in The Complete Collection of the Four Treasuries (Siku quanshu 四庫全書). 1500 Vols. Taipei.
Karashima, N. and Sen, T. 2009. Appendix II: Chinese Texts Describing or Referring to the Chola Kingdom as Zhu-nian, in: Kulke, H., Kesavapany, K. And Sakhuja, V. (eds), *Nagappattinam to Suvarnadwipa: Reflections on Chola Naval Expeditions*, New Delhi, 292–315.

Keay, J. 2000. *India: A History*, New York.
Liu, X. 1985. *Early Commercial and Cultural Exchanges Between India and China, 1st–6th Centuries AD*. PhD dissertation: University of Pennsylvania.
Liu, X. 1988. *Ancient India and Ancient China: Trade and Religious Exchanges*, Delhi.
Luo X. 1986. Songdai zhulianguo shi salisanwen ju hua xingcheng kao 宋代注輦國使娑 里三文入華行程考, *Songshi yanjiu ji* 宋史研究集, Taipei.
Manguin, P. 1989. *A Bibliography for Sriwijayan Studies*, Jakarta.
Mathews, R.H. 1931. *Mathews' Chinese-English Dictionary*, Shanghai.
Mullaney, S. 1983. Strange Things, Gross Terms, Curious Customs: The Rehearsal of Cultures in the Late Renaissance, *Representations* 3, 40–67.
Nienhauser, W.H. Jr. (ed.). 1986. *The Indiana Companion to Traditional Chinese Literature*, Bloomington.
Pang Yuanying 龐元英 (fl. 1078–1082). 2006. *Miscellaneous Records of the Wenchang Ministry* (*Wenchang Zalu* 文昌雜錄) in Zhu, Yi'an (ed.) *Quan Song Biji* 全宋筆記. Ser. 2, Vol. 4, Zhengzhou.
Peng Cheng 彭乘. 1916. *Xu Moke huixi* 續墨客揮犀, *Hanfen Lou Biji*, Shanghai.
Rafael, V.L. 1993. *Contracting Colonialism: Translation and Christian Conversion in Tagalog Society under Early Spanish Rule*, Durham.
Rossabi, M. (ed.). 1983. *China among Equals: The Middle Kingdom and its Neighbors, 10th–14th Centuries*, Berkeley.
Schafer, E.H. 1963. *The Golden Peaches of Samarkand: A Study of T'ang Exotics*, Berkeley.
Schafer, E.H. 1967. *The Vermilion Bird: T'ang Images of the South*, Berkeley.
Serruys, H. 1967. *Tribute System and Diplomatic Missions: 1400–1600*. Mélanges Chinois et Bouddhiques vol. 14, Brussels.
Shen Gua 沈括 (1031–1095). 1961. *Dream Pool Essays* (*Mengxi Bitan* 夢溪筆談), Taipei.
Spencer, G.W. 1983. *The Politics of Expansion: The Chola Conquest of Sri Lanka and Sri Vaya*, Madras.
Su Shi 蘇軾. 1936. *Jizhu fenlei Dongpo xiansheng shi* 集註分類東坡先生詩, Shanghai.
Su Shi 蘇軾. 1967. *Jizhu fenlei Dongpo shi* 集注分類東坡詩, Taipei.
Sen, T. 2009. The Military Campaigns of Rajendra Chola and the Chola-Srivijya-China Triangle, in: Kulke, H., Kesavapany, K. And Sakhuja, V. (eds), *Nagappattinam to Suvarnadwipa: Reflections on Chola Naval Expeditions to Southeast Asia*, New Delhi, 61–75.
Thapar, R. 1966. *A History of India*, vol. 1, Middlesex.
Toghto 脫脫 *et al.* 1983. *Songshi* 宋史, Beijing.
van der Veer, P. and Gibson, T. 2016. *The Value of Comparison*, Durham.
White, H. 1978. *Tropics of Discourse: Essays in Cultural Criticism*, Baltimore.
WikiCommons, Gregors. 2011. Rajendra Chola's Territories *c.* 1030 CE. https://en.wikipedia.org/wiki/Rajendra_Chola_I#/media/File:Rajendra_map_new.svg (accessed February 2020).
Wolters, O.W. 1967. *Early Indonesian Commerce: A Study of the Origins of Srivijaya*, Ithaca.
Wright, A.F. 1965. *Buddhism in Chinese History*, New York.
Xu Song 徐松 *et al.* 1964. *Songhuiyao jigao* 宋會要輯稿, Taipei.
Xue Jixuan 薛季宣 (1134–1173) *et al.* annotated by Zhang Liangquan 張良權 2003. *Xue Jixuan ji* 薛季宣集, Shanghai.
Ye Mengde 葉夢得 (1077–1148). 2006. *Dinnertime Discussions of the Stone Forest* (*Shilin Yanyu* 石林燕語) in Zhu, Yi'an (ed.) Quan Song Biji 全宋筆記. Ser. 2, Vol. 10, Zhengzhou.

Chapter 15

'Flying Cash': Credit Instruments on the Silk Roads[1]

Dan Du

In the Geniza Archive, which preserved tens of thousands of fragments documenting the daily lives of Jewish communities in Old Cairo, Egypt, during the Fatimid Period (909–1171), Shelomo Dov Goitein discovered considerable records of promissory notes.[2] A promissory note is a promise in writing to repay a debt in the future. These notes of '3, 5, or 7 dirhems should not surprise the reader,' Goitein remarked, 'Silver money was scarce. Day-to-day transactions...were largely made by paper.'[3] Similarly, modern readers should not find that credit instruments excavated on the overland Silk Roads are unexpected. The three promissory notes unearthed in Khotan (in present southwestern China) record the trade and debt relationships between Chinese and Khotanese dealers during the 7th to early 9th centuries.[4] In 9th-century Tibet, owing to the rarity of cash, travelers resorted to drafts to convert into local supplies and cash. Bearing the expression *dus-sum-gyi-dpan-grya* ('attested signature of the three times'), which possibly indicates a way of sealing contracts by drawing finger joints, the drafts resemble bills of exchange or travelers' checks, two types of credit instruments that represent orders of payment in different locations.[5]

Promissory notes, bills of exchange, and checks, are the three major credit instruments of modern finance. Even though scholars often associate them with capitalism and European credit techniques, evidence on pre-modern, non-European credit instruments abounds in the history of the Silk Roads, a network of overland and maritime routes that connected Eurasia and Africa from the 2nd century BCE through the 15th century CE.[6] However, most research on the Silk Roads tends to highlight commodity monies, such as spice and silk, or hard money, such as gold and silver.[7] Given the cost and risk of transporting cash long distances, cashless payments were no less critical in facilitating the trade on the overland and maritime Silk Roads. Helen Wang has examined the credit economy of eastern Central Asia, Shelomo Dov

Goitein has explored paper bills of the Jewish communities in Egypt, and Peng Xinwei 彭信威 has elaborated on the development of paper money in China.[8] Building on their findings, this paper explores well-documented credit instruments in China, India, and the Islamic Empire, the three major powers on the Silk Roads, and revisits the financial structure of pre-modern world trade.

The preoccupation with commodity and hard monies also sets the nature of world trade during the Silk Road period apart from that in the European-dominated era, which featured a credit economy. Most research on credit instruments is Eurocentric, even though it includes Jewish and Islamic credit techniques.[9] By portraying credit instruments as an invention of modern times and industrialism, such works easily reinforce the linear monetary theory, which, hypothesized by the Germans, chronologizes a progression from primitive commodity money and hard money to nominal currencies, such as paper money, and ultimately credit money.[10] Pre-modern societies, thus, appear to be at the bottom of the pyramid. Expanding on the Needham Questions and the 'Great Divergence' hypothesis, some even posit a 'financial divergence' theory to depict a robust West and a stagnant East in terms of financial innovations, including the invention of credit instruments.[11] However, continuity rather than divergence better describes the financial structure of world trade during the Silk Road Era and modern times. The credit instruments employed as fiscal and commercial devices in China, India, and the Islamic Empire highlight the sophistication of credit techniques in the East. Even if the West has been playing a leading role in modern finance, European imperial expansion after the Age of Exploration was no less critical than military, cultural, geographical, mathematical, or any contingent factors in shaping the different paces of development.[12]

Credit Instruments

The phrases 'credit instruments' and 'negotiable instruments' have been in frequent use in the English-speaking world since the late 18th century.[13] Modern credit instruments primarily refer to promissory notes, bills of exchange, and checks.[14] A promissory note is a *promise* to pay on a future date. A bill of exchange or a check represents an *order* to pay at a different location.[15] In Europe, checks developed out of bills of exchange around 1400.[16] While most bills of exchange save the cost and risk of shipping cash in international trade, inland bills and checks usually circulate between different locations, and sometimes in the same place, within the issuing country.[17] Debtors *write* promissory notes to pay their creditors or the bearers of their notes, so that a note initiates a credit relationship between two parties. By contrast, debtors *draw* bills or checks from the drawees to pay their payees, who are possibly the creditors or related parties; therefore, bills and checks involve three parties in the initial credit relationship.[18] While the drawee of a bill can be anyone, the drawee of a check is ordinarily a bank or similar institution.[19] Despite the differences in format and function, they all facilitate cashless transactions.

These credit instruments are also negotiable instruments, a quality that enhances their ability to lubricate money-scarce economy. Negotiability refers to the transferability of the instruments from one person to another with or without endorsement. It is a special type of transfer, in which the transferee becomes the temporary holder.[20] This quality allows credit instruments to circulate like currency until they are cashed. The primary function of currency includes means of payment, medium of exchange (the settlement of a debt relationship), unit of account, and store of value.[21] While precious metals and commodities, due to their intrinsic value, serve all four of these purposes, credit instruments usually play the roles of means of payment and media of exchange.[22] Through transference to different holders, negotiability allows credit instruments to perform the monetary functions for people beyond the original credit relationships and, therefore, expands the scope of their circulation.

However, these standardized formats and terminology of credit instruments were established by continental European and Anglo-American legal traditions, thus often failing to describe the credit techniques on the Silk Roads. Modern distinctions between a promissory note, a bill of exchange, and a check are too clear-cut for ancient credit instruments, the format and function of which were shaped by local traditions. Thus, it is necessary to clarify the concepts of credit and credit instruments before discussing the non-European techniques. Scholars have defined credit as a transfer of commodity or service without an actual transfer of an equivalent on the spot.[23] Therefore, the gap of time in transactions generates either commercial credit (the sale of commodities without an immediate transfer of monetary objects) or monetary credit (the lending or remittance of money).[24] The nature of credit even led some medieval Europeans, such as William of Auxerre, the French theologian and official in the Roman Catholic Church, to attack usury, not only owing to the high interest it generated, but also because a usurer 'sells time, which,' he asserted, 'is common to all creatures'.[25] This understanding of credit as a means to sell and buy time is sophisticated. Credit instruments are the devices that bridge the gap of time, and often distance, in transactions. A promissory note links the past with the future through a documented promise. A bill of exchange and a check, orders to offer payment in different locations, mitigate the cost and risk of transporting cash and thereby add a spatial dimension. With these credit instruments in commercial and financial transactions, cash is able to 'fly' over time and space.

The techniques of credit instruments, including the concept of negotiability, began in antiquity. Scholars found the earliest extant evidence in Mesopotamia. The Murasu Archive discovered in Nippur, an ancient city in modern Iraq, boasts a considerable number of promissory notes produced in the Achaemenid Empire (550–330 BCE); the large number of female holders reveals women's role in ancient trade and debt relationships.[26] The cuneiform tablets discovered in Borsippa (modern Birs-Nimrud, Iraq) record credit relationships during the Neo-Babylonian period (ca. 620–539 BCE) and early Persian rule (ca. 538–484 BCE).[27] A clay tablet held by the Oriental Institute at the University of Chicago pushes the dates of promissory notes in Mesopotamia further back to the Old-Babylonian Period (1830–1531 BCE).[28] A.H. Pruessner's article published in 1928 introduces 18 Babylonian tablets, the credit instruments that document credit transactions in grain, silver, sheep, and oil as early as 2105 BCE.[29]

The Babylonian tablets, thus, help to elucidate the origin of credit instruments. A promissory note, the earliest form of credit instruments, is a loan contract by nature. The Babylonian silver loan contract inscribed on a clay tablet, dating to the seventh month of the year 1820 BCE, demonstrates their shared features:

> One and one-sixth shekels silver (*i.e.*, 9.33 grams), to which the standard interest is to be added, Ilshu-bani, the son of Nabi-ilishu, received from (the god) Shamash and from Sin-tajjar. At harvest time he will repay the silver and the interest. Before five witnesses (their names are listed). In month seven of the year that Apil-Sin built the temple of Inanna of Elip.[30]

By clarifying the maker, payee, amount of the loan, duration of the debt, rate of interest, and promise of a future payment, the document resembles the format of a modern promissory note.[31] Bills of exchange, scholars agree, originated from the order of payment that accompanied the loan contract.[32] Arthur Robert Burns, a 20th-century economist, recognized some of the Babylonian clay tablets as bills of exchange. From the 9th and 8th centuries BCE, he noticed, people in Babylonia and Chalda inscribed an amount of money, expressed in wrights of silver and copper, onto clay bricks in one place, while the holder of the brick withdrew the sum in a different place. This practice lowered the cost of transporting money and provided security for traders against robbers in the caravan trade.[33] Although evidence of these notes and bills being sold or traded is absent, some Babylonian loan tablets document the repayment to a third party; scholars also discovered a loan record in the archive of an individual unmentioned in the text. These pieces of information indicate the transferability or negotiability of the tablet credit instruments.[34]

Credit Instruments on the Silk Roads

If the evidence of negotiability in Babylonia is dubious, scholars agree that a fully negotiable credit instrument emerged in ancient China. That was paper money, which Marco Polo, the well-known Silk Road traveler, famously accounted.[35] Other well-documented credit instruments in India and the Islamic Empire, serving as fiscal and commercial devices, also facilitated the internal and international commerce of these regions, thus undergirding the credit economy of the Silk Roads.

From Commercial Device to Legal Tender: Paper Money in China

Despite being regarded as the origin of paper money, *feiqian* 飛錢 (literally meaning 'Flying Cash'), or *bianhuan* 便換 ('Convenience Note') was still a bill of exchange, the format of which developed out of traditional Chinese loan contracts. Introduced during the reign of Emperor Xianzong (r. 805–820) of the Tang Dynasty (618–907), *feiqian* facilitated merchants' remittance of coins from the capital city to the circuits (the administrative unit of Tang China).[36] After a merchant deposited coins with the *jinzouyuan* 進奏院, the Capital Liaison Office of each local circuit, he would receive a paper document as the receipt. The merchant held half of the receipt, while the *jinzouyuan* sent the other half to his local government and afforded the fees for the remittance. After returning to the home circuit, the merchant could cash the receipt with the local government if the two halves matched. Traditional Chinese loan contracts shaped the format of the *feiqian*, because the creditor and debtor usually held half of the loan agreement, and, when the loan came due, matched both halves to clear the debt.[37] The function of a *feiqian* resembled an inland bill of exchange, as it was an order drawn by the *jinzouyuan* on a local government to pay a merchant. However, the *jinzouyuan* never monopolized the business, as military commanders, the *hubu* 戶部 (the Ministry of Revenue), and merchant houses also issued *feiqian* to lubricate the money-scarce economy, so it remained a commercial device.[38]

The practice continued into the Song dynasty (960–1279). As a popular means of payment, the *feiqian* functioned as the equivalent of a modern check. Ordinary Song Chinese could draw on the *zuocangku* 左藏庫 (National Treasury) and later a more specialized bureau, *bianqianwu* 便錢務 (Convenience Note Bureau), to obtain *feiqian*. The value of the *feiqian* in circulation soared to 1.7 million *guan* (a unit of currency, 1 *guan* equaled 1000 coins) in 997 CE and 2.8 million *guan* in 1021 CE.[39] The Imperial Court of the Song mandated local bureaus to redeem all holders' *feiqian* on sight, indicating the considerable volume and significance of this instrument in society. The trouble that the 'convenience cash' created in intraregional commerce further demonstrates its popularity. During the years 1027–1028, the merchants who remitted *feiqian* to purchase gold from the miners in Enzhou 恩州 (near present-day Guangzhou) had become fierce competitors of the government in the gold market, so Fan Yong 范庸 persuaded the central government to place a ban on the flow of *feiqian* to Enzhou.[40]

However, the *feiqian* was only one of the various paper credit instruments that boosted the commercial prosperity of the Song. Even *sengdao dudie* 僧道度牒, the officially certified identification documents of Buddhist and Daoist monks, became popular credit instruments during this period, because holders could cash them with coins, silver, or gold. Sold originally at 130,000 *guan* by the central government, a piece of *sengdao dudie* commanded as much as 800,000 *guan* under speculation.[41] More famous credit instruments during this period were *jiaozi* 交子, *guanzi* 關子, *qianyin* 錢引, *huizi* 會子, *yanyin* 鹽引, and *chayin* 茶引. Emerging in chronological order, *jiaozi*, *guanzi*, *qianyin*, and *huizi*, all played a similar role to that of the *feiqian*.[42] *Jiao*, *guan*, and *hui*, the names of *jiaozi*, *guanzi*, and *huizi*, all signify 'match' in Chinese, akin to the idea that the holder and issuer held half of the *feiqian*.[43] The Song government printed new versions of *jiaozi* in 1105 as *qianyin*, a name which either indicated its co-circulation with coins (*qian*) or its function as the permit (*yin*) to obtain paper bills from government departments.[44] In addition to the permit to exchange currency, the official permits to trade salt (*yan*) and tea (*cha*), *yanyin* and *chayin*, also grew to be popular credit instruments. According to the *Song History* (*Songshi* 宋史), Song merchants usually invested half of their proceeds in purchasing the salt permits.[45] *Yanyin* thus acquired the name *yanchao* 鹽鈔 (literally, 'salt money'), a term ultimately used to name Chinese paper money.[46]

Therefore, from these credit instruments, Chinese paper money acquired its format and name. These devices were initially in constant contest with one another; however, the Imperial Court of the Song gradually transformed the *jiaozi*, *guanzi*, and *huizi* from ordinary commercial bills to fiscal devices. After the collapse of the Tang dynasty, China dissolved into various political and monetary districts. In Sichuan, where iron coins circulated as the primary currency, *jiaozi* emerged during the years of 1008–1016 as a privately and freely issued note to resolve the problem of money inadequacy and inconvenience. By the early 11th century, however, 16 prominent merchant houses had dominated its issuance and circulation.[47] When financially embarrassed merchants were unable to redeem their *jiaozi*, the Song government shouldered the responsibility to issue, redeem, and manage the paper notes. Bearing hidden insignia, signature, and seal, the official *jiaozi* inherited the format and function of the private notes in iron-coin districts.[48] During the Southern Song period (1127–1279), *guanzi* and *huizi* circulated more widely in northwest and southern China.[49] Moreover, in the late 12th century, the *hubu* (Ministry of Revenue) allowed commoners to use the *huizi* to pay taxes.[50] In the 1170s, the government used centrally-issued *huizi* to purchase local *huizi*. Even though the local *huizi* was allowed to recirculate later, this attempt demonstrated the government's attempt to elevate the status of the official *huizi* over the other credit instruments as a fiscal device, an action which possibly enhanced the negotiability of the *huizi* in the monetary market.[51]

However, it was the Jurchens of the Jin Regime (1115–1234) who transformed the fiscal device into a legal tender. The centrally issued *jiaozi* and *huizi* of the Song were by no means paper money in modern sense, for they had expiration dates and the central government redeemed them every two or three years.[52] Thus, a *huizi* functioned as not only a bill of exchange and a check, but also a promissory note, which represented the government's promise to pay the noteholder two or three years later. However, not only did the government pay no interest, but noteholders had to pay a three percent commission fee for their redemption.[53] Jurchen rulers, nevertheless, issued unexpired paper notes, namely *jiaochao* 交鈔.[54] *Chao* 鈔, originally referring to the *yanyin*, became the designation for Chinese paper money.[55] Emperor Zhangzong of the Jin, a staunch supporter of the credit money policy, continued to elevate its status over other credit instruments. During 1197–1198, he mandated the purchase of the *yanyin* with *jiaochao* at a discount, so the official notes commanded higher prices than the salt money in the market. He also paid his soldiers with *jiaochao*, thus saving hard money for the national treasury.[56] In 1207, he even encouraged commoners to use the *jiaochao* in daily transactions involving as little as one *guan* in value.[57] Although the Jin economy was plagued by inflation, the monetary policy revolutionized the nature of the credit instrument into a legal tender, which, in addition to being a medium of exchange and a means of payment, assumed such roles as a unit of account and a store of value. Readily transferred between holders, these paper notes were fully negotiable.

The life of paper money also crystalized the influence of the Silk Roads in China. Qiu Shiyu 邱石玉, a Chinese scholar who specializes in the history of the Jurchens, argues that Jewish traders influenced the design of the *jiaochao* during the Jin period, because the leaf motif on the edges of the note was unique to Jewish culture.[58] This connection, if substantiated by further research, would fortify the significance of the Silk Roads in shaping Chinese monetary history. Moreover, under the influence of the Song Chinese and Jurchens, Mongol rulers also issued paper money backed by silk or silver.[59] When Marco Polo reported on the use of paper money in China, it had been transformed from an ordinary credit instrument to a fiat currency. The vast territory of the Mongol Empire enabled the concept of paper money to travel to Persia. Although Persian rulers were Mongols, they inscribed Chinese characters on their notes.[60] Even though the idea of credit instruments had emerged in Europe, scholars believe that Polo's acclaim for Chinese paper money inspired European goldsmiths to issue promissory notes in exchange for deposits.[61] These paper bills in Europe evolved into banknotes when issuers became banks and eventually a form of legal tender when the Bank of England monopolized the issuing privilege in 1833.[62]

In addition to the transmission of ideas along the Silk Roads, foreign merchants also utilized the device to ease their trade with China. The *Song History* mentions in passing that in the year 1250 'owing to the low prices of the *huizi*, Lizong of the Southern Song reissued the ban on the export [of hard money] via maritime routes (*Yi huijia dijian, fuyanshenxiahaizhijin* 以會價低減, 復申嚴下海之禁),' thus suggesting the use of the paper bill in China's foreign trade.[63] The declining demand for the *huizi*, Emperor Lizong assumed, resulted from the increasing use and outflow of hard money. Indeed, contemporary rulers in the area harbored similar fears. The Jurchen Emperor, Zhangzong of the Jin, also ordered in 1198 that anyone who used coins to trade with foreign merchants would be sentenced to death.[64] In 1212, to curb the smuggling of hard money outside Southern Song along the Yangzi River, especially into the Jurgen Jin regime, Emperor Ningzong mandated his officials to use *huizi* to purchase copper coins that circulated in Jiangbei 江北 (north of the lower reaches of the Yangzi River) and Jingxi 京西 (west of present-day Hangzhou), thus tightening the government's control over hard money and promoting the circulation of the paper currency.[65] Song rulers also prohibited the outflow of hard money via the southern port cities active in the maritime Silk Road trade, such as Quanzhou 泉州 and Guangzhou 廣州, in 1182, 1208, 1234, 1244, 1248, and 1250.[66] In 1248, Chen Qiulu 陳求魯, Monitor Censor 監察御史 of the Song, complained to the emperor that 'the mountainous vessels from foreign countries have braved the wind and waves and penetrated into the interior of China to trade luxurious but useless items for our wealth (*fanbo jusou, xing ruo shanyue, cheng feng jia lang, shen ru ya zou. Fan yu Zhongguo zhe jie fumi wuyong zhi yiwu, er xie yu waiyi zhe nai guojia fugui zhi caobing* 蕃舶巨艘，形若山嶽，乘風駕浪，深入遐陬。販於中國者皆浮靡無用之異物，

而泄於外夷者乃國家富貴之操柄),' primarily in the form of hard money.[67] Therefore, it is tenable that paper devices played a critical role in China's Silk Road trade to avoid the outflow of bullion and coins.

After the Mongols conquered Eurasia, including Song China, and established the Yuan Dynasty (1279–1368) in East Asia, historical documents record the use of paper money in the Silk Roads trade. The Florentine businessman, Francesco Balducci Pegolotti, learned that foreign traders had to deposit their silver in the treasury of the Yuan in exchange for *balishi*, a piece of 'yellow paper, stamped with the seal' of Mongol rulers, to buy silk and other merchandise. Four pieces of *balishi* was worth one sommo (a Byzantine unit of currency) of silver. According to Pegolotti, people in the country were expected to accept *balishi* at par.[68] The term *balishi* possibly derived from the Chinese name for the Yuan paper money, *zhongtong yuanbaochao* 中統元寶鈔, which was commonly known as *baochao* 寶鈔. *Baochao* was also important for traders from Central and South Asia, as extant historical documents record the exchange rates of *baochao* for local currencies. For example, in the Kingdom of Lavo (450–1388), present-day Thailand and Vietnam, 24 *liang* (Chinese unit of weight and currency) *baochao* equaled 10,000 local currency.[69]

After driving the Mongols out of Chinese provinces, the Imperial Court of the Ming Dynasty (1368–1644) monopolized legal foreign trade and conceptualized this commercial intercourse as a tributary relationship. Commodities and paper money thus became tributes and gifts between foreign rulers and Ming emperors.[70] During Zheng He's voyages to the Indian Ocean and Africa from 1405 to 1433, for instance, Emperor Yongle (r. 1402–1424) granted paper money to not only his soldiers, but also the envoys from Calicut and Quilon in India, Semudera and Aru on Sumatra, Malacca, Hormuz in Africa, as well as unspecified 'other countries'.[71] Emperor Renzong (r. 1424–1425) and Xianzong (r. 1464–1487) of the Ming also presented *Da Ming baochao* 大明寶鈔 as a gift to many countries that offered goods to the court, such as Java and Malacca on the maritime Silk Road.[72] In 1571, the use of *Da Ming baochao* extended to the tea-horse trade, which was traditionally a barter economy, as Emperor Muzong (r.1567–1572) gave paper money to those who offered fine horses to the Ming court.[73] Paper money was so prevalent in these diplomatic and commercial contacts with China that Fan Ji 范濟 claimed, by 1473, 'the treasure notes of our glorious Ming have been circulating in all the tributary countries (*Da Ming baochao, Huayi zhuguo mo bu fengxing*大明寶鈔, 華夷諸國莫不奉行).'[74] However, Fan possibly exaggerated the widespread popularity of *daming baochao*; in all likelihood foreign ambassadors had exchanged it for Chinese goods before returning to their own country.[75]

After the mid-15th century, the inflation-ridden Ming resumed the use of hard money, particularly silver, as primary currency.[76] The abolition of paper money, nevertheless, by no means eliminated credit instruments in China; instead, this policy only paved the way for private promissory notes, bills of exchange, and checks issued by merchants and moneyshops.[77] In the year 1626, a person named Zhu Zuwen 朱祖文 bribed officials and social elites in the capital city with *huipiao* 會票, issued by merchant firms or moneyshops, to rescue his imprisoned friend.[78] In 1880, a store titled *Boling yinji* 博陵寅記店 drew on a silk commission house, *Shunhao sibaozhan* 順號絲寶棧, to pay a firm named *Chunsheng baozhuang* 椿生寶莊; this bill of exchange reveals the close commercial ties among these merchant houses.[79] In 1899, the storeowner of *Wanchuntang* 萬春堂 wrote a check, drawn on a moneyshop or a merchant firm *Fushundedianhao* 福順德店號, to pay a person named Ren Jinbao 任進寶.[80] In addition to these cases, scholars have found miscellaneous notes and bills – namely *huiquan* 會券, *qipiao* 期票, *duipiao* 兌票, *xinpiao* 信票, and *liangtie* 糧帖 – issued by moneyshops, merchant houses, and traders during the Ming and Qing dynasties.[81] During the Qing period (1644–1912), merchant firms in Laiyang 萊陽 County, Shandong Province, started to use *qiantie* 钱帖 or *piaohui* 票回 in de facto transactions.[82] Bearing part of a Chinese character as insignia on each half of a paper bill, many of these instruments were able to preclude counterfeiting and spread widely in daily transactions.[83] Thus, scholar Xu Mei 許楣 (1797–1870) lamented, 'The majority choose *qianpiao* 錢票 (bank notes) and *huipiao* (bills of exchange and checks) in our society, but all would reduce to waste paper once the issuing money shops failed (*Minjian duo yong qianpiao huipiao, mei yu qianzhuang xiebi, quan gui wu yong* 民間多用錢票會票,每遇錢莊歇閉,全歸無用).'[84] A staunch supporter of hard-money policy as Xu was, his stance might compromise the validity of his remarks, but cannot refute the prevalence of these private paper instruments in daily transactions.

Private commercial bills also continued to facilitate China's foreign trade. In 1814, a Chinese merchant, Pan Kun 潘崑 (better known as Consequa to American merchants), petitioned President James Madison regarding American merchants' defaults on the large amount of debt owed to him.[85] Consequa's loans to his American counterparts totaled $1,000,000, at least $632,952.58 of which was extended through promissory notes, but he was only one of the dozens of Chinese merchants who attempted to collect debt from their American commercial partners.[86] These private credit instruments filled the breach left by paper money in the credit economy of China, even though some scholars, preoccupied with the large volume of silver flowing into China, tended to portray the foreign trade of Ming and Qing as a silverized economy with 'no credit system, no paper money'.[87]

From Fiscal Device to Commercial Bill: Hundis in India

Similar to modern associations of China with silver, India was well known for the use of cowrie shells as currency.[88] However, it is likely that a form of credit machinery had operated since the 6th or 5th century BCE, when money lenders began to fund India's commerce with the Persian Gulf.[89] Early Buddhist literature, such as the *Jatakas*, recounts such credit instruments as letters of credit, I.O.U.s, or debt sheets (*ina-pannani*) issued by wealthy merchants.[90] The most famous and mature credit instrument in India, however, was the *hundi*, also named *hoondee*, *hundika*, *hoondy*, *hondavi*, or *hundwi*.[91] While some regard the term as a Persian corruption of the Indian word Hindi or Hindu, modern scholars believe that it derives from the Sanskrit root *hund*, meaning 'to collect'.[92] While the Persian origin theory of the term suggests how commercial connections between India and Persia influenced the language of business, the Sanskrit verb root reveals that the *hundi* was a credit instrument by nature. The *Oxford Hindi-English Dictionary* defines it as a promissory note, a bill of exchange, and a draft (bank check).[93] With multiple functions and formats, *hundis* circulated in the Indian subcontinent from ancient times to raise money, remit funds, and finance inland or international trade.[94]

While the relationship between the *hundi* and the credit instruments recorded in the *Jatakas* is unclear, scholars have identified its earlier forms as a bill of exchange, namely *hundika* or *adesa*.[95] The earliest extant historical reference to the use of *hundika* is *Rāj Huṉḍikā Yathā* (*A Royal Treasury Draft*) in the *Lekhpaddhati*, a collection of model documents in pre-Sultanate Gujarat from the 8th to 15th centuries.[96] In 744–745 CE, an officer referred to as *rājaputra* was instructed to 'collect and pay' eight foot soldiers '8 dra.[*drammsas*] per day towards undressed provisions until the account is clear,' and then 'send (the sum) to the treasury'.[97] While this 8th-century *hundika* still reads like a letter, the format of the royal *hundika* issued during 1475–1476 resembles a modern bill.

> By the order of the King, on the word of X of territory A, a *huṉḍikā* is issued to Y [as follows]: You should pay to *rājaputra* X, son of Y, *dra*. 124, in words, one hundred and twenty-four *drammas*, when the *huṉḍikā* becomes due, according to the usage of the *huṉḍikā*. The period (of payment) is 15 days. For each day beyond the limit, you should pay 1 *dramma* ka 2.[98]

With the drawer, drawee, payee, amount of money, and period of payment clarified in the documents, the *hundika* functioned like a bill of exchange. Ranjit Sitaram Pandit, editor of *Rajatarangini* (*The River of Kings*), a 12th-century chronicle of Kashmir, asserts that an ancient and simpler form of the *hundika* was *adesa*, 'an order to a third person to pay the bearer a sum of money on behalf of the sender', which is the very definition of a bill of exchange.[99]

Both Gujarati and Kashmiri kings employed the *hundi* as a fiscal device, suggesting the wide distribution of this practice in medieval India. By the 10th century, kings in Kashmir had commonly employed this credit technique to pay their soldiers. 'In that realm whose kings had occupied Kanauj and other territories the livelihood of kings depended upon the delivery of *Hundis* to the Tantrin,' recounts the *Rajatarangini*.[100] Stories have it that Vastupala, the prime minister of Gujarat in the 13th century, drew a *hundi* of 10 crores (one crore denotes 10 million) on a city banker of Ahmadabad in Gujarat to build the temple of Dilwara in the northern state of Rajasthan.[101] The above-mentioned royal bill issued in 1475–1476 shows that Gujarati rulers continued to use *hundika* as a means of payment into the 15th century. However, it is unclear when the term transitioned from *hundika* to *hundi*.

It is certain that the *hundi* or *hundika* served as a promissory note, a bill of exchange, or a check for commoners to settle transactions or remit money, thus facilitating private businesses on the subcontinent. *Lokaprakasha*, a lexicon compiled by the

11th-century Kashmiri poet, Ksemendra, includes various terms regarding endowments to temples, bonds, and credit documents.[102] It shows that Kashmiri villagers wrote *hundika* ranging in value from 300 to 10,000 dinnars as promises to repay their debts in cash (*dinnar hundika*), rice (*dhanya hundika*), or wheat and barley (*yavaghodupi hundika*). They also employed *kriyakar hundika* and *gotika hundika* to purchase wine and horses.[103] This bill remained a popular instrument during the Mughal period (1526–1857). Abul Fazl, a historian in the 16th century, commended the *hundi* as a marvelous instrument for remitting money.[104] 'He took out a *hundi* for 500 rupees, And started laying in a store of cloth', Banarasidas, a Jain businessman and poet active in the 17th century, describes how he used the *hundi* to purchase cloth in his autobiography *Ardhakathānaka* (*A Half Story*).[105] The *hundi* was still popular in the mid-18th century, noticed a Gujarati historian, for people used this paper bill, instead of cash, for commercial payments.[106]

The *hundi* was a negotiable instrument, and the Indian concept of negotiability rendered it more acceptable than cash. Om Prakash has traced the negotiability of the *hundi* to the 17th century. In the late 1630s, he argues, when *hundi* holders redeemed their notes with *sarrafs* (an Arabic term for professional money changers), they could either receive cash with a half percent additional charge or endorse the *hundi* to their creditors to clear the debt obligations.[107] Endorsement indicates negotiability, so Prakash reasons. In fact, the *hundika* issued six centuries ago has demonstrated the negotiability of this instrument. In 11th-century Kashmir, after loaning out three *kharis* (a unit of measurement) of produce, a villager wrote an interest-bearing *hundika*, prompting the debtor to repay 'five *kharis* of good, healthy, and dry rice' at the beginning of the current winter season. 'Not observing the time-limit for realisation or remitting all the profits (then) in whosoever's hand this *hundika* it will be like this,' the villager declares in the *hundika*.[108] This line suggests that the interest rate would apply to any holder of the *hundika* when it came due. Transferability, thus, indicates negotiability. Moreover, different from the Euro-American practice, where liability was exempted after the instrument was transferred to another holder, anyone who had discounted a *hundi* became liable for the debt if the drawee failed to cash the bill. This unique practice in India enhanced the negotiability of the credit instrument, so a *hundi* discounted by *sarrafs* or reputable merchants became more acceptable in the market.[109] Thus, in the transactions at Agra, the *hundi* was one percent higher in value than cash in 1652.[110]

While early literature indicates the *hundi* as a fiscal device, it appears frequently as a commercial bill in more recent historical sources, as *sarrafs* in India orchestrated the remittance of the *hundi* through their local representatives.[111] At the core of the credit system, small and large *sarrafs* constructed a vast *hundi* network to serve fiscal and commercial demands. In 1717, for example, Keshav Rai Kishanchand, whose headquarters were located at Agra, helped to redeem the Surman embassy's Rs. 59,000 bills through his agents at Delhi, Kora, and Allahabad.[112] The Jagat Seth family, the most prominent *sarraf* establishment in the first half of the 18th century, transferred over 10 million rupees annually for the imperial exchequer of the Mughal Empire from their headquarters in Bengal to Delhi.[113] Indian *sarrafs*' expansive *hundi* network also facilitated Dutch and British businesses in India. The Director of the Dutch East India Company, for instance, remitted *hundis*, at the value of 60,000 rupees, to the Dutchman van Hoorn in Patna in 1669, arranged by Indian *sarrafs* in the two cities.[114] The British East India Company also accepted 'hoodian on bills granted by the Shroffs' [the English term derived from *sarrafs*] for delivery of cash, which were later payable in silver from the Company treasury'.[115] Relying on native bankers and their *hundis*, the Dutch and British raised capital, distributed resources, and expanded their imperialism in India.[116] Amazed by the reliability of the *hundi* system, Charles Northcote Cooke remarked in 1868, 'The dishonoring of a hoondee is an event of rare occurrence with [*sarrafs*], and without the aid of which, indeed, an English house of business would scarcely be secure.'[117]

Under the operation of Indian *sarrafs* and their agents, *hundis* also facilitated international trade on both the maritime and overland Silk Roads. According to Jean-Baptiste Tavernier, the French gem merchant who traveled to Persia and India in 1630–1668, the credit network of Jain and Bania merchant-bankers stretched from Surat and Agra eastward to Bantam, Achin, and the Philippines through the Strait of Malacca and westward to Basra, Hormuz, and Mokha across the Arabian Sea.[118] Without ready cash, a merchant at every major commercial center in India and any port on the Indian Ocean could have drawn a *hundi* on a Bania *sarraf* of Surat to cover the cost of remittance, exchange, and even insurance.[119] Therefore, stable money markets for the *hundi* developed on the maritime Silk Roads. Although the exchange rates varied locally, they were steadily standardized in each region.[120] The *hundi* system had also spread to Central Asia by the 18th century. Due to the danger-ridden routes of the overland Silk Roads, *sarrafs* usually sent agents, namely the *gumasthas*, to the vast area spanning from Khorrassan, Iran to Turkestan, thus transferring theirs and their clients' funds exclusively in the form of *hundis*.[121] The credit system of the *hundi* obviated the risk of long-distance trade and connected India with other regions on the Silk Roads.

Fiscal and Commercial Devices in Islamic Credit Culture

Credit instruments in the Islamic world have also received considerable attention.[122] Some scholars identify the Arabic devices as the prototype of European credit instruments, whereas others refute Arabic influence by highlighting distinctions between these techniques.[123] The Islamic forms of a promissory note (*ruq'a*), a bill of exchange (*saftaji*), and a check (*sakk*) had been in circulation in early Islamic history. The European term 'check' was derived from *sakka* or *sakk*.[124] The French word *aval*, meaning 'endorsement,' originated from the Arabic *hawala*, which was an arrangement that purchasers or debtors made with courts, notaries, or bankers to transfer payment obligations to their debtors.[125] However, Islamic credit instruments were indeed distinctive from their European counterparts. Unlike the European concept of endorsement, *hawala* was both a legal doctrine and a concrete instrument.[126] While Islamic legal texts only record *saftaji* and *hawala*, the *sakk* and *ruq'a* documented in historical texts were the actual practices and proofs of *hawala*.[127] Moreover, different from the European promissory note, a *ruq'a* played the dual role of a promise of repayment and an order of payment.[128] A *saftaji*, compared with a modern bill of exchange, was not only non-negotiable, but also different in the kind of currency for payment.[129] The holders of *suftaja* (plural of *saftaji*) expected to receive the same type of money that they had deposited with the issuing bankers, rather than local currency.[130] However, European monetary history also saw the use of inland bills of exchange at Medieval fairs and in modern England.[131] Islamic *suftaja*, thus, functioned like inland bills of exchange but employed in international trade. Therefore, it is difficult to determine whether these different practices refute the Islamic origin of modern credit instruments, or they merely demonstrate that the development of credit instruments was a historical process; after all, credit techniques were constantly evolving in accordance with the ever-increasing demand for convenience.[132]

It is certain that the creation of credit instruments started in the early Islamic economy. Arab poets before and during the Mohammedan period hint at the existence of credit documents to bridge the temporal gap of transactions, as in their poetry, the Arabic word دين, which originally meant 'debt,' refers to a bill of sale.[133] In Muhammad's lifetime, the outside influence from Egypt, Syria, Persia as well as the caravan trade to the north and east prompted Arabs to document their debt relations.[134] Therefore, a chapter in the *Koran*, ascribed to Muhammad's Medinan period (622–629), repeatedly accentuates the importance of recording debts: 'When you contract a debt for a fixed period, put it in writing.' The sura continues, 'Let a scribe write it down for you with fairness,' 'be they small or large, together with the date of payment.'[135] With documentation, even God would side with the creditors whose debtors defaulted, so 'let the debtor dictate, fearing God his Lord and not diminishing the sum he owes,' and 'do not fail to put your debt in writing'.[136]

Although the format of a bill of sale or a debt contract is unclear, the *sakk* and *safatij* emerged fully formed under the earliest caliphates. Witnessed and sealed at the bottom of the instrument, *sukuk* (plural of *sakk*) served as Arab caliphs' checks, drawn on the treasury, to pay for such governmental expenses as rations, stipends, military provisions, and merchants' supplies.[137] Umar I (r. 634–644), the senior companion of Muhammad and the second caliph, arguably initiated the use of the *sakk* in the form of grain vouchers in 641. Drawn on the treasury, these vouchers entitled the locals of Medina to a particular amount of grain shipped from Egypt.[138] Moreover, a secondary money market formed at Al Djar, a port on the Red Sea, where grain merchants doubled their profits by buying and selling the *sukuk*. The volume of their dealings was too large to eliminate until his successor, Uthmān (r. 644–656), banned the trade.[139] Al-Hajjaj, the governor of the Umayyad Caliphate (661–750), also drew *sukuk* on the treasury to pay military stipends and provisions in Iraq.[140] The practice continued into the 10th century in the Abbasid Empire (750–1258), where soldiers had to display *sikaks*, the certificate from the *diwan* (the central financial department) that bore the ruler's official mark, to obtain allowances and actual payment.[141] The *Safatij* also emerged in the 7th century to save hard money when transferring funds in different locations. Ibn Abbas (619–687), the uncle of Muhammad, and Ibn az-Zubayr (r. 683–692), the leader of the caliphate in Mecca, took cash and wrote *suftaja* to merchants, who would then cash them in Basra and Kufa.[142] In the 10th century, Abbasid caliphs employed *suftaja* to collect provincial tax revenues.[143] In 916, the public exchequer in Baghdad recorded *suftaja* from Ahwaz, Isfahan, Fars and the Eastern provinces, while the financial inspector of Egypt and Syria, Ali b. Isa, sent 147,000 dinars of taxes via *suftaja* from the West. With special messengers carrying *suftaja* to Baghdad, this credit instrument simplified the financial administration of the Abbasid government.[144]

Even though Islamic leaders used the *sakk* and *safatij* to distribute government expense and revenue, these credit instruments remained commercial devices, as court merchant-bankers, *jahbadh*, managed the financial operations. The *jahbadh* emerged in the 9th century to resolve the difficulties in money changing posed by the transition from silver to gold standard. As such, they performed three important functions: administering funds for government officials; transferring funds through cash and *suftaja* across the empire; and, when taxes could not sustain governmental expenses, extending interest-bearing loans to the government.[145] Rather than simply anchored in an administrative apparatus of the government, the *jahbadh* established banks and relied on a network of merchants and money-changers to fulfill their financial function.[146] For example, two high-ranking Muslim officials, Ibn al-Furat and Muhassin b. al-Furat drew *sakuk* on the same *jahbadh*, Aaron b. Amram, to pay a certain Ali b. Isa 3,000 dinars.[147] In 913, lacking the necessary funds for payment, the above-mentioned inspector of Egypt and Syria, Ali b. Isa, deposited unpaid *suftaja* from the provinces as security with Aaron b. Amram and Joseph b. Phineas, another prominent *jahbadh*, to obtain a loan of 10,000 dinars for a period of 16 years.[148] The *jahbadh* also cashed *suftaja* for the government and charged commission fees for their service, thus constituting an important source of income for these court merchant-bankers.[149]

Undergirded by the operations of private merchant-bankers, Islamic credit instruments also promoted the flow of private capital in internal and intraregional transactions. As the line between commerce and finance was fluid in Islamic practice, commercial and financial activities went hand in hand.[150] The internal and intraregional trade, thus, became a critical source of funds for these merchant-bankers; in turn, their credit network woven by the credit instruments buttressed commercial activities. In the 10th century, for example, merchants in the Abbasid Empire drew *sukuk* on the money-changers that they had deposited cash with to pay for local transactions.[151] The Persian traveler Nasir-I Khosrau recognized the *sakk* as a common means of payment at Basra in the mid-11th century.[152] Every trader paid only in *sakk*, drawn on his bank account, in the bazaar, thus making merchant-bankers an indispensable part of Basra's economy. Similarly, about 200 banks existed in the banker's bazaar in Isfahan.[153] Around the same period, while specie were no longer sufficient for Jewish communities in Fatimid Egypt, people presented *ruq'as* to pay at government offices, such as toll stations, or to buy grapes from grocers, who accumulated 5, 10, or 20 orders to demand payment altogether.[154] Nahray b. Nissim, a merchant-banker in the mid-11th century, redeemed a 44-dinar *safatij* with several *ruq'as* and only paid the remaining sum of less than 2 dinars in specie.[155] The accounts between Abu Zikri Kohen, a prominent India trader in Old Cairo, and his banker, Khiyar b. Nissim, testify to the considerable amount of promissory notes (*ruq'as*) that circulated between Abu Zikri and his business partners in the 1140s.[156] Issued by, or drawn on, higher-standing bankers, *suftaja* also eased daily lives in Islamic empires.[157] It is not surprising that a traveler chose to embark on a long journey with only two servants, a guide, and 5,000-dinar *safatij*; donors conveyed *suftaja* from Kairouan, a city in Tunisia's inland desert, to Sura and Pumbadita; and people used *suftaja* for bribery.[158] The *suftaja* kept by the Jewish communities connected the areas between Fatimid Egypt and its neighboring regions to the east, such as Ramle, Jerusalem, Tyre, Damascus, Baghdad, and Ascalon.[159] Therefore, in the early 11th century, when an old man traveled from Northern Iran to Egypt, he obtained a *safatij* of 100 dinars drawn on the Tustari brothers, the leading bankers in Fustat, from their Baghdad agent.[160] *Suftaja* with small sums of 10–19 ½ dinars sent from Old Cairo to Jerusalem also frequently appear in the Geniza Archives.[161] The traces that the *sakk*, *ruq'a*, and *saftij* left in the documents reflect the vast reach of the Islamic credit culture.

Modernization of the Credit Instruments

The fall of Constantinople to Ottoman Sultan Mehmed II in 1453 disrupted European overland routes to the East, and thereby partially led to the Age of Exploration and the expansion of European colonialism. Initially, European traders employed local credit instruments to distribute their resources in Asia. In Southeast Asia, where the Chinese mercantile diaspora held considerable influence, Europeans used Chinese promissory notes extensively to facilitate their local trade. In the subcontinent, Indian bankers transferred British revenue in Bengal, Bihar, and Banaras to Western India through the *hundi* operations, thus assisting the British imperial expansion in the interior.[162] As their influence in Asia grew, however, European credit instruments gradually replaced the ancient forms.

In British India, Asian traditions and European practices came into direct contact and conflict. By the mid-19th century, Great Britain had tightened its grip over India, including the monetary system.[163] Not only did William Bentinck fix the ratio of silver rupee to British gold in 1835, but English lawmakers also attempted to sublimate India's credit instruments into their legal framework.[164] The Negotiable Instruments Act of 1881 stipulated promissory notes, bills of exchange, and checks as standardized credit instruments, thus excluding the *hundi* from this category.[165] When imposing taxes on credit instruments, however, the Indian Stamp Act of 1899 recognized not only those 'defined by the Negotiable Instruments Act, 1881 but also a *hundi*'.[166] More paradoxically, it clarified that 'A Bill of Exchange may include a '*hundi*' but a '*hundi*' does not include a Bill of Exchange,' although both a bill of exchange and a promissory note could be written on a piece of paper bearing the designation 'hundi'.[167] In 1930, the Provincial Banking Enquiry Committee in Burma reserved 14 pages in its report to the definition of the *hundi*, particularly whether 'a bill of exchange is really a particular kind of *hundi*,' or whether a *hundi* is really a type of a bill of exchange.[168] The five-decade long wrangling over the subset relation between Indian and English credit instruments reflects the power struggle between Asian and European credit techniques.

As Europe began to dominate world trade, European negotiable instrument laws came to define the technique and terminology of modern finance, even though divergence in commercial customs, legal traditions, and matters of detail, such as whether a check is a credit instrument or merely a means of payment, had divided the European camp into Anglo-American and Continental European systems.[169] The English Bills of Exchange Act of 1882 gradually encompassed, with few variations, the British dependencies, such as India, Siam, Costa Rica, and the United States, while the US Negotiable Instruments Act of 1896 spread the Anglo-American system further to Alaska, Hawaii, Puerto Rico, the Philippines, Panama, and Colombia by the turn of the century.[170] The Continental European System, based on German and French theories, began to influence world credit instruments laws in the 20th century. The Hague Convention of 1910 and 1912 were the first occasions when 'all civilized nations' gathered to reconcile various credit instruments laws, despite the failed attempts to synthesize Anglo-American laws into the Continental European system.[171] The Uniform Law of the Convention encompassed all continental European countries, Central and South American states, and Japan.[172] Influenced by Japan, China passed its first Law of Negotiable Instruments in 1929 in accordance with the Continental European system.[173] With the passage of the Geneva Uniform Laws for Bills of Exchange in 1930 and 1931, the League of Nations helped to maintain the influence of the Continental framework in 25 countries and regions in Europe, Asia, and South America.[174] From 1968, the United Nations continued to 'minimize departures from the content of the two existing principal legal systems'.[175] Endorsing the United Nations Convention on International Bills of Exchange and International Promissory Notes, countries in Asia and Africa, such as China, India, Iran, Iraq, and Egypt, have been collaborating with those in Anglo-American and Continental European systems to promote the vocabulary of Euro-American credit instruments laws worldwide.[176]

While credit instruments have become a feature of modern global trade, it is hard to gauge the scale of pre-modern credit economy. Given the transportation and communication techniques during the Silk Road era, nevertheless, credit instruments were no less critical to the long-distance trade than they are today. Goitein describes the medieval Islamic economy in Fatimid Egypt as 'a paper economy'.[177]

After investigating the *hundis* circulating in the Mughal Empire, Habib Irfan asserts that 'there is little reason to believe that at that time the *hundis* was [*sic*] any less used in Indian commerce than bills in England'.[178] The long history of paper money and private commercial bills crystalized the scale of China's credit economy. In addition to these well-documented devices, scholars also discovered various forms of credit instruments on the Silk Roads.[179] Therefore, monetary development was never linear, and the Silk Road trade was by no means a cumbersome hard-money system. The quest for a convenient means of payment and remittance may not be a modern advantage, but a characteristic of human commercial activities. However, the credit instruments employed on the Silk Roads is another forgotten story. The attempts to standardize and modernize these devices with European traditions reflect another round of struggles between the declining and rising powers in the globalization process.

Notes

1 Special thanks to Michael Bates, Nick Guoth, and Raisur Rahman for providing me with insightful comments. I wish to acknowledge Qiong Zhang, Xin Wu, and Yunli Shi for pointing out important sources. I also want to thank the participants of the Silk Roads Conference for their helpful remarks.

2 The Geniza Archives, originally a lumber room at the Ben Ezra Synagogue in Old Cairo, Egypt, were discovered in the 19th century. Many scholars, such as Goitein, have employed the tens of thousands of documents to conduct important research on the Jewish communities in Fatimid Egypt (909–1171). For more details, see Greif 1989, 859–860.

3 Goitein 1973, 300.

4 See Wang 2004, 99 table 43.

5 According to Wang, the phrase of 'attested signature of the three times' may refer to the practice of 'drawing finger joints', which means, while signing a contract, the two parties placed their fingers on the document and drew out their finger joints on it (2004, 109).

6 For example, Lapavitsas assumes that banknotes, bills of exchange, and 'all forms of commercial IOU' are 'more advanced [than commodity money and fiat money] and ... specifically capitalist forms of money' (2016, 35–38). For a definition of the 'Silk Road', see Liu 2012, 1. However, this paper regards the fall of Constantinople in 1453, instead of the 13th century, as the declining event of the Silk Roads, for it triggered the Age of Exploration and the beginning of a new world trade dominated by Europeans.

7 See Frank 1998; Alram 2001, 271–291; Hansen 2011; Hansen and Rong 2013; Yang 2018.

8 Goitein 1967; Wang 2004; Peng 2015. Written more than six decades ago, Peng's work remains one of the most brilliant on Chinese monetary history.

9 It is easy to find works on European and American credit instruments in the early 20th century, such as Ogden 1909; Johnson 1911; Moulton 1926; Burns 1927. For recent works, see Holden 1955; Los Angeles Center for Medieval and Renaissance Studies University of 1979; Rogers 2004; Quinn and Roberds 2008; Geva 2011; Lapavitsas 2016.

10 For works that portray the techniques of credit instruments as a modern invention, see Ogden 1909, esp. 8; Burns 1927, esp. 284–285. On linear monetary theory, see Moulton 1926, 16; von Glahn 2005, 65–66.

11 See Goetzmann and Rouwenhorst 2005, 10. By showing multiple possible consequences of financial evolutions, the authors refute the claim that Western capitalist economies are the inevitable and sole equilibrium of the outcome. However, they paint a stagnated East, particularly

China, as opposed to a robust West in terms of financial innovation. Oriented by the chapters in this edited volume, their major focus is the abolition of paper money in China. In doing so they neglect to examine how private commercial institutions in China contributed to financial innovations.

12 Goetzmann and Rouwenhorst 2005, 10. Their essay, or more precisely, the articles in this volume, include military, cultural, geographical, and mathematical explanations of the divergence, but overlook how European imperialism created the institutional environment that was conducive for the spread of Western financial innovations.

13 A search of the two phrases in Google Ngram reveals that 'negotiable instruments' emerged in English literature in the last decade of the 18th century, while 'credit instruments' appeared in the last decade of the 19th century.

14 Ogden 1909, 1. However, in continental Europe and the countries influenced by continental European laws of credit instruments, lawmakers regard a check merely as a means of payment, not a credit instrument. By contrast, the Anglo-American laws of credit instruments include the check. Effros 1971, 115. The third section of this paper discusses the two mainstream systems of negotiable instrument laws: Continental European and Anglo-American.

15 Quinn and Roberds 2008, 2–5. For the standard forms of a promissory note, bill of exchange, and check, see Ogden 1909, 1, 5; 'Text', in United Nations 1988, Article 3, 2.

16 For the relationship between a bill and a check in Europe, see Quinn and Roberds 2008, 3.

17 Some of the earliest bills in Europe were inland bills, used for payments in the same location, such as medieval fairs. See Quinn and Roberds 2008, 3 n. 4.

18 Ogden 1909, 5; Johnson 1911, 58.

19 Effros 1971, 115.

20 White 1975, 225.

21 Ogden 1909, 9; Quinn and Roberds 2008, 6. As to the roles of money, see Pamuk 2000, 1–2. As to the meaning of 'exchange', see Johnson 1911, 58.

22 The nature of a check is more complicated than such documents in modern finance. In continental Europe and the countries influenced by continental European laws of credit instruments, a check functions only as a means of payment, not a credit instrument. However, in the Anglo-American legal framework a check is a credit instrument. See Effros 1971, 114.

23 Moulton 1926, 28–29. Similarly, Ray defines credit as 'any transfer of the ownership of a property and its risk of loss not against an immediately-perceived countervalue, but against an obligation to provide such countervalue at a later time' (1997, 49).

24 Lapavitsas 2016, 24.

25 Cited in Goetzmann and Rouwenhorst 2005, 5.

26 See Dandamyev 2004, 141; Tolini 2012.

27 Zadok and Zadok 2004, 44; Still 2019, esp. chapter 3.

28 At least two tablets in the collection of the Oriental Institute have been identified as promissory notes: an Old-Babylonian text (Tablet A152) and a Neo-Babylonian one (Tablet A1558).

29 Pruessner 1928, 88–107.

30 van de Mieroop 2005, 21.

31 For the format of a modern promissory note, see Ogden 1909, 1–5; 'Text', in United Nations 1988, Article 3, 2.

32 For the origin of bills, see Ashtor 1972, 553.

33 Burns 1927, 284–285.

34 Goetzmann and Rouwenhorst 2005, 23.

35 Goetzmann and Rouwenhorst 2005, 10.

36 Peng 2015, 291 n. 3.

37 Peng 2015, 78, 290.

38 Peng 2015, 290.

39 Peng 2015, 396–397.
40 *Treatise on Food and Commodities* (Shihuo 食貨), *j*. 34, chapter 15, in Xu 1862.
41 Peng 2015, 322 n. 35, quoting earlier work by Wang, noted that Wang did not clarify the unit of currency; since coins were still the primary currency of the Song, the unit might have been the *guan*.
42 Huyan 1763, vol. 3, 222; 'Treatise on Food and Commodities, Part II, Section 3 (*Shihuo xia san* 食貨下三)' in Tuotuo and Alutu 1345, *j*. 181, 4403–4404.
43 Peng 2015, 317.
44 On the former theory regarding the *qianyin*, 'Treatise on Politics (*Zhengzhi zhi* 政治志)', in Wang 1935, *j*. 3, 532. On the latter, Peng 2015, 317.
45 'Treatise on Food and Commodities, Part II, Section 2 (*Shihuo xia san* 食貨下二)', in Tuotuo and Alutu 1345, *j*. 180, 4391.
46 'Treatise on Food and Commodities (*Shihuo zhi* 食貨志)', in Yang and Zhong 1932, *j*. 2, 207; Peng 2015, 318.
47 Peng 2015, 315. Citing Hansen 2015, Gortzmann and Koll depict merchants' issuing *jiaozi* as a privilege granted and later somehow retracted by the Song government. However, Peng and von Glahn have shown that the emergence of *jiaozi* was a bottom-up response to the scarcity and inconvenience of iron coins. Thus, *jiaozi* was originally a commercial bill. See Goetzmann and Koll 2005, 93; von Glahn 2005, 68; Peng 2015, 314–315.
48 von Glahn 2005, 68; Peng 2015, 315, 317.
49 Peng 2015, 316–320.
50 Peng 2015, 319.
51 Peng 2015, 320.
52 von Glahn 2005, 69; Peng 2015, 317, 319.
53 von Glahn 2005, 69.
54 'Treatise on Food and Commodities III (Shihuo san 食貨二)', in Tuotuo 1345, j. 48, 1069–1073; Peng 2015, 405.
55 For some references to *chao* as *yanyin*, see Tuotuo and Alutu 1345, *j*. 180, 4391; *j*. 181, 4403, 4424–4426; vol. 182, 4445–4447. Chinese historical texts document the changing meaning of *chao* from *yanyin* to paper money. For example, 'The term *Chao*, which first appears in the *Song History*, refers to today's *yanyin* (*Chao zhi ming shi jian Songshi, gai ji jin zhi yanyin ye* 鈔之名始見《宋史》，蓋即今鹽引也)', Lu 1436–1494, *j*. 8, 35. 'The paper notes printed during the Jin, Yuan, and Ming dynasties all adopted the name, *Chao*, which were initially regarded as *yanyin*, *chayin*, and other permits. However, the notes circulated temporarily with coins for so long that people started to treat them as currency (*Jin Yuan Ming xiangyan yinxing tong ming zhi yue chao. Qishi yi yancha quanyin zhi shu shi zhi, er zan yi quan qian xing zhi, ji jiu nai ji yidang qian lun zhe* 金元明相沿印行統名之曰鈔. 其始以鹽茶券引之屬視之，而暫以權錢行之，既久乃即以當錢論者)', 'Treatise on Food and Commodities (*Shihuo zhi* 食貨志)', in Yang and Zhong 1932, *j*. 2, 207. Goetzmann and Koll also examine the meaning of *chao* (Gortzmann and Koll 2005, 93–96).
56 'Treatise on Food and Commodities III (*Shihuo san* 食貨三)', in Tuotuo 1345, *j*. 48, 1075–1076.
57 'Annals XII (Benji dishier 本紀第十二)', in Tuotuo 1345, vol. 12, 281.
58 Han and Ji 2000, 007. The image of the printing plate of the *jiaochao* 交鈔 is available at the web page of the Shaanxi History Museum, http://www.sxhm.com/index.php?ac=article&at=read&did=10801 (accessed 27 May 2019).
59 Peng 2015, 409.
60 Peng 2015, 414–415.
61 Orrell and Chlupatý 2016, 64–65.
62 Quinn and Roberds 2008, 8.
63 'Treatise on Food and Commodities, Part II, Section 2 (*Shihuo xia san* 食貨下二)', in Tuotuo and Alutu 1345, *j*. 180, 4399.

64 'Treatise on Food and Commodities III (*Shihuo san* 食貨三)', in Tuotuo 1345, *j*. 48, 1076.
65 'Treatise on Food and Commodities, Part II, Section 2 (*Shihuo xia san* 食貨下二)', in Tuotuo and Alutu 1345, *j*. 180, 4398.
66 'Treatise on Food and Commodities, Part II, Section 2 (*Shihuo xia san* 食貨下二)', in Tuotuo and Alutu 1345, *j*. 180, 4396–99.
67 'Treatise on Food and Commodities, Part II, Section 2 (*Shihuo xia san* 食貨下二)', in Tuotuo and Alutu 1345, *j*. 180, 4399.
68 Pegolotti 1916.
69 Wang Dayuan 汪大淵, *A Brief Account of Foreign Islanders* (*Daoyi zhilue* 島夷志略), cited in Peng 2015, 414.
70 For more details about the concept, Tributary System, see Fairbank and Ch'en 1968.
71 Dreyer 2007, 58–59, 91, 138, 162.
72 Peng 2015, 466.
73 'The emperor granted paper money to those who presented tribute horses outside the horse market (*Shi yi wai you gong ma zhe, yi chaobi jia ci zhi* 市易外有貢馬者，以鈔幣加賜之)', Zhang 1739, *j*. 80, 2277.
74 Cited in Peng 2015, 466.
75 Peng 2015, 466.
76 Peng 2015, 483–84, 554–555.
77 Peng (2015, 396–397, 554–555) argues that, due to the convenience and prevalence of paper money, the instruments for money-changing and remittance issued by money shops were in decline. Consequently, banking in imperial China was less developed than that in medieval Europe.
78 Liu and Xie 2006, 3.
79 Yu 1994, 38–39.
80 Yu 1994, 38–39.
81 For details about these credit instruments, see Liu and Xie 2006. Yu 1995.
82 Wang 1935, 532.
83 The characters on each paper bill could be combined into a phrase, which constituted the series number for a set of bills. For details and images of these insignia, see Yu 2007.
84 Dai 1995, 9.
85 Consequa 1814.
86 Grant Jr. 1987, 49; Du 2018, 183–186.
87 Howe 1900, 120. For major works on the conventional perception of the China trade, see Dennett 1963; Downs 1997. For more details about the debate on the silverized economy of China, see Deng 2008.
88 Yang 2011, esp. 2, 7; Frank 1998, 136–137.
89 Jain 1929, 6–7.
90 Jain 1929, 7.
91 For the variants of the term *hundi*, see Yule 1903, 425; Bayly 1983, 477; Habib 1990, 394; Martin 2009, 911, 915–916.
92 For the Persian origin theory, see Cooke 1863, 21; Martin 2012, 49. For the Sanskrit origin theory, see Prasad 1989, 129 n. 1; Jost and Sandhu 2000, 15.
93 McGregor 1993, 1076.
94 Many types of *hundis* exist. The two broadest categories include demand and fixed-term *hundis*; the former was payable at sight or within a few days, and the latter between 30 and 120 days. Within these categories, individual merchant communities classified their *hundis* into several sub-categories in terms of the bearers, drawees, or goods dispatched. For details, see Jain 1929, 73–80.

95 According to Pandit (1935, 178 n. 266) and Ray (1995, 461), the *hundi* is the modern form of the *hundika*, even though the *Oxford Hindi-English Dictionary* (see McGregor 1993, 1076) does not distinguish the two terms and it is unclear when the transition occurred.
96 Although *Lekhapaddhati* was compiled in the 15th century, the documents can be dated from 744–745 to 1475–1476 CE (Prasad 1989, 124).
97 Prasad 1989, 124–125; 2007, 75;.
98 Prasad translates '*hundika*' as bills of exchange. See Prasad 2007, 75.
99 Pandit 1935, 178. The 1988 United Nations Convention on International Bills of Exchange and International Promissory Notes defines a bill of exchange as 'an unconditional order whereby the drawer directs the drawee to pay a definite sum of money to the payee or to his order' (United Nations 1988, 2).
100 Pandit 1935, 178; Ray 1995, 458.
101 Jain 1929, 70–71.
102 Mazumdar 1960, 224; Prasad 1989, 124.
103 Mazumdar 1960, 223–224; Gopal 1965, 171; Prasad 1989, 126–129.
104 Ray 1995, 461–62. It is unclear whether the term had transitioned from *hundika* to *hundi* by the 16th century or the translators of these sources just used the better-known term, *hundi*, to name these credit instruments.
105 Chowdhury 2009, verse 494.
106 Habib 1964, 401.
107 Prakash 1991, 486–487.
108 Prasad 1989, 127.
109 Habib 1990, 393–394.
110 Habib 1990, 394.
111 Prakash 1998, 161–163.
112 Habib 1990, 391.
113 Prakash 1998, 162.
114 Prakash 1991, 482.
115 Accountant Generals' Office Correspondence to Earl Cornwallis K.G., Governor General in Council in the Public Department of Fort William, 11th June 1789, in: Martin 2009, 917. Although Martin assumes that this statement reveals that the *hundi* was a promissory note, it appears more like a bill of exchange, because as an order drawn by the Indian shroffs (bankers) on the Company, instead of a promise by the Company to pay the holder, it involved three parties: the holder, the shroff, and the Company. However, the *hundi* could be a promissory note and a bill of exchange, since the boundary for these instruments, a European practice, had not been established during that time. On the terms, 'shroff' and 'saraf', see *Oxford English Dictionary Online*.
116 For details about the Dutch, see Prakash 1991, 481–490. For the British, see Bayly 1983, 241; Ray 1995, 493–501.
117 Cooke 1863, 15.
118 On *Travels in India (1605-1689)* by Jean Baptiste Tavernier, see Ray 1995, 462.
119 Ray 1995, 461–462. For more details on the credit network of Indian merchants, Habib 1990, 391–392.
120 For detailed exchange rates, see Ray 1995, 462.
121 Markovits 2000, 158–160. *Gumashtas* is a Persian term for agents, see Habib 1990, 390.
122 See Fischel 1933, 569–603; Goitein 1967; Udovotch 1981; Geva 2011, esp. chapter 6; Ray 1997, 43–90.
123 For the debate over the Islamic or European origins of the credit instruments, see Ashtor 1972, 553–554, 555, 558. Ashtor and Geva disbelieve the influence of Islamic credit instruments on Europe, whereas Ray and Quinn and Roberds argue that medieval Islamic credit techniques

was introduced to Europe and influenced the development of credit instruments there. See Ray 1997, 80; Quinn and Roberds 2008, 2.

124 Ashtor 1972, 555; Geva 2011, 277–280.

125 Goitein 1967, 241–42; Bloom 2001, 137–138; Geva 2011, 259.

126 Discrepancies exist in different portrayals of *hawala*. Geva insists that *hawala* (*hawale*) is not a distinct instrument but a principle, concept, or legal doctrine, but he also notices that people named all cashless payments *hawale* (2011, 258–60). Ray and Abū ʿAbdallāh al-Khwārazmī treat *hawala* as a financial instrument. *Hawala* is a concrete instrument, such as a bill or check according to the *Encyclopedia of Islam* (1927 and 1961). See Bosworth 1969, 140; Ray 1997, 60, 65, 73; Martin 2009, 913–914. It is unclear whether Geva projected the European concept of endorsement onto his interpretation of *hawala*. A nuanced study of this concept in an Islamic historical context would greatly help to elucidate its function.

127 Ray 1997, 60, 73.

128 Goitein 1967, 245; Geva 2011, 277–278.

129 Goitein found that the Geniza Archives contain abundant negotiable *sukuk* and *ruq'as*, but nothing to indicate the transferability of *suftaja* (1967, 245). For an overview on how the four primary Islamic legal schools – Hanafi, Maliki, Shafi'I, and Hanbali – discussed the transferability of the *sakk*, *ruq'a*, and especially *safatji* based on the principle of *hawala*, see Geva 2011, 252–306, esp. 279–284.

130 Goitein 1967, 242.

131 Quinn and Roberds 2008, 3 n. 4; Johnson 1911, 57.

132 Pruessner 1928, 90 brought up this argument when he inquired into the oriental origins of negotiable instruments.

133 Chester 1896, xlvi.

134 Chester 1896, xlvi.

135 *Koran* 2: 282–284, in: Koehler 2014, 111–112.

136 *Koran* 2: 282–284, in: Koehler 2014, 111–112.

137 Morony 1993, 711.

138 While Morony (1993, 719 n. 278) depicts the grain vouchers as *sukuk* (checks), Koehler (2014, 112) categorizes them as *suftaja* (bills of exchange). Since *suftaja* usually circulated in different locations but these vouchers were used locally, this paper regards them as *sukuk*.

139 Koehler 2014, 112.

140 Morony 1993, 711.

141 Bosworth 1969, 125–126.

142 Morony 1993, 711.

143 Morony 1993, 711.

144 Fischel 1937, 19.

145 Ray 1997, 69. For the definition of *jahbad*, see also Fischel 1937, 2–10; Geva 2011, 254. For the important role of *jahbads* in managing monetary and fiscal affairs of the Abbasid Empire, see Fischel 1933, 339–352.

146 Fischel 1937, 29; Ray 1997, 70.

147 Ray 1997, 72 considers these orders as *sukuk*, instead of *suftaja* as Fischel 1937, 20–21, 24.

148 Fischel 1937, 24; Ray 1997, 70.

149 Fischel 1937, 21.

150 Wealthy merchants also played the role of bankers, so for purposes of this article I have adopted the term merchant-bankers. For details, see Fischel 1933, 584; Udovitch 1981, esp. 269; Koehler 2014, 113.

151 Morony 1993, 711.

152 Ashtor 1972, 555.

153 Ray 1997, 72–73.
154 Goitein 1966, 28; 1967, 240;.
155 Goitein 1967, 246.
156 Goitein 1973, 299–304. India trade refers to commercial activities stretching from the Red Sea to Sumatra, Indonesia, Goitein and Friedman 2008, 6.
157 Goitein 1967, 243.
158 Fischel 1937, 17–18.
159 Goitein 1967, 242–245. Goitein (1966, 29; 1967, 245) noticed that in the Geniza records, credit instruments rarely appeared in the trade between Egypt and areas to the west. Purses of gold and silver remained the major means of payment and remittance. Ray (1997, 64, 78) assumes that the dominance of the Maliki school of law in the Muslim West precluded the circulation of *suftaja* in the area. Unlike the other three schools, the Maliki legal school forbade the use of *suftaja* except under extreme necessity. A *sakk* was sent from Tripoli to Fustat as an example of the use of Maliki *hawala* for long-distance transfer of funds between Egypt and its areas to the west.
160 Goitein 1967, 244.
161 Goitein 1967, 244.
162 Ray 1995, 470.
163 Ray 1995, 495.
164 Ray 1995, 495.
165 Negotiable Instruments Act 1881.
166 Basu 1933, 12.
167 Basu 1933, 14, 63, 109, 372.
168 Martin 2012, 56.
169 Effros 1971, 115.
170 Yntema 1951, 178.
171 American Bar Association 1911, 724; Lorenzen 1916, 137.
172 The French group included Argentine Republic, Bolivia, Brazil, Chile, Columbia, Ecuador, Egypt, France, Greece, Guatemala, Hayti, Luxemburg, Monaco, Mexico, Netherlands, Nicaragua, Panama, Paraguay, Polish Russia, Servia, Turkey, and Uruguay. The countries belonging to the German group included Austria-Hungary, Belgaria, Denmark, Germany, Italy, Japan, Norway, Peru, Portugal, Rumania, Russia (exclusive of Polish Russia), Salvador, Sweden, Switzerland, Venezuela. In addition, the countries that stood between the French and German groups included Belgium, Cuba, Honduras, Malta, the Philippines, Porto Rico, and Spain. See Lorenzen 1916, 137–138.
173 Effros 1971, 114; Sommers and Phillips 1998, 318 n. 1.
174 These regions include Austria, Belgium, Denmark, Finland, France, Germany, Greece, Italy, Monaco, Netherlands, Norway, Poland, Portugal, Sweden, Switzerland, Spain, Turkey, Soviet, and Yugoslavia in Eurasia; Japan in East Asia; and Brazil, Colombia, Czechoslovakia, Ecuador, Peru in South America. United Nations 1930.
175 'Explanatory Note' in United Nations 1988, 42. For all sessions held by the United National Commission on International Trade Law held each year, see https://uncitral.un.org/commission.
176 United Nations 1987; 1988.
177 Goitein 1967, 240; 1973, 300;.
178 Habib 1971, 300.
179 In addition to the ones mentioned at the beginning of this paper, Helen Liu also describes various unidentified credit payments, such as Khotanese vouchers and Chinese wood slips. For details, see Wang 2004, 47–58, 97.

Bibliography

Alram, M. 2001. Coins and the Silk Road, in: Juliano, A.L., Lerner, J.A. and Alram, M. (eds), *Monks and Merchants: Silk Road Treasures from Northwest China Gansu and Ningxia 4th–7th Century*, New York, 271–291.

American Bar Association. 1911. *The Thirty-Fourth Annual Meeting of American Bar Association Held at Boston, Massachusetts*, Baltimore.

Ashtor, E. 1972. Banking Instruments Between the Muslim East and the Christian West, *Journal of European Economic History* 1, 553–573.

Basu, M.N. 1933. *The Indian Stamp Act*, Calcutta.

Bayly, C.A. 1983. *Rulers Townsmen and Bazaars: North Indian Society in the Age of British Expansion, 1770–1870*, New Delhi.

Bloom, J. 2001. *Paper before Print: The History and Impact of Paper in the Islamic World*, New Haven.

Bosworth, C.E. 1969. Abū ʿAbdallāh Al-Khwārazmī on the Technical Terms of the Secretary's Art: A Contribution to the Administrative History of Mediaeval Islam, *Journal of the Economic and Social History of the Orient* 12, 113–164.

Burns, A.R. 1927. *Money and Monetary Policy in Early Times*, New York.

Chester, F.D. 1896. On Early Moslem Promissory Notes, *Journal of the American Oriental Society* 16, xliii–xlvii.

Chowdhury, R. 2009. *Banarasidas Ardhakathanak (A Half Story)*, London.

Consequa. 1814. To James Madison from Consequa [10 February 1814], *Founders Online*, National Archives. https://founders.archives.gov/documents/Madison/03-07-02-0242 (accessed 22 February 2016).

Cooke, C.N. 1863. *The Rise, Progress, and Present Condition of Banking in India*, Calcutta.

Dai Xuewen 戴学文. 1995. Qingdai huipiao xilun 清代会票析论, *Zhongguo qianbi* 中国钱币 4, 9–13.

Dandamyev, M. 2004. Ethnic Minorities in Babylonia, in: Schnegg, K. (ed.), *Commerce and Monetary Systems in the Ancient World: Means of Transmission and Cultural Interaction: Proceedings of the Fifth Annual Symposium of the Assyrian and Babylonian Intellectual Heritage Project, Held in Innsbruck, Austria, October 3rd–8th 2002*, Stuttgart, 137–151.

Deng, K.G. 2008. Miracle or Mirage? Foreign Silver, China's Economy and Globalization from the Sixteenth to the Nineteenth Centuries, *Pacific Economic Review* 13.3, 320–357.

Dennett, T. 1963. *Americans in Eastern Asia: A Critical Study of United States' Policy in the Far East in the Nineteenth Century*, New York.

Downs, J.M. 1997. *The Golden Ghetto: The American Commercial Community at Canton and the Shaping of American China Policy, 1784–1844*, Bethlehem.

Dreyer, E.L. 2007. *Zheng He: China and the Oceans in the Early Ming Dynasty, 1405–1433*, New York.

Du, D. 2018. Green Gold and Paper Gold: Seeking Independence through the Chinese-American Tea Trade, 1784–1815, *Early American Studies* 16.1, 151–191.

Effros, R.C. 1971. The Problem of Postdated Checks in the Republic of China, *International Monetary Fund, Staff Papers* 18.1, 113–135.

Fairbank, J.K. (ed.) 1968. *The Chinese World Order; Traditional China's Foreign Relations*, Cambridge.

Fischel, W. 1933. The Origin of Banking in Mediaeval Islam: A Contribution to the Economic History of the Jews of Baghdad in the Tenth Century, *Journal of the Royal Asiatic Society of Great Britain and Ireland* 3, 569–603.

Fischel, W. 1937. *Jews in the Economic and Political Life of Mediaeval Islam*, v. 22, Hertford.

Frank, A.G. 1998. *ReOrient: Global Economy in the Asian Age*, Berkeley.

Geva, B. 2011. *The Payment Order of Antiquity and the Middle Ages: A Legal History*, London.

Goetzmann, W.N. and Koll, E. 2005. Paying in Paper: A Government Voucher from the Southern Song, in: Goetzmann and Rouwenhorst 2005, 91–103.

Goetzmann, W.N. and Rouwenhorst, K.G. (eds) 2005. *The Origins of Value: The Financial Innovations That Created Modern Capital Markets*, Oxford/New York.

Goetzmann, W.N. and Rouwenhorst, K.G. 2005. Introduction: Financial Innovations in History, in: Goetzmann and Rouwenhorst 2005, 3–16.

Goitein, S.D. 1966. Bankers Accounts from the Eleventh Century AD, *Journal of the Economic and Social History of the Orient* 9, 28–66.

Goitein, S.D. 1967. *A Mediterranean Society: the Jewish Communities of the Arab World as Portrayed in the Documents of the Cairo Geniza*, Berkeley.

Goitein, S.D. 1973. *Letters of Medieval Jewish Traders*, Princeton.

Goitein, S.D. and Friedman, M.A. 2008. *India Traders of the Middle Ages: Documents from the Cairo Geniza 'India Book,'* Leiden.

Gopal, L. 1965. *Economic Life of Northern India: AD 700–1200*, Delhi.

Grant, F.D., Jr. 1987. Hong Merchant Litigation in the American Courts, *Proceedings of the Massachusetts Historical Society* 99, 49.

Greif, A. 1989. Reputation and Coalitions in Medieval Trade: Evidence on the Maghribi Traders, *The Journal of Economic History* 49.4, 857–882.

Habib, I. 1964. Usury in Medieval India, *Comparative Studies in Society and History* 6.4, 393–419.

Habib, I. 1971. The System of Bills of Exchange ('Hundis') in the Mughal Empire, *Proceedings of the Indian History Congress* 33, 290–303.

Habib, I. 1990. Merchant Communities in Precolonial India, in: Tracy, J.D. (ed.), *The Rise of Merchant Empires: Long-Distance Trade in the Early Modern World, 1350–1750*, Cambridge.

Han Xuegeng 韩学耕 and Ji Jianhua 姬建华. 2000. Youtairen zhu zhongguo zao 'jiaozi' 犹太人助中国造'交子, *Renmin ribao (haiwai ban)* 《人民日报》海外版, November 22, 2007.

Hansen, V. 2011. The Place of Coins and Their Alternatives in the Silk Road Trade, in: Shanghai B. (ed.), *Sichou zhilu guguo qianbi ji silu wenhua guoji xueshu taolunhui wenji*, Shanghai, 83–113.

Hansen, V. 2013. Rong, Xinjiang. How the Residents of Turfan Used Textiles as Money, 273–796 CE, *Journal of the Royal Asiatic Society of Great Britain and Ireland* 23, 281–305.

Hansen, V. 2015. *The Open Empire: A History of China to 1800*, 2nd ed. New York.

Holden, J.M. 1955. *The History of Negotiable Instruments in English Law*, London.

Howe, J.W. 1900. *Masterpieces of American Eloquence*, New York.

Huyan Huaguo 呼延華國. 1763. *Didaozhou zhi* 狄道州志, *Airusheng Zhongguo fangzhiku* 愛如生中國方志庫 (accessed 9 July 2019).

İnalcık, H. 2012. Ḥawāla, in: Bearman, P., Bianquis, T., Bosworth, C. E., van Donzel, E. and Heinrichs, W. P. (eds), *Encyclopaedia of Islam*, 2nd ed. Leiden. http://dx.doi.org/10.1163/1573-3912_islam_SIM_2807 (accessed 29 November 2019).

Jain, L.C. 1929. *Indigenous Banking in India*, London.

Johnson, J.W. 1911. *Promissory Notes, Bills of Exchange, Cheques and Other Negotiable Instruments: A Reference Book for Business Men and a Text Book for Commercial Students*, Belleville.

Jost, P.M. and Sandhu, H.S. 2000. *The Hawala Alternative Remittance System and Its Role in Money Laundering*, Lyon.

Koehler, B. 2014. *Early Islam and the Birth of Capitalism*, Lanham.

Lapavitsas, C. 2016. *Marxist Monetary Theory: Collected Papers*, Leiden.

Liu Qiugen 刘秋根 and Xie Xiuli 谢秀丽. 2006. Mingqing minjian shangye xinyong piaoju de chubu fazhan – yi huipiao, huidui wei zhongxin 明清民间商业信用票据化的初步发展 – 以汇票、汇兑为中心, *Zhongguo qianbi* 中国钱币 1, 3–11, 79.

Liu, X. 2012. *The Silk Roads: A Brief History with Documents*, Boston.

Lorenzen, E. 1916. Hague Convention of 1912 Relating to Bills of Exchange and Promissory Notes: A Comparison with Anglo-American Law (Part 1), *Illinois Law Review* 11.3, 137–158.

Los Angeles Center for Medieval and Renaissance Studies University of California (ed.) 1979. *The Dawn of Modern Banking*, New Haven.

Lu Rong 陸容. n.d. *Shuyuan zaji* 菽園雜記, *Zhongguo zhexue dianzihua jihua* 中國哲學書電子化計劃 (accessed 6 March 2019).

Markovits, C. 2000. *The Global World of Indian Merchants, 1750–1947: Traders of Sind from Bukhara to Panama*, Cambridge.

Martin, M. 2009. Hundi/Hawala: The Problem of Definition. *Modern Asian Studies* 43, 909–937.

Martin, M. 2012. *An Economic History of Hundi, 1858–1978*. PhD Thesis, The London School of Economics and Political Science, London. http://etheses.lse.ac.uk/315/.

Mazumdar, B.P. 1960. *Socio-Economic History of Northern India (1030–1194 AD)*, Calcutta.

McGregor, R.S. (ed.) 1993. Hundi, *The Oxford Hindi-English Dictionary*, Oxford.

Morony, M. 1993. Commerce in Early Islamic Iraq, *Asien, Afrika, Lateinamerika* 20, 699–720.

Moulton, H.G. 1926. *Principles of Money and Banking: A Series of Selected Materials, with Explanatory Introductions*, Chicago.

Negotiable Instruments Act. 1881. http://devgan.in/negotiable_instruments_act/ (accessed 8 October 2019).

Ogden, J.M. 1909. *The Law of Negotiable Instruments, Including Promissory Notes, Bills of Exchange, Bank Checks and Other Commercial Paper, with the Negotiable Instruments Law Annotated and Forms of Pleading, Trial Evidence and Comparative Tables Arranged Alphabetically by States*, Chicago.

Orrell, D. and Chlupatý, R. 2016. *The Evolution of Money*, New York.

Oxford English Dictionary Online. n.d. Oxford. https://www.oed.com/ (accessed 23 October 2019).

Pamuk, Ş. 2000. *A Monetary History of the Ottoman Empire*, Cambridge.

Pandit, R.S. 1935. *River of Kings (rajatarangini)*, Allailabad.

Peglotti, F.B. 1916. Pegolotti's Merchant Handbook. Excerpted from the partial translation, in: Yule, H. and Cordier, H. (transl. and ed.), *Cathay and the Way Thither, Being a Collection of Medieval Notices of China*, vol. 3, London, 143–171. https://depts.washington.edu/silkroad/texts/pegol.html (accessed 21 September 2019).

Peng Xinwei 彭信威. 2015 (1954). *Zhongguo huobishi* 中國貨幣史, Shanghai.

Prakash, O. 1991. Sarrafs, Financial Intermediation and Credit Network in Mughal India, in: van Cauwenberghe, E.H.G. (ed.), *Money, Coins and Commerce: Essays in the Monetary History of Asia and Europe from Antiquity to Modern Times*, Leuven, 473–390.

Prakash, O. 1998. *European Commercial Enterprise in Pre-Colonial India*, Cambridge.

Prasad, P. 1989. Hundika Documents from Lokaprakasa, *Proceedings of the Indian History Congress* 50, 124–131.

Prasad, P. 2007. *Lekhapaddhati: Documents of State and Everyday Life from Ancient and Medieval Gujarat*, Oxford.

Pruessner, A.H. 1928. The Earliest Traces of Negotiable Instruments, *The American Journal of Semitic Languages and Literatures* 44.2, 88–107.

Quinn, S. and Roberds, W. 2008. The Evolution of the Check as a Means of Payment: A Historical Survey, *Economic Review* 93, 1–28.

Ray, N.D. 1997. The Medieval Islamic System of Credit and Banking: Legal and Historical Considerations, *Arab Law Quarterly* 12, 43–90.

Ray, R.K. 1995. Asian Capital in the Age of European Domination: The Rise of the Bazaar, 1800–1914, *Modern Asian Studies* 29, 449–554.

Rogers, J.S. 2004. *The Early History of the Law of Bills and Notes: A Study of the Origins of Anglo-American Commercial Law*, Cambridge.

Still, B. 2019. *The Social World of the Babylonian Priest*, Leiden.

Sommers, A.L. 1998. Phillips, Kara L. The P.R.C.'s Negotiable Instruments Law: An Instrument for Facilitating Private Economic Activity or Monetary Control?, *Houston Journal of International Law* 20, 317–351.

Tablet A152. The Oriental Institute, The University of Chicago, https://oi-idb.uchicago.edu/id/44795732-5314-4103-98ff-f19857b81bd3 (accessed 10 July 2019)

Tablet A1558. The Oriental Institute, The University of Chicago, https://oi-idb.uchicago.edu/id/5f0e20bb-ee53-41e9-be76-4d268e290381(accessed 10 July 2019).

Tolini, G. 2012. Women and Family Solidarities in the Murašû Archive (Nippur-Fifth century BC), *Neo-Assyr./Neo-Babyl./Achaemenid, Workshop-1* (Nanterre 11, 2012), https://refema.hypotheses.org/227 (accessed 3 February 2019).

Tuotuo 脱脱. 1345. *Jinshi* 金史, *Scripta Sinica* (*Hanji dianzi wenxian ziliaoku* 漢籍電子文獻資料庫) (accessed 4 February 2019).

Tuotuo 脱脱 and Alutu 阿魯圖. 1345. *Songshi* 宋史, *Scripta Sinica* (*Hanji dianzi wenxian ziliaoku* 漢籍電子文獻資料庫) (accessed 4 February 2019).

Udovotch, A.L. 1981. *Bankers without Banks: Commerce, Banking, and Society in the Islamic World of the Middle Ages*, Princeton.

United Nations. 1930. Convention Providing a Uniform Law for Bills of Exchange and Promissory Notes, Geneva, 7 June, https://treaties.un.org/Pages/LONViewDetails.aspx?src=LON&id=552&chapter=30&clang=_en (accessed 28 March 2019).

United Nations. 1987. Report of the Working Group on International Negotiable Instruments on the Work of Its Fifteenth Session, New York, 17–27 February, https://undocs.org/en/A/CN.9/288 (accessed 8 October 2019).

United Nations. 1988. United Nations Convention on International Bills of Exchange and International Promissory Notes, New York, https://uncitral.un.org/en/texts/payments/conventions/bills_of_exchange (accessed 26 April 2019).

van de Mieroop, M. 2005. The Invention of Interest: Sumerian Loans, in: Goetzmann and Rouwenhorst 2005, 17–30.

von Glahn, R. 2005. The Origins of Paper Money in China, in: Goetzmann and Rouwenhorst 2005, 65–89.

Wang, H. 2004. *Money on the Silk Road: The Evidence from Eastern Central Asia to c. AD 800*, London.

Wang Pixu 王丕煦. 1935. *Laiyang Xianzhi* 萊陽縣志, *Airusheng Zhongguo fangzhiku* 愛如生中國方志庫 (accessed 9 July 2019).

White, H. 1975. The Convention on the Uniform Law of International Bills of Exchange and International Promissory Notes: A Comparison to the Uniform Commercial Code, *Georgia Journal of International & Comparative Law* 5:1, 216–247.

Xu Song 徐松. 1862 (?). *Songhuiyao jigao* 宋會要輯稿, *Scripta Sinica* (*Hanji dianziwenxian ziliaoku* 漢籍電子文獻資料庫) (accessed 4 February 2019).

Yang, B. 2011. The Rise and Fall of Cowrie Shells: The Asian Story, *Journal of World History* 22.1, 1–25.

Yang, B. 2018. *Cowrie Shells and Cowrie Money: A Global History*, New York.

Yang, Weizhong 楊維中. 1932. Zhong, Zhengmao 鍾正懋 (ed.). *Quxianzhi* 渠縣志, *Airusheng Zhongguo fangzhiku* 愛如生中國方志庫 (accessed 9 July 2019).

Yntema, H.E. 1951. Unification of the Laws Respecting Negotiable Instruments, *The International Law Quarterly* 4, 178–191.

Yu Hongchang 俞鸿昌. 1994. The *Huipiao* of the Guangxu Period (Guangxu huipiao 光绪会票), *Chinese Money* (*Zhongguo qianbi*中国钱币) 4, 38–39.

Yu Hongchang 俞鸿昌. 1995. *Qingdai huipiao gailun* 清代会票概论, *Zhongguo qianbi* 中国钱币 4, 13, 19–28.

Yu Hongchang 俞鸿昌. 2007. Qingdai huipiao de hanzi miya 清代会票的汉字密押, *Zhongguo qianbi* 中国钱币 99.

Yule, H. 1903. Burnell, Arthur Coke. *Hobson-Jobson: A Glossary of Colloquial Anglo-Indian Words and Phrases, and of Kindred Terms, Etymological, Historical, Geographical and Discursive*, London.

Zadok, R. and Zadok, T. 2004. A Promissory Note from the Atkuppu Archive, *Nouvelles Assyriologiques Brèves et Utilitaires*, 44, http://www.achemenet.com/pdf/nabu/Nabu2004-044.pdf (accessed 3 February 2019).

Zhang Tingyu 張廷玉 (ed.) 1739. *Mingshi*明史, *Scripta Sinica* (*Hanji dianziwenxian ziliaoku* 漢籍電子文獻資料庫) (accessed 4 February 2019).

Section Five

Long-Distance Commodity Trade

Chapter 16

The Case for Shipwrecked Indians in Germany

Jeffrey D. Lerner

Three versions recorded by different authors document a peculiar incident involving a group of strange looking men whose ship had been caught in a storm and dashed onto the shores of Germany. All three authors attribute their accounts to Cornelius Nepos (ca. 110–24 BCE), who wrote that in 62 BCE the survivors, identified as Indians, were captured and presented as a gift by a king from Germany to the Proconsul of Gallia Cisalpina (Northern Italy), Quintus Caecilius Metellus Celer.

The earliest account to mention this passage appears almost three quarters of a century later from the Roman geographer Pomponius Mela, who in 43 or 44 CE published a work entitled, *de Chrorographia.* He writes (3.5.44–45):

> **44.** Ultra Caspium sinum quidnam esset ambiguum aliquamdiu fuit, idemne Oceanus an tellus infesta frigoribus sine ambitu ac sine fine proiecta. **45.** Sed praeter physicos Homerumque qui uniuersum orbem mari circumfusum esse dixerunt, Cornelius Nepos ut recentior, auctoritate sic certior; testem autem rei Quintum Metellum Celerem adicit, eumque ita rettulisse commemorat; cum Galliae pro consule praeesset, Indos quosdam a rege Boiorum dono sibi datos; unde in eas terras deuenissent requirendo cognosse ui tempestatium ex Indicis aequoribus abreptos, emensosque quae intererant, tandem in Germaniae litora exisse. Restat ergo pelagus, sed reliqua lateris eiusdem adsiduo gelu durantur et ideo deserta sunt.
>
> **44.** Quite a long time ago one wondered what was beyond the Caspian Gulf, the same Ocean or a land that the cold makes hostile and that extends without circumference and without end. **45.** But besides the natural philosophers and Homer, who have all said that the whole entire known world was surrounded by the sea, there is the authority of Cornelius Nepos, a more recent author, consequently more worthy of faith. However, Nepos produces on the testimony of Quintus Metellus Celer and quotes from him the following account: when he was proconsul in Gaul, the king of the Boii had given him the gift of some Indians; asking them where they had gone to reach these countries, he learned that a storm had carried them out of Indian waters, and that after having crossed the distance separating them they had finally landed on the coasts of Germany. The sea is therefore continuous, but what remains on this side of the Earth is hardened by the persistent cold and therefore deserted.

Almost 35 years later in 77–79 CE, the Roman author and naturalist philosopher, Pliny the Elder, also referenced this event in his *Naturalis Historia*. Pliny's account, however, differs slightly from that of Mela (2.170):[1]

> Idem Nepos de septentrionali circuitu tradit Quinto Metello Celeri, Afrani in consulatu collegae, sed turn Galliae proconsuli, Indos a rege Sueborum dono datos, qui ex India commercii causa navigantes tempestatibus essent in Germaniam abrepti. sic maria circumfusa undique dividuo globo partem orbis auferunt nobis, nec inde huc nec hinc illo pervio tractu. quae contemplatio apta detegendae mortalium vanitati poscere videtur, ut totum hoc, quicquid est, in quo singulis nihil satis est, ceu subiectum oculis quantum sit ostendam.
>
> The same Cornelius Nepos, when speaking of the northern circumnavigation, tells us that Quintus Metellus Celer, the colleague of Lucius Afranius in the consulship, and later a proconsul in Gaul, had a present made to him by the king of the Suevi, of certain Indians, who, sailing from India for the purpose of commerce, had been driven by tempests into Germany. Thus it appears, that the seas, which flow completely round the globe, and divide it, as it were, into two parts, exclude us from one part of it, as there is no way open to it on either side. And as the contemplation of these things is adapted to detect the vanity of mortals, it seems incumbent on me to display, and lay open to our eyes, the whole of it, whatever it be, in which there is nothing which can satisfy the desires of certain individuals.

The last to reference this strange episode is Martianus Minneus Felix Capella (fl. ca. 411–428 CE), a 5th-century Latin writer, who in his *De Nuptiis Philologiae et Mercurii* ('*Marriage of Philology and Mercury*') states simply (6.621):

> Idem Cornelius post captos Indos per Germaniam nauigauit.
> The same Cornelius, after capturing Indians, sailed by Germany.

Unfortunately, the work on which these accounts are based is lost. It may have appeared in Nepos' three volume *Chronica*,[2] devoted to the history of the world purportedly written in the late 40s or early 50s BCE following the annalistic tradition, or in his *Exempla* ('*Models*'),[3] a compendium of oddities and moralizing anecdotes related to geography, history, and physics in at least five books, published at some point after 43 BCE. There is also considerable debate as to whether this account formed part of another work that focused on geography, in which Nepos discussed peripheral areas of Europe with special emphasis placed on those regions in which the Celts resided,[4] or if it appeared in his aforementioned *Exempla*.[5]

We learn from Mela and Pliny that the source of Nepos' account was based on the testimony of Metellus Celer himself when he served as proconsul in Gallia Cisalpina in 62 BCE. There are some differences, however, in the versions supplied by Mela and Pliny. The former states that the king of the Boii gave the Indians to Metellus, while Pliny claims it was the king of the Suebi. Both are in agreement that a storm carried the ship out of Indian waters ultimately bringing it to the coast of Germany, although Pliny supplies the additional detail that the survivors were merchants. The reason that Mela and Pliny relay the episode is that it is used to support an argument that Nepos had advanced and that they were following: a continuous river Ocean flows around the entire earth, and it is thus possible to circumnavigate it. As proof Mela and Pliny by referencing Nepos use the incident to illustrate that it is indeed possible to sail from India through a northwest or northeast passage, which is how these hapless Indians overtaken by a storm at sea came to find themselves suddenly in Germany, either by way of the Atlantic Ocean or the Baltic Sea.[6] This incident is less about the appearance of Indian merchants in Germany as an unexpected phenomenon as it is an attempt to offer a reasonable explanation of how they got there in the first place. The tradition of describing the habitable world that both authors follow stems from Nepos,[7] who appeared to have knowledge of only the contours of the coastal regions of the three continents bordering the river Ocean at the expense of the interiors of the countries of which they were part. Consequently, Mela and Pliny relied strictly on Nepos who formed his conception of the Afro-Eurasian land mass from the earlier works of Homer and the natural philosophers.

By the time this episode reached Martianus Capella the tradition, or his understanding of it, had deteriorated so greatly that he completely misunderstood the original account. Nepos was not on a voyage when this incident occurred, and he certainly did not capture these unfortunates nor did he sail past Germany. Yet, it was this version that led Otto of Freisingen, the uncle of Frederick Barbarossa, to update the episode so as to occur 'in the time of the German emperors' and write that 'an Indian ship with Indian merchants was taken on the coast of Germany which evidently had been driven there from the east,' as recorded by Aeneas Sylvius Piccolomini (Pope Pius II) in his 1477 work, *Historia rerum ubique gestarum*.[8] For this reason, we may completely disregard this account as ahistorical.

By all accounts the event occurred in the year 62 BCE during Metellus' proconsulship (62–59 BCE) in Gallia Cisalpina, and the hapless survivors had perhaps been captured a year earlier.[9] Nepos, Mela, and Pliny were each convinced that it was possible to sail from India to Germany through a northern route, and thus the survivors about whom they wrote originated from India and were thus Indians, and did not come from elsewhere. At this point, the explanation of the circumstances that led to this episode takes a most unexpected turn.

The prevailing opinion among a number of researchers is that these shipwrecked individuals neither hailed from India nor were they Indian, precisely because a journey through a northern passage in antiquity was impossible. Instead it is commonly accepted that these individuals represent the first known pre-Columbian contact between the Old World and the New World as the only viable explanation for the identity of these individuals, and thus from whence they came, and how they could possibly have arrived on the coast of Germany. Henning, for example, argued that it would have been impossible for Indian merchants to have sailed to Germany from the Indian Ocean, thereby making these survivors 'imaginary Indians,' who were really Native Americans or Eskimos from either North America or Greenland. He, therefore, credited Nepos – albeit unwittingly – as having recorded the first written testimony of a people from the New World who appeared in the Old World, some

1,500 years prior to Columbus' voyage,[10] and having committed the same mistake as the famed navigator by having been confused by their identity. Henning's analysis is based on the notion that 'India' as a term had by this time become a catchall for all things *exotica* and would remain as such well into the Middle Ages.[11] In support of this thesis, Henning, like Plischke before him,[12] cited numerous examples of Eskimo kayaks, fishing boats, and sometimes Eskimos themselves (dead or alive) as having washed up on the shores of England, Germany, Norway, the Orkney Islands and elsewhere with various museums and cathedrals housing an assortment of objects that had appeared on their respective coasts.[13] Such was the case in Rouen, France in 1507 or 1509 when a boat appeared on the shore with one Eskimo still alive, while six of his comrades had died. Other incidents of a similar nature also occurred, including one in 1577 on the coast of the Netherlands.[14]

Although this idea remains attractive, the majority of researchers have come to regard these 'Indians' as having originated closer to Germany itself, like Ditten who was prompted to argue that the survivors may have simply been Germans or Slavs, although he did not completely rule out the possibility that they may indeed have been Eskimos.[15]

Tausend[16] likewise shares the view that these 'Indians' were not from India, but were identified as such by the Romans. At least they were regarded as such publicly, because he posits without foundation that Metellus remained Pompey's agent in Gallia Cisalpina after having served him as his legate in Armenia and had also been a member of the general's staff.[17] Tausend builds on an idea initially proposed by Bengston[18] that Pompey's campaign in the East had aroused Roman interest about India, especially the possibility of developing relations with the country. He, therefore, saw in these survivors what he took to be Indians.[19] He overlooks that fact that while Cicero had assisted Metellus with obtaining command against the Catilinarians and the province of Gallia Cisalpina, he subsequently turned against Cicero,[20] and in 60 as consul fell out with Pompey on account of the general having divorced his sister Mucia Tertia.[21]

Nonetheless, Tausend follows Pliny's assignment of the king who handed the captives to Metellus as rex Sueborum, whom he identifies as none other than Ariovist, the famed German king mentioned by Caesar in his *Commentarii*, who dwelt near the Main River, but at the time of this incident presumably inhabited the region on the Rhine. Tausend further maintains that the so-called *Boti* (from rex Botorum) mentioned by Pomponius Mela is a corruption of Goti, or Goths, based on a similar mistake committed by Strabo.[22] This scenario would have it that the Goths (Gutons or Gotones), who would have already been well-ensconced in southern Sweden, sent the so-called 'Indians' as a goodwill gesture to their ally Ariovist. For his part, the German protagonist was attempting to promote closer relations with Roman officials, by way of the Harudes or Lugier, who were his allies and served in his army. They occupied the upper Oder and Vistula river region and in this transaction would have played the role of mediator. Consequently, it is possible that these 'Indians,' who appeared to them as a 'strange-looking people' were probably Finno-Ugric tribesmen living east of the Gulf of Bothnia. They were thus Lapps, known to Tacitus as Fenni, and Ptolemy as Φιννοι,[23] who might have come from further north in the continent, and could have been Mordvins or Permians, or Permyaks.[24] This interpretation depends on the notion that these 'Indians' were in reality a Finno-Ugrian people from the southeastern or eastern Baltic region, or somewhere further north, and that the resolution of Boti as Goths, who at this time occupied southern Sweden, is correct.

While not ruling out the possibility that these survivors came from the eastern Baltic region, Podossinov rightly notes that this reconstruction suffers from the fact that it relies on too many assumptions.[25] For example, if these were Baltic peoples who had drifted southward to Germany, they could not have been Scandinavians, because as Germans they would not have been mistaken as Indians. He also argues that this idea relies on the notion that the name of the Boti, extrapolated from Ptolemy's localized Goths, were situated in southern Sweden, but he goes too far in maintaining Tacitus' statement (*Germania* 48) that the Lapps would have been too primitive to have been able to sail to the coast of Germany. Nonetheless, he conjectures that the *Indi* of Mela and Pliny is actually a corruption of Vindi (Vinedi, Venedi), a presumed Slavonic tribe of the southeastern Baltic coast east of the Vistula River mentioned in the Verona copy of Roman provinces (*Nomina provinciarum omnium*) of 'barbarian peoples' of northern Europe produced in the late 3rd–early 4th century CE in which a people called the *Indii* appear. He concluded that the Indians of Mela and Pliny bear no relation to the Indians of South Asia, rather they were either Eskimos from the New World who were carried by storms and currents to the shores of northern Europe, or were inhabitants of the Baltic Sea, the Slavonic eastern neighbors of the Germans mistakenly called Venedi, or *Indians*. In either case, their language, presumably unknown to either Germans or Romans, would have made them appear as 'Indians'.[26] The problem with this line of reasoning is that the identification of the Venedi as a tribe of early Slavs, or even their ancestors, is unsubstantiated and remains mere conjecture. Indeed there is nothing about the name Venedi to suggest an association of it as Slavic. Although afterward the Germanic Wends was used to indicate Slavs in general, they do not enter recognizably into the historical record until the 6th century in Slovakia and Moravia. In actuality, the identity of the ancient Venedi remains at present unknown.[27]

It is apparent that the analysis of these passages falls into one of two camps: had the 'Indians' come from the East, they could not have arrived in Germany by way of a northeastern or northwestern trans-Atlantic route, and, conversely, they could not have come by way of the south Atlantic. Nepos reflected the dominate view of geographic knowledge, which held that the river 'Ocean' encircled all the world's lands. Therefore, there were two possible oceanic routes by which one could have sailed from India to Germany: the northern, which was known to be impossible as it would mean a voyage through the Artic, and the southern, for which all known attempts were the subject of failure.[28] For example, a route that might have been taken along the western African coast can be dismissed, due to the insurmountable difficulties that such a journey would have entailed, including the direction of the winds and ocean currents as well as the lack of natural harbors making access to food and water scarce, especially along the Saharan coast.[29] It is worth noting that we are in agreement with André[30] that these individuals could not have been African from either the Sudan or north Africa, as Jullian proposed without explanation, simply because Metellus or anyone of his associates would not have made such a mistake, especially as people from these regions were well-known to the Romans.

For Nepos, and later Mela and Pliny, this episode serves as proof that in fact it is possible to sail around the world on the river Ocean and that it is possible to do so using a northern route. In doing so, ancient concepts of geography dating from Homer and the natural philosophers are upheld as true, as evidenced by Metellus' report that was purported by Nepos.[31] It is little wonder, then, that Biffi[32] has called into question the veracity of the report itself, arguing that this episode stands alone in reporting that a group of Indians arrived in the West well before their supposed first venture when Indian ambassadors would later visit Augustus.[33] Since his proconsulship lacked anything outstanding, he holds that Metellus either invented or greatly embellished a story about these purported Indian merchants in an attempt to win the recognition of the *populus Romanus*, especially the Roman merchant class, as an important person. In this regard, Nepos embraced the account, because it supported his conceptualization that the Ocean encircles the whole of the *oikoumene*. On the other hand, it is difficult to agree with Biffi that since Strabo (ca. 64/3 BCE–24 CE) remains silent about this incident, it therefore did not occur; especially as Luisi[34] proposed that if in fact Nepos had written a universal geography in 59 BCE, then we are compelled to call into question Strabo's originality, and not that of Nepos'.

Aside from Biffi, who is convinced that the episode never took place, the majority of researchers are in agreement that whoever these shipwrecked individuals were, they could not have been Indians from India. On the other hand, Bengston[35] long ago advanced the notion that a group of Indians from India may well have been shipwrecked and washed up on the shores of Germany. Subsequently, they were captured and surrendered to Metellus, but the route that they took deviates sharply from the original report stated by Nepos. Bengston had the route that these merchants follow correspond to an account noted by Marcus Terentius Varro an officer in Pompey's army in 65/64 BCE during the Third Mithridatic War (73–63 BCE), and preserved in a passage by Pliny.[36] Varro learned that it took a seven days' march from India to reach the river Icarus, a tributary of the Oxos (Amu Daria) in Baktria (northern Afghanistan). From the Icarus merchants sailed along the Oxos and by way of the Kelif Uzboi entered the Caspian Sea where they would have made their way up the Kuros River (modern Kura) and after a five-days' portage through the Surami Pass they would have arrived at the Phasis (modern Rioni). Undoubtedly, they would have rested at Phasis, the principal city of Kolkhis, situated at the confluence of the Phasis and the Black Sea. At this point Bengston's reconstruction significantly departs from the source tradition, for he has these merchants march overland across the southern

Russian steppes into eastern and central Europe. In order to avoid various hostile tribes of Skythians, Sarmatians, Thrakians, and Germans, they then detoured northward to the Vistula River, skirting the Dacian realm of Burbista, and upon reaching the Baltic Sea boarded another ship only to encounter the fateful storm, which dashed them ashore. Ultimately, they were captured and subsequently brought to Metellus. As Pompey's legate during the eastern campaigns of 66–63 BCE, Metellus Celer would have been cognizant of Varro's report, especially since he participated in the war against the Caucasian Albanians in Lesser Armenia.[37] Bengston further supposes that Metellus would have known of Pompey's desire to emulate Alexander the Great by undertaking an expedition into India, leading to a general desire among the Romans of all things Indian. Metellus was thus able to capitalize on this curiosity by publicizing his captives.

Given the propensity by researchers to dismiss the possibility that these survivors could have been Indian, it comes as little surprise that Bengston's interpretation has won no supporters. Certainly, the idea of a journey upcountry from the city of Phasis to the Baltic Sea has rightly been condemned as mere invention. There is no indication in the accounts of Mela or Pliny[38] that these individuals had ever undertaken an overland trek upcountry to the Baltic Sea where they would have obtained another ship and set sail for open waters. There is also no evidence that the Greeks or Romans of the 1st century BCE were familiar with the lands and the peoples living inland of the Black Sea, the Danube, or the Rhine.

At the same time, Bengston's critics have been too quick to reject his reconstruction entirely. Although evidence is lacking to support a presumed overland journey from Phasis to the Baltic Sea, there is merit in considering the feasibility that a group of Indian merchants could have sailed along inland waterways to reach Phasis from Baktria. In recent years the existence of such a route has found a small but growing group of adherents.[39] The first notice of the existence of such a waterway connecting Baktria with the Black Sea dates back to a report by Patrokles, the admiral of Seleukos I and Antiochos I during their co-regency in ca. 292–281 BCE, when he was charged with exploring the Caspian Sea.[40] Having described the same route as would later Varro, Strabo[41] notes that the polis of Phasis was regularly visited by merchants traveling between India and Europe. Generally, opinion is divided between those who maintain that this route was so little used that it was inconsequential,[42] and those who argue that by the time of Varro's report it was no longer in existence.[43] Both camps, however, overlook another report supplied by Pseudo-Skymnos in his treatise, *Περίοδος (Periodos)*,[44] dedicated to Nikomedes II Epiphanes (149–127 BCE) or his son Nikomedes III Euergetes (127–97 BCE).[45] He informs us that different peoples speaking a variety of languages from some 60 nations converged at Phasis, including those from Baktria and India. Evidently, Phasis was a well-known international entrepôt where merchants plied their wares. This waterway became such a popular highway that by 1600 a Mughal-Indian merchant population had established itself in an Indian community at the southwestern Russian Caspian port city of Astrakhan, formerly Khadzhi-Tarkhan, situated in the delta of the Volga River some 60 miles from the Caspian Sea. From there they acted as intermediaries, selling goods between Russia and Iran, much like their counterparts on the Absheron Peninsula in Azerbaijan, some 30 miles north of Baku.[46] It is, therefore, entirely conceivable that from here merchants, including those from India, would have easily been able to sail to the Black Sea, even making their way into the Mediterranean.

There is another route that these sailors might have taken on a voyage from India to Germany. Molchanov[47] has proposed that Indian merchants were already by the 3rd century BCE sailing to various Ptolemaic ports in the Red Sea. He argues that it is entirely possible that from there they could have marched across the Eastern Desert to Alexandreia just as others would later do under Roman rule. He further suggested that Nepos' Indians could have made such a journey to Alexandreia where they could have obtained a ship allowing them to sail into the Atlantic Ocean. A journey of this sort would thus account for how they encountered a storm and found themselves cast upon the shores of Germany.

Evidence of trade in the western Indian Ocean, including the Persian Gulf and the Red Sea, dates to the third millennium BCE between the Indus Valley, Mesopotamia, and Egypt.[48] In the Hellenistic Period (323–30 BCE) trade in this region included the Mediterranean world under the Seleukid and Ptolemaic dynasties, successors of Alexander the Great, who controlled the Persian Gulf and Red Sea, respectively.[49] By the end of the 2nd century BCE the Characenians, vassals of the Parthian kingdom,[50] replaced the Seleukids in the Persian Gulf. Although the Ptolemies established additional routes, waystations, and ports along the Eastern Desert and the Red Sea coast, it was not until the Roman conquest of Egypt in 30 BCE that trade between the Mediterranean and the Indian Ocean became more regularized and profitable, perhaps due at least in part to the free market economy that replaced the Ptolemaic monopolies on imported eastern goods.[51]

While merchants from Ptolemaic Egypt set sail on voyages from the Red Sea in the late 2nd to early 1st centuries BCE using the Southwest Monsoon to reach India, Indian sailors had for centuries – if not for millennia – freely operated throughout the Indian Ocean, having developed highly advanced seafaring technology for undertaking such maritime activities as fishing and commercial ventures. The first notice in classical literature of a Greek to undertake ocean voyages to India occurred between ca. 116 and 107 BCE[52] by Eudoxos of Kyzikos.[53] In the course of his retelling of how Eudoxos came to sail to India, Strabo mentions how a shipwrecked Indian sailor was rescued by the Ptolemaic coast guard and brought to the royal court. There, after learning Greek, he agreed to act as a guide on an expedition to India with Euxodos selected by Ptolemy VIII as its leader. The venture succeeded and when Euxodos returned with a cargo filled with aromatics and precious stones the king seized it for himself. Following the king's death, his wife Kleopatra III commissioned another expedition, which again Euxodos led only to lose this cargo to her son. The account of a different Indian Ocean crossing by a Greek is recorded by Pliny and appears in the *Periplus*,[54] where credit is given to a certain Hippalos, who explained that the Southwest Monsoon enabled sailors to complete a voyage across the ocean[55] to India as opposed to hugging the coastline as had been the norm.[56] Like Bengston's idea, this reconstruction has also been dismissed as implausible, although without explanation.[57]

It is apparent that the only proof of the origin of our shipwrecked Indians in Germany, and the events that landed them in the hands of Metellus is circumstantial. There is one point, however, on which all modern explanations rest: the account of Nepos involving an ocean voyage must be true, even as they acknowledge that his account is far from trustworthy. For example, Pliny, who often relies on Nepos as an important source, states that Nepos reported that Eudoxos quit Egypt to avoid capture from Ptolemy Lathyros and sailed from the Red Sea to Gades.[58] This version of the episode stands as a corruption of Posidonios' original account, and thereby calls into question the veracity of the report that these Indian sailors arrived in Germany by way of the Atlantic. The sole reason that Nepos recounted the story was to show that it was possible to circumnavigate the *oikoumene*, and thereby to travel from India to western Europe. In other words, this whole affair is used as a proof: 'Ergo, the sea is continuous....'[59] If the original account of this event as reported by Nepos resembles Mela's version of it, then Nepos held no interest in examining the interiors of Germany, the Alpine, and Danubian provinces, while Dacia is entirely omitted, not to mention the interior of Asia.[60] A similar pattern is discernible in Pliny with regard to Nepos' account of these Indians. We would be hard pressed to discover a better example of the unrestrained manner in which Pliny has 'raked together his materials, or the total want of critical judgement, or even common accuracy with which he made use of them.'[61] Although the shoreline of northern Europe had been explored during Augustus' reign, rather than drawing on this acquired knowledge Pliny opted to embrace as fact popular Roman imagination of this part of the world as it still persisted. This also explains his contorted account of Eudoxos, which he obtained from Nepos. The primary distinction between Mela and Pliny on the one hand, and Strabo[62] on the other is that the latter employed Posidonios' account of Eudoxos' voyages to argue that they amount to no more than a flight of fancy.

While there is no reason to suspect that the main premise of this account is fictional, namely that a ship bearing Indians was involved in a storm and washed up on *a coast* in Germany, we may call into doubt that this incident occurred as the result of an ocean voyage, either in the Atlantic Ocean or in the Baltic Sea. There is no need to invent a shipwreck involving a group of travelers who were not Indian and who had not come from India. How then might these Indians have arrived in Germany? The most probable explanation is that they came by way of an inland waterway that would have begun in Baktria and led to the city of Phasis (Fig. 16.1). Although there is a distinct possibility of a ship bearing Indian merchants that might have made its way from Alexandreia in Egypt, there is no such record in the pre-Augustan period

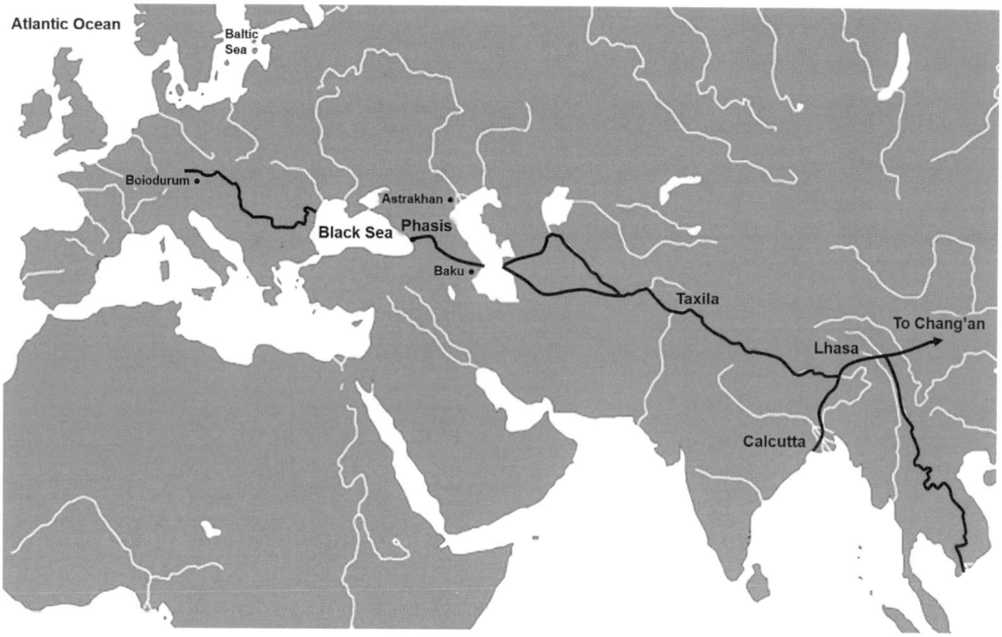

Fig. 16.1. Network of trade routes linking China to Western Europe using inland waterways.

of any such voyage into either the Black Sea or the Atlantic, let alone into the Mediterranean itself.

The problem of a storm that could have resulted in a shipwreck with Indian passengers is easily resolved if we allow that it occurred on a river like the Danube, and not in the Atlantic. In this regard, the ethnicity of the travelers as Indian is maintained without the necessity of inventing some other group of survivors who appeared in kayaks having originated from somewhere in the New World, Africa, the Baltic Sea, or from somewhere else. It is likewise not difficult to conceive that these Indians could have learned Greek in Baktria or at Phasis, and thereby have been able to communicate with Metellus. Finally, it is entirely plausible that Nepos in his desire to offer as proof the existence of a route from India to Germany sailing the river Ocean would either have not understood the system of inland waterways that connected the Black Sea to Baktria, or would not have accepted its existence because such a voyage would have conflicted with his worldview.

This leaves the equally sticky issue of the identification of the people responsible for 'capturing' these merchants and presenting them to Metellus. Unfortunately, the texts are in disagreement: Mela states that it was *rege Bootorum*, resolved as the king of the Bootes (or Boii), against Pliny's *rex Sueborum*, the king of the Svevi (Suebi).[63] André[64] attributes this apparent confusion to Nepos' text, which he conjectures must have appeared as something like *a rege Sueborum seu Boiorum*, and therefore discounts the possibility that Pliny substituted Sueborum for Boiorum, or that Mela committed the same error in reverse. Consequently, each author chose the name that he saw fit as an emendation to Nepos' text. In this regard, Tausend[65] sees in Mela's *rex Bootorum* the name of the *Boti*, which he takes as a corruption of the name *Gotorum* or Goths of southern Sweden who conveyed the Indians through the *Harudes* whom he identifies as the neighbors of these presumed Goths and allies of the Suebian king Ariovist. André, too, considers the Germanic Ariovist to have been the Suebian king who was responsible for conveying the Indians to the Romans. Both maintain that in 62 BCE the Suebians, under Ariovist, occupied the area of the rivers Neckar and Main, south of the *Vbii*. They see him, and not the Boii, as the party responsible for turning these Indians over to the Romans, because he sought favorable relations with Rome based on his *subsequent* interactions with Caesar.[66] André, however, confines the Boii north of the Danube in Bohemia, prior to their expulsion by the Dacians when they were driven to Noricum and there joined with the migrating Helvetii. Metellus was wintering with his legions in the Alpian city of Aquileia readying to face an onslaught of invasions from the east.

One other possibility exists. By faithfully copying Nepos' text, Mela would have retained the original name of the tribe who handed the Indian captives to the Romans as the Keltic *Bootorum/Boiorum* (*i.e.*, the *Boti*). Pliny, on the other hand, writing almost a century and a half after Nepos emended the identification of this tribe by updating it to reflect the demographic realities of his day as the Germanic (and non-Germanic) generic Svevi (Suebi).[67] Pliny was able to link the Suebi with Caesar's campaigns a few years later in Gaul and thereby reflected a similar demographic makeup of the tribes beyond the *limes* that appear in Tacitus' ethnographic work, *Germania*, written some three decades later. In this manner, the tribes Pliny identifies as responsible for gifting Metellus the Indians would have changed according to the period in which each author wrote.

In 62 BCE the Boii, who were in the process of migrating westward from Bohemia (*Boiohaemum*), could very well have been in possession of the settlement of Boioduron in modern Passau at the confluence of the Inn and Danube,[68] from where later they would join the Helvetians on their migration to Gaul in 58 BCE where Caesar[69] encountered them. The unknown king of this Keltic tribe was then the one responsible for bringing these captives to Metellus, perhaps while he wintered at his headquarters at Aquileia, and not as later tradition would have it, the Suebian Ariovist. As for what the Indians were doing in this part of the world, we may look no further than to see in them merchants, who sought new sources of exotica, like amber, with which to trade elsewhere.

Almost 1100 years after this shipwreck, a nun between 40 and 60 years of age died, and her corpse was buried in a cemetery near the monastery where she had lived in Dalheim, south of Frankfurt.[70] When scientists examined her skull they came upon a most unusual discovery: they found flecks of blue that were embedded and preserved in the fossilized dental plaque (calculus) of the nun's teeth. It so happened that this ultramarine substance was in actuality a costly and rare painting material made from semi-precious lapis lazuli. The reason that she had this pigment was because she had been a illustrator of lavish biblical manuscripts, a profession thought to have been a task reserved only for monks. As to why the pigment was in her mouth, the researchers concluded that she must have licked the end of her paintbrushes as she carried out her meticulous work.

Adding to the importance of this discovery is that lapis lazuli used to create illuminated manuscripts could only have come to medieval Europe through a long-distant trading network that had as its origin the Badakhshan Mountains of Afghanistan and Tajikistan some 3,000 miles to the east. It may well have been that the merchants who brought this bright blue metamorphic rock followed the same inland waterway system to Europe as their Indian predecessors who more than a millennium earlier had washed up on the German side of the Danube, and unwittingly entered into Roman history.

Notes

1 The only other reference to Nepos occurs at 3.9.90. Pomponius' *De chorographia* stands as the first extant work on geography written in Latin and the only known work of its kind devoted exclusively to the subject (Duff 1964, 101).
2 It was to this work that prompted Catullus (1.5–7) to tease Nepos, see Wissowa 1901, 1411; Detlefsen 1908, 56; Elder 1966, 143–149; Quinn 1970, 90.
3 Schanz 1909, 151–152; cf. Charisius, 1.146K; Gellius, *Attic Nights* 6.18.11.
4 Pliny *Naturalis Historia* 3.4, 3.132, 4.77; note 5.4 in which he passes an unfavorable judgement on the work; cf. Detlefsen 1908, 56.
5 A concise overview of the controversy crediting Cornelius Nepos with having composed a work on geography, see Stem 2012, 88–95 with ample bibliography.
6 Thomson 1948, 198–200.
7 See Bunbury 1959, ii, for an overview of the problems with Mela's geography, *De chorographia* 352–357, and with Pliny's reconstruction, 371–384, esp. 383.
8 Tarducci 1893, 280; cf. Plischke 1916, 94.
9 Bengston 1954, 229–231; André 1982, 46–48 with bibliography.
10 Henning 1944, 298–299; Plischke 1916, 95.
11 Henning 1944, 289–292; André 1982, 53–55; André and Filliozat 1986, 343ff.; contra Silberman 1988, 278.
12 Plischke 1916, 93; 1921, 125; Henning 1944, 298–299.
13 *e.g.* MacRitchie 1911, 284–286; Plischke 1921, 125.
14 Henning 1944, 298–299; Contra André 1982, 53–55; André and Fillozat 1986, 343–344.
15 Ditten 1988, 547–548 in his comments on Pomponius Mela *De chorographia* 3.45.
16 Tausend 1999, 116–120.
17 Plutarch *Pompey* 34.2; Cassius Dio *Roman History* 36.54.1 and 37.4.4.
18 Bengston 1954, 234.
19 Tausend 1999, 116–117.
20 We are told that he did so in favor of his adopted brother Quintus Caecilius Metellus Nepos (Cicero *Epistulae ad familiares* 5.1–2).
21 For an overview, Gruen 1974, 58, 130–131, 219–220, 396 et passim.
22 Strabo *Geography* 7.1.3 (292).

23 Tacitus *Germania* 46.3; Ptolemy 2.11.16.
24 Tausend 1999, 120–124.
25 Podossinov 2014, 137–138.
26 Podossinov 2014, 138–144; cf. Подосинов 2002, 101, 104–105 for a translation of the text and a discussion of the term.
27 Barford 2001, 86–103; 2014, 19–27.
28 For a succinct overview of such failed ventures, see Biffi 2003, 147–148.
29 This is the same route that the Carthiginian admiral Hanno attempted, *Periplus of Hanno* 18; Mela *De chorographia* 3.90 with n. 61. It seems that ships which ventured beyond the wadi Loukkos never returned (Strabo *Geography* 2.3.4 [98–100]).
30 See André 1982, 51–52, 54–55. Cf. Thomson 1948, 199 with n. 2, wherein he cites examples of later explorers, like Gilbert, who literally believed this account and sought 'a northwest passage round Canada', or Frobisher, who brought home 'an Eskimo whose Tartar features seemed to prove that Asia had actually been reached'.
31 Podossinov 2014, 140–144.
32 Biffi 2003, 150–151, 153.
33 *Res Gestae* 31.1. Two embassies are attested in Augustus' lifetime: one in 25 BCE at Tarragone (Orosius 6.21.19), the other in Samos in 20 BCE (Cassius Dio *Roman History* 54.9). Strabo (*Geography* 15.1.4 [686]) reports on one the missions (15.1.73 [719]) whose details he drew from Nikolaos of Damascus, having reported that he met these ambassadors at Antiocheia near Daphne, who probably reached Augustus in Samos; cf. Florus 2.34.62; Suetonius *Divus Augustus* 21.6.
34 Luisi 1988, 41–51.
35 Bengston 1954–1955, 229–236.
36 Pliny *Naturalis Historia* 6.19 (52).
37 Cassius Dio *Roman History* 34.54, 36.54, 37.4.4; cf. Plutarch *Pompey* 34.2.
38 André 1982, 82; Tausend 1999, 117; Podossinov 2014, 136–137.
39 Rtveladze 2010, 80–96; Ртвеладзе 2012 (for a review, Lerner 2014a); Olbrycht 2013; Lerner 2014b, 155–174.
40 Strabo *Geography* 2.1.15 (73), 11.6.1 (507), 11.7.1 (508), 11.7.3 (509), 11.11.5–6 (518); cf. Curtius 4.4.19.
41 Strabo *Geography* 11.2.17 (507).
42 Гуленков 1999, 95–101; Lordkipanidze 2000, 30; Гуленков 2001, 169–175; Дзарасов 2003, 27; cf. Gardiner-Garden 1987, 39–48; Lordkipanidze 1991, 135; Lordkipanidze 1996, 113–120; Braund 1999, 39–40.
43 Braund 1994, 40–42; 2002, 291–295; cf. Pearson 1960, 164.
44 *Ad Nicomedem regem*, 934 [F20]. A variety of titles have been proposed for the work composed in iambic trimeters, including *Perigesis* (*GGM* 1.196–237), *Perigesis ad Nicomedem regem* (Diller 1952, 165–176); *περίοδος γῆς* (Marcotte 2000, lv–lxxii; cf. Marcotte 2002), *περίοδος γῆς ἐν κωμικῷ μέτρῳ εἰς Νικομήδην βασιλέα* (Korenjak 2003). For a succinct overview, Ryan 2007, 137–143.
45 On the date of the work's composition, see Bianchetti 1990, 23–35; Marcotte 2000, 7–16; Boshnakov 2004, 4–6, 70–78; cf. Bravo 2009, xiii–xiv, xvii–xviii. Several attempts have also been made to identify the author, including Pausanias of Damascus (Diller 1955, 276–278), and Apollodorus of Athens (Marcotte 2000, 35–46, who in addition provides an overview of previous attempts at identifying the author). For an overview of the historical and geographical tradition of the *Periodos*, Hunter 2006, 123–140.
46 Dale 1994, 101–124, especially 59 n. 63, 108 n. 24, 111, and 128.
47 Молчанов 2006, 137–138.

48 Wainwright 1947, 143–144; Curtin 1984, 71–73; Ray 1994, 12–17; Vogt 1996, 126–127; Moore and Lewis 1999, 52–58; Shaw 2000, 316–317; Warburton 2007, 9–21; Smith 2009, 32–36, 41–45; Beaujard 2015, 15; Gaur and Sundaresh 2016, 199; Salles 2016, 137.
49 Desanges 1978, 243–305; Salles 1987, 85–88; 1995, 125, 134; 1996a, 302–304; 2016, 149–150; Sherwin-White and Kuhrt 1993, 65–66; Ray 1994, 55; Parker 2002, 70–71; Tomber 2008, 112.
50 Salles 1995, 117; Potts 1996, 282; Gregoratti 2011, 214–225.
51 Raschke 1975, 244; Sidebotham 1991, 15; 1996, 287–288; Whittaker 2004, 163, 167; Tomber 2008, 18, 71; McLaughlin 2010, 29, 169; Sidebotham 2011, 5; Gurukkal 2016, 20, 139; Tomber 2017, 537.
52 Sidebotham 2011, 17; Habicht 2013, 199.
53 Strabo *Geography* 2.3.4–5 (98–102).
54 Pliny, *Naturalis Historia* 6.26.100, 104–105; *Periplus* 57.
55 Opinion is divided between the name Hippalos as a name commonly found in Egypt (Habicht 2013, 202) and that it is the name of a wind called Hippalus (Mazzarino 1997, 72–79).
56 Herodotos *Histories* 4.44. Romm 1992, 84–85; Hourani 1995, 25; Tchernia 1995b, 991, 995–996; 1997b, 256; 2016, 232; Salles 1996b, 254; Parker 2008, 14–18.
57 Podossinov 2014, 137.
58 Pliny *Naturalis Historia* 6.169: *Nepos Cornelius auctor est Eudoxum quendam sua aetate, cum Lathyrum regum fugeret, Arabico sinu egressum Gades usque pervectum.*
59 Mela *De chorographia* 3.45 translation by Romer, who points out that even Pliny *Naturalis Historia* 5.4 was well aware that Nepos enjoyed a reputation for gullibility (114 n.24); cf. Silberman's critique of the assorted difficulties connected with this passage (1988, 277–278 ns. 1–3).
60 Bunbury 1959, ii, 355. Medians, Bactrians, and Arians are known by name only, while countries such as Drangiana, Margiana, and Sogdiana pass without notice.
61 Bunbury 1959, ii, 383; cf. 186–191.
62 Strabo *Geography* 2.3.5 (100–102).
63 Bengston 1954, 234–235 with n. 3 for the earlier bibliography.
64 André 1982, 49.
65 Tausend 1999, 119–124.
66 Caesar *De Bello Gallico* 1.40, 42. The entirety of the interactions between Casear and Ariovist occurs in 1.31–50. André 1982, 49–50; Tausend 1999, 119–121.
67 An assessment of archaeological sites in Germany dated to this period reveal that it is impossible to discern a specific group of people who can be identified as 'Suebian'; instead the results indicate that there numerous tribes-German and Keltic-that shared some customs with a considerable amount of variation (*e.g.* Godlowski 1992; Gropengiesser 1992). Similarly, the name of the Suebians indicates that it merely refers to a member of a collective that differentiated itself from other groups, like Kelts and Skythians (Neumann 1992), while a linguistic analysis of ethnic names likewise proves unhelpful (Schmid 1992). The depictions by Tacitus and Caesar of the Suebians are essentially useless in reconstructing a trustworthy ethnography (Timpe 1992a; 1992b).
68 Krause 2006, 225.
69 Thus Caesar *De Bello Gallico* 1.5.4, 29.2 et passim, ultimately settling in the region of the Aedui, 1.28.5.
70 I am grateful to my colleague and good friend, Ulrike Wiethaus, for bringing this study to my attention: Radini *et al.* 2019.

Bibliography

André, J. 1982. Des Indiens en Germanie?, *Journal des Savants* 1982, 45–55.
André, J. and Filliozat, J. 1986. *L'Inde vue par de Rome: Textes latins de l'Antiquité relatifs à l'Inde*, Paris.

Barford, P.M. 2001. *The Early Slavs*, London/Ithaca.
Barford, P.M. 2014. From 'Proto-Slavs' to Proto-state, in: Gleason, A. (ed.), *A Companion to Russian History*, Chichester, 17–33.
Beaujard, P. 2015. The Worlds of the Indian Ocean, in: Pearson, M. (ed.), *Trade, Circulation and Flow in the Indian Ocean World*, London, 15–26.
Bengtson, H. 1954–1955. Q.Caecilius Metellus Celer (cos 60) und die Inder, *Historia* 3, 229–236.
Bianchetti, S. 1990. *Πλωτὰ καὶ πορευτά. Sulle tracce di una periegesi anonima*, Florence.
Biffi, N. 2003. Melello e gli Indiani, *Geographia Antiqua* 12, 146–153.
Boshnakov, K. 2004. *Pseudo-Skymnos (Semos von Delos?)*, Palingenesia 82, Stuttgart.
Braund, D. 1994. *Georgia in Antiquity: A History of Colchis and Transcaucasian Iberia: 550 BC–AD 562*, Oxford/New York.
Braund, D. 1999. Indian Traders at Phasis, in: Lordkipanidze, O. (ed.), *Trade on the Black Sea in the Archaic and Classical Periods: Historical Perspective of the Silk Road*, Tbilisi, 39–40.
Braund, D. 2002. Indian Traders at Phasis: Neglected Texts on Ancient Georgia, in: Faudot, M., Fraysse, A. and Geny, E. (eds), *Pont-Euxin et commerce: la genèse de la Route de la soie: actes du IXe Symposium de Vani, Colchide, 1999*, Besanon/Paris, 287–295.
Bravo, B. 2009. *La Chronique d'Apollodore et le Pseudo-Skymnos: érudition antiquaire et littérature géographique dans la seconde moitié du IIe siècle av. J.-C.*, Studia Hellenistica 46, Leuven.
Bunbury, E.H. 1959. *A History of Ancient Geography Among the Greeks and Romans from the Earliest Ages till the Fall of the Roman Empire*, 2 vols. 2nd ed., New York.
Curtin, P.D. 1984. *Cross-cultural Trade in World History*, Cambridge.
Dale, S.F. 1994. *Indian Merchants and Eurasian Trade, 1600–1750*. Cambridge studies in Islamic Civilization, Cambridge.
Desanges, J. 1978. *Recherches sur l'activité des Méditerranéens aux confins de l'Afrique (vi^e siècle av. J.-C.-ive siècle ap. J.-C.)*, Rome.
Detlefsen, D. 1908. *Die Geographie Afrikas bei Plinus und Mela und ihre Quellen*, Berlin.
Diller, A. 1952. *The Tradition of the Minor Greek Geographers*, New York.
Diller, A. 1955. The Authors Named Pausanias, *Transactions and proceedings of the American Philological Association* 86, 268–279.
Ditten, H. 1988. Pomponius Mela: 296–302, 542–551, in: Herrmann, J. (Hrsg.) *Griechische und lateinische Quellen zur Frühgeschichte Mitteleuropas bis zur Mitte des 1. Jahrtausends u.Z. Teil 1. Von Homer bis Plutarch (8. Jh. v. u.Z. bis 1. Jh. u.Z.)*. Schriften und Quellen der alten Welt Bd. 37.1, Berlin.
Elder, J.P. 1966. Catullus 1, His Poetic Creed, and Nepos, *Harvard Studies in Classical Philology* 52, 143–149.
Gardiner-Garden, J.R. 1987. *Greek Conceptions of Inner Asian Geography and Ethnography from Ephoros to Eratosthanes*. Papers on Inner Asia no. 9, Bloomington.
Gaur, A.S. and Sundaresh. 2016. Ancient technology of jetties and anchorage system along the Saurashtra coast, India, in: Boussac, M.-F., Salles, J.-F. and Yon, J.-B. (eds), *Athens, Aden, Arikamedu: Essays on the Interrelations Between India, Arabia and the Eastern Mediterranean*, New Delhi, 199–215.
Godłowski, K. 1992. Die Przeworsk-Kultur, in: Neumann and Seemann (eds), 9–90.
Gregoratti, L. 2011. A Parthian port on the Persian Gulf: Characene and its Trade, *Anabasis: Studia Classica et Orientalia*, 2, 209–229.
Gropengiesser, E. 1992. Die Neckarsueben, in: Neumann and Seemann (eds), 91–123.
Gruen, E. 1974. *The Last Generation of the Roman Republic*, Berkeley/Los Angeles/London.
Gurukkal, R. 2016. *Rethinking Classical Indo-Roman Trade: Political Economy of Eastern Mediterranean Exchange Relations*, New Delhi.
Habicht, C. 2013. Eudoxus of Cyzicus and Ptolemaic Exploration of the Sea Route to India, in: Buraselis, K., Stefanou, M. and Thompson, D.J. (eds), *The Ptolemies, the Sea and the Nile: Studies in Water Borne Power*, Cambridge, 197–206.

Hennig, R. 1944. *Terrae incognitae 1: Altertum bis Ptolemäus,* 2nd ed., Leiden.
Hourani, G.F. 1995. *Arab Seafaring in the Indian Ocean in Ancient and Medieval Times. Revised and Expanded ed. J. Carswell*, Princeton.
Hunter, R. 2006. The Prologue of the Periodos to Nicomedes ('Pseudo-Scymnus'), in: Harder, M.A., Regtuit, R.F. and Wakker, G.C. (eds), *Groningen Workshops on Hellenistic Poetry: Beyond the Canon.* Hellenistica Groninga 11, Leuven, 123–140.
Korenjak, M. 2003. *Die Welt-Rundreise eines anonymen griechischen Autors („Pseudo-Skymnos'),* Hildesheim.
Krause, R. 2006. Boii and the Celts in Bohemia, in: Koch, J.T. (ed.), *Celtic Culture: A Historical Encyclopedia*, vol. 1, Santa Barbara/Denver/Oxford, 222–226.
Lerner, J.D. 2014a. Review: Ртвеладзе, Е.В. 2012. *Великий индийский путь: из истории важнейших торговых дорог Евразии.* Санкт-Петербург in *Anabasis: Studia Classica et Orientalia* 5, 209–214.
Lerner, J.D. 2014b. On the Inland Water Ways from Europe to Central Asia, *Ancient West and East* 13, 155–174.
Lordkipanidze, O. 1991. *Archäologie in Georgien: von der Altsteinzeit zum Mittelalter, Quellen und Forschungen zur prähistorischen und provinzialrömischen Archäologie*, Bd. 5, Weinheim.
Lordkipanidze, O. 1996. *Das alte Georgien (Kolchis und Iberien) in Strabons Geographie*, Neue Scholien, Deutsch von N. Begiaschwili, Amsterdam.
Lordkipanidze, O. 2000. *Phasis: the River and City in Colchis, Geographica Historica*, Bd. 15, Stuttgart.
Luisi, A. 1988. Cornelio Nepote geografo, in: Sordi, M. (ed.), *Geografia e storiografia nel mondo classic*, Milan, 41–51.
MacRitchie, D. 1911. Der Kajak im nördlichen Europa, *Petermanns Geographische Mitteilungen*, 57, 284–286, Tafel 51.
Marcotte, D. 2000. *Les geographes grecs*, vol. 1, Paris.
Marcotte, D. 2002. *Pseudo-Scymnos, Circuit de la Terre*, Paris.
Mazzarino, S. 1997. On the name of the *Hipalus (Hippalus)* wind in Pliny, in: De Romanis, F. and Tchernia, A. (eds), *Crossings: Early Mediterranean Contacts with India*, New Delhi, 72–79.
McLaughlin, R. 2010. *Rome and the Distant East: Trade Routes to the Ancient Lands of Arabia, India and China*, London.
Mela, P. 1998. [*De chorographia*] *Pomponius Mela's description of the world*, F.E. Romer, Ann Arbor.
Moore, K. and Lewis, D. 1999. *Birth of the Multinational: 2000 Years of Ancient Business History – From Ashur to Augustus*, Copenhagen.
Müllenhoff, K. 1892. Über den Anhang zu dem Provinzialverzeichnis von 297, in: Müllenhoff, K. (Hrsg.), *Deutsche Altertumskunde*, vol. 3, Berlin, 311–325.
Müller, C. (ed.) 1855–1861. *GGM. Geographici Graeci Minores*, 2 vols, Paris.
Neumann, G. 1992. Der Name der Sweben, in: Neumann and Seemann (eds), 153–166.
Neumann, G. and Seemann, H. (eds). 1992. *Beiträge zum Verständnis. Beiträge zum Verständnis der Germania des Tacitus: Bericht über die Kolloquien der Kommission für die Altertumskunde Nord- und Mitteleuropas im Jahre 1986 und 1987.* Teil 2. Abhandlungen der Akademie der Wissenschaften in Göttingen, Philogisch-Historische Klasse, Folge 3, Nr. 195, Göttingen.
Olbrycht, M.J. 2013. *Die Geschichte der Seidenstraße in antiker Zeit: eine Einführung*, in: *Die Krim goldene Insel im Schwarzen Meer: Griechen-Skythen-Goten. Begleitbuch zur Ausstellung*, Darmstadt, 67–87.
Parker, G. 2002. Ex Oriente Luxuria: Indian Commodities and Roman Experience, *The Journal of the Economic and Social History of the Orient* 45.1, 40–95.
Parker, G. 2008. *The Making of Roman India*, Cambridge.
Pearson, L.I.C. 1960. *Lost Histories of Alexander the Great*, New York.
Plischke, H. 1916. Verschlagungen von Bewohnern Amerikas nach Europa im Altertum und Mittelalter, *Petermanns Geographische Mitteilungen* 1916, 93–95.
Plischke, H. 1921. Eine Verschlagung von Eskimos nach Holland aus dem Jahr 1577, *Petermanns Geographische Mitteilungen* 1921, 125.

Podossinov, A.V. 2014. The Indians in Northern Europe? On the Ancient Roman Notion of the Configuration of Eurasia, in: Podossinov, A.V. (ed.), *The Periphery of the Classical World in Ancient Geography and Cartography*. Colloquia Antiqua 12, Leuven/Paris/Wapole, MA., 133–145.

Potts, D.T. 1996. The Parthian Presence in the Arabian Gulf, in: Reade, J. (ed.), *The Indian Ocean in Antiquity*, London, 269–285.

Quinn, K. 1970. *Catullus: The Poems*, London.

Radini, A., Tromp, M., Beach, A., Tong, E., Speller, C., McCormick, M., Dudgeon, J.V., Collins, M.J., Rühli, F., Kröger, R. And Warinner, C. 2019. Medieval women's early involvement in manuscript production suggested by lapis lazuli identification in dental calculus, *Science Advances* 5, eaau7126.

Raschke, M.G. 1975. Papyrological Evidence for Ptolemaic and Roman Trade, in: *Proceedings of the XIV International Congress of Papyrologists: Oxford, 24-31 July 1974*, Oxford, 241–246.

Ray, H.P. 1994. *The Winds of Change: Buddhism and the Maritime Links of Early South Asia*, Oxford.

Romm, J.S. 1992. *The Edges of the Earth in Ancient Thought: Geography, Exploration and Fiction*, Princeton.

Rtveladze, E.V. 2010. The Great Indian Road: India-Central Asia-Transcaucasia, *Anabasis: Studia Classica et Orientalia* 1, 80–96.

Ryan, F.X. 2007. Der sogenannte Pseudo-Skymnos, *Quaderni Urbinati di Cultura Classica*, 87/3 n.s., 137–143.

Salles, J.-F. 1987. The Arab-Persian Gulf under the Seleucids, in: Kuhrt, A. and Sherwin-White, S. (eds), *Hellenism in the East: The Interaction of Greek and Non-Greek Civilizations from Syria to Central Asia after Alexander*, London, 75–109.

Salles, J.-F. 1995. The *Periplus of the Erythrean Sea* and the Arab-Persian Gulf, in: Boussac, M.-F., Salles, J.-F. and Yon, J.-B. (eds), *Athens, Aden, Arikamedu: Essays on the interrelations between India, Arabia and the eastern Mediterranean*, New Delhi, 115–146.

Salles, J.-F. 1996a. Hellenistic Seafaring in the Indian Ocean. A Perspective from Arabia, in: Ray, H.P. and Salles, J.-F. (eds), *Tradition and Archaeology: Early Maritime Contacts in the Indian Ocean*, New Delhi, 293–309.

Salles, J.-F. 1996b. Achaemenid and Hellenistic Trade in the Indian Ocean, in: Reade, J. (ed.), *The Indian Ocean in Antiquity*, London, 251–267.

Salles, J.-F. 2016. Towards a Geography of the Harbours in the Persian Gulf in Antiquity (Sixth Century BC–Sixth Century AD), in: Boussac, M.-F., Salles, J.-F. and Yon, J.-B. (eds), *Athens, Aden, Arikamedu: Essays on the Interrelations Between India, Arabia and the Eastern Mediterranean*, New Delhi, 137–161.

Schanz, M. 1909. *Geschichte der römischen Litteratur bis zum Gesetzgebungswerk des Kaisers Justinian*, vol. 1.2, Aufl. 3, München.

Schmid, W.P. 1992. Der Namenhorizont im germanischen Osten: Suebi und Veneti, in: Neumann and Seemann (eds), 190–202.

Schoff, W.H. (trans.) 1912. *Periplus of Hanno: A Voyage of Discovery Down the West African Coast by a Carthaginian Admiral of the 5th Century BC*, Philadelphia.

Shaw, I. 2000. Egypt and the Outside World, in: Shaw, I. (ed.), *The Oxford History of Ancient Egypt*, Oxford, 308–323.

Sherwin-White, S. and Kuhrt, A. 1993. *From Samarkhand to Sardis: A New Approach to the Seleucid Empire*, London.

Sidebotham, S.E. 1991. Ports of the Red Sea and the Arabia-India trade, in: Begley, V. and Puma, R.D. (eds), *Rome and India. The Ancient Sea Trade*, Madison, 12–38.

Sidebotham, S.E. 1996. Roman Interests in the Red Sea and Indian Ocean, in: Reade, J. (ed.), *The Indian Ocean in Antiquity*, London, 287–308.

Sidebotham, S.E. 2011. *Berenike and the Ancient Maritime Spice Route*, London.

Smith, R.I. 2009. *Trade in World History*, London.

Stem, S.R. 2012. *The Political Biographies of Cornelius Nepos*, Ann Arbor.

Tarducci, F. 1893. *John and Sebastian Cabot*. Transl. H.F. Brownson, Detroit.

Tausend, K. 1999. Inder in Germanien, *Orbis Terrarum* 5, 115–125.

Tchernia, A. 1995. Moussons et monnaies: les voies du commerce entre le Monde Gréco-Romain et l'Inde, *Annales Histoire, Sciences Sociales* 50.5, 991–1009.
Tchernia, A. 1997. Winds and Coins: From the Supposed Discovery of the Monsoon to the *Denarii* of Tiberius, in: De Romanis, F. and Tchernia, A. (eds), *Crossings: Early Mediterranean Contacts with India*, New Delhi, 250–276.
Tchernia, A. 2016. *The Romans and Trade.* Transl. J. Grieve with E. Minchin, Oxford.
Thomson, J.O. 1948. *History of Ancient Geography*, Cambridge.
Timpe, D. 1992a. Die Landesnatur der Germania nach Tacitus, in: Neumann and Seemann (eds), 258–277.
Timpe, D. 1992b. Der Sueben-Begriff bei Tacitus, in: Neumann and Seemann (eds), 278–310.
Tomber, R. 2008. *Indo-Roman trade: From Pots to Pepper*, London.
Tomber, R. 2017. Egypt and Eastern Commerce During the Second Century AD and Later, in: Wilson, A. and Bowman, A. (eds), *Trade, Commerce and the State in the Roman World*, Cambridge, 531–555.
Vogt, B. 1996. Bronze Age Maritime Trade in the Indian Ocean: Harappan Trits on the Oman Peninsula, in: Reade, J. (ed.), *The Indian Ocean in Antiquity*, London, 107–132.
Wainwright, G.A. 1947. Early Foreign Trade in East Africa, *Man* 47, 143–148.
Warburton, D.A. 2007. What Happened in the Near East ca. 2000 BC?, in: E.H. Seland (ed.), *The Indian Ocean in the Ancient Period: Definite Places, Translocal Exchange*, Oxford, 9–21.
Whittaker, C.R. 2004. *Rome and its Frontiers: The Dynamics of Empire*, London.
Wissowa, G. 1901. *Real Encyclopädie*, s.v. Cornelius Nepos, 4.1, 1411–1416.
Гуленков, К.Л. 1999. «Каспийский» торговый путь во времена Митридата VI Евпатора, в: *Античность: события и исследователи*, Казань, 95–101.
Гуленков К.Л. 2001. « Северный » торговый путь во времена Митридата VI Евпатора, *Проблемы истории, филологии, культуры* Вып. 10, 169–175.
Дзарасов, А.А. 2003. К вопросу о роли аорсов и алан в системе торговли и обмена на северном кавказе в первые века н.э., *Известия вузов. Северо-Кавказский регион. Общественные науки. Приложение* 2003. № 8, 27–34.
Молчанов А.А. 2006. Индийцы на севере Европы в 62 г. до н.е.: реконслрукция маршрута путешествия при эго сокращенном описания в источниках, *Восточная Европа в древности и средневековье* 18, 135–138.
Подосинов, А.В. 2002. *Восточная Европа в римской картографической традиции. Тексты, перевод, комментарии.* Древнейшие источники по истории Восточной Европы, Москва.
Подосинов, А.В. 2005. Индийцы на Севере Европы? (Несколько замечаний к Mela, III, 45), в: Чтения памяти В.Т. Пашуто, *Восточная Европа в древности и средневековье. Проблемы источниковедения.*, Москва, 29–33.
Ртвеладзе, Е.В. 2012. *Великий индийский путь:* из *истории важнейших торговых дорог Евразии*, Санкт-Петербург.

Chapter 17

Samuel Shaw's 'Maritime Silk Road' from American Independence towards Monopoly, 1784–1794

John A. Ruddiman

After the Revolutionary War, John Adams wrote his fellow American diplomat John Jay: 'There is no better advice to be given to the merchants of the United States, than to push their commerce to the East Indies as fast and as far as it will go.'[1] This proved easier said than done, despite celebratory narratives later published in histories of America's success in the 'Old China Trade'. The career and writings of Samuel Shaw, a merchant who helped facilitate trading voyages to China in the late 1780s and early 1790s and was appointed first diplomatic consul of the United States in Canton, show Americans' uneven efforts to master the mechanics of the trade and their expectations about the relationships between private merchants and government policy.

The stories of American merchants' entrance into the tea trade at Canton – and particularly the inaugural voyage made by the ship *Empress of China* – are well known. Historians have considered these first American trading expeditions to China in celebratory patriotic terms, as part of studies of capitalism or the material culture of tea, or to weigh the era's changing understandings of race and national identity. Studies have usefully emphasized the evolution of the trade across time and American merchants' eventual innovations.[2] Rather than focusing on the remarkable successes of American merchants that unfolded in the period running up to 1842, this essay considers Americans' initial search for expertise; a case-study consideration of Samuel Shaw casts new light on expectations and relationships in the first moment of American participation in the trade with China.

Examination of the correspondence of Samuel Shaw, the merchant supercargo on the first American-flagged voyage to Canton, reveals that these first Americans in China were not innovators. They were trying to sail with a mercantile convoy that stretched back across centuries. In Canton, they sought to join a specific trade system that had been evolving for 80 years. Shaw's voyages and his writings reveal two processes intertwining: a search for relevant expertise and attempts to shape government policy. First, American merchants had to figure out the relationships and mechanics of the China trade itself; in the first voyage of the *Empress of China*, they got lucky. Thereafter, as Shaw gained fluency, he lobbied from within the new national government to reorganize the American side of that trade under a government-sponsored monopoly to match 18th-century traditions and expectations.

Hunting for Opportunity and Expertise

Late in 1783 Samuel Shaw faced an economic crisis. Born in 1754, this elite son of New England had dedicated his entire youth to the military struggle against Britain. Shaw's own war almost started early, in 1774, when the teenaged youth challenged a British lieutenant to a duel to avenge the officer's slanders of the American colonials as 'cowards and rebels'.[3] Shaw then served across the eight years of the War of American Independence as an artillery officer, ending as aide-de-camp to General Henry Knox, one of George Washington's key lieutenants. During the long war, Shaw had sensed precious time and economic opportunities were slipping away while he served in the army. In 1779, when he was about 25 years old, he wrote home: 'I have made a discovery lately.... I am old and growing older every day! ... I hope we shall have peace before we grow much older....' He asked his brother, 'Don't you sometimes think of these matters, and cast about how a family is to be maintained? I'm afraid if this unlucky war should continue I shall be the batchelor of the family.'[4] When the war did end in 1783, Shaw found himself broke and unemployed. There was no room for him to join his brothers in the already shaky family business of land speculation in frontier Maine. But what Shaw lacked in cash or a civilian vocation, he more than made up with his government connections forged in the Continental Army. General Washington himself praised the young veteran in a letter of recommendation: Shaw had 'distinguished himself in everything which could entitle him to the character of an intelligent, active and brave officer'.[5] Knox, Shaw's patron, promisingly continued as head of the War Department, which apart from the Congress itself practically comprised the rest of the loose 'national' government of the United States under the Articles of Confederation. Shaw was well-situated to rise, but his next steps were uncertain. The young veteran would have to figure out what to do with himself.

Just as Shaw faced a path forward that seemed equally promising and rocky, the elite merchants of the young United States had to learn how to navigate in their new and uncertain position independent from Britain. Shut out of British markets, the newly-independent United States faced an economic challenge in 1783; they needed to access new trading partners to bring currency and specie into a catastrophically disrupted and depressed economy. Distressingly, as one American merchant reported in 1783, Spanish authorities in Cuba also 'caused a stop to be put to our trade... and thereby cutt us off from the brightest prospects'.[6] The 1778 trade treaty between France and the United States proved inadequate to absorb the American agricultural exports; building relationships to access those markets fully would take years. Consequently, American merchants had to look beyond familiar ports. Rays of hope shone from the far east: the navigation regulations of British empire had barred American colonials from directly trading with India or China; independence ostensibly opened those routes. American merchants, however, still had to learn how to conduct this expensive and challenging trade to the other side of the globe.

First, Americans looking to join the Indies trade needed to secure commodities in demand in Asia to make up for their inadequate supply of specie silver and their limited access to credit. They expected animal furs taken from cold Pacific coasts might serve. In the early 1780s, John Ledyard, a sailor on Captain Cook's last pacific voyage, concocted a plan to organize an expedition to the Pacific Northwest that could then carry valuable pelts across the wide ocean to China. He pitched the idea to an interested Philadelphia merchant, Robert Morris, who was also the Superintendent of Finance for the United States. This plan, which proposed sending one ship east to China via Africa's Cape of Good Hope and at least two others south and west around South America's Cape Horn, proved too complicated. Morris and his partners chose instead to ship ginseng east to China. Dug from their Appalachian mountains in Virginia, the medicinal root was more accessible and more easily transported as east-bound cargo. (Daniel Boone would join other frontier prospectors seeking this commodity, digging in Kentucky for 'The Sang' as he called the medicinal root, sending it to coastal merchants.) They would send one ship east in 1784.[7]

Much of this first American mercantile expedition to China was jury-rigged, repurposed, and showed the lack of experience burdening these merchants. Robert Morris and his partners purchased a newly constructed ship, designed during the war to sail fast and light as a privateer that would prey on British traders. The war's end meant her repurposing as a merchant vessel; in an act of public relations, the investors in this scheme rechristened the ship the *Empress of China*.[8] This marketing was over-optimistic. Key advice from one of Morris's collaborators in Paris arrived too late to be useful: he warned that because all ships paid a flat, heavy fee at Canton 'no persons send any [vessels] but those of 800 or 1000 Tons. If a Ship only of 300 is sent, so considerable a duty makes a great addition to the cost of the cargo'. The *Empress of China* was only 360 tons.[9] At least finding a ship captain proved straightforward; Morris selected navy veteran John Green. Though he had never sailed east of the Atlantic, as was true for most American mariners, Green was an experienced and confident seaman. The investing partners also grabbed for experience where they could find it: one of the ship's officers, Frederick Molineaux, was likely hired as the captain's clerk because he had earlier worked for the British East India Company.[10] Hopefully he would have advice on the mechanics of the trade. Unsure what details might be required in the 'Sea Letters' that acted as a sort of passport, Morris obtained a catch-all document from the Congress impractically addressed to all nations of the world and their 'Most Serene, Serene, most puissant, puissant, high, illustrious, noble, honorable, venerable, wise and prudent Emperors, Kings, Republicks, Princes, Dukes,

Earls, Barons, Lords, Burgomasters, Councillors....'[11] Captain Green acknowledged their struggle to overcome their inexperience: 'this is a fittaguing Business and new to us all. I hope our Next outfit will be more Fammiller.... [and] Hope all will End in harmony & to our Satisfaktison as well to open a New doore for Commerce at a futour day.'[12]

The risks posed by their inexperience were heavy. Troublingly, the investors had struggled to secure insurance for the voyage from French firms because of the novelty of the Americans' undertaking.[13] Morris and his partners had pulled together almost $120,000 to buy the ship and fill it with dried ginseng and some silver to be exchanged for Chinese tea, silk, and porcelain. This staggering sum was 10 times the cost of financing a European trading voyage. The capital outlay equaled the cash reserves of one of the handful of American banks; alternately, such a sum could purchase 24,000 acres of good frontier land or a large slave plantation in the rice-growing South Carolina low country.[14] (Comparatively speaking, this investment considered as a share of US gross domestic product (GDP) today would total 13 billion dollars.[15]) These investors already knew they were up against stiff competition from vessels sailing from Britain, France, and Portugal. The partners accordingly ordered their captain to depart and sail as quickly as possible, avoiding putting into other ports: 'It being of very great Consequence to arrive in Canton before any other Vessel which can go from Europe with fresh Ginseng' – to say nothing of more easily exchanged silver.[16] When one of the investing partners declined to sail east himself, Morris and his colleagues had to find another supercargo to be responsible for the contents of the hold and represent their interests. Government-connected Robert Morris thus hired government-connected Major Samuel Shaw, who otherwise had little experience to recommend him. Shaw wrote his brother on Christmas Eve, 1783, just days after leaving the army: 'I shall sail for China...the terms on which I go promise something clever, and I hope to shake you by the hand in less than two years.'[17] The ship sailed from New York harbor in February 1784 to fanfare. The investors incorrectly expected their vessel would reach China in just three months; in fact, it took six.[18] In their selected cargo, composition of the crew, the instructions, and the ship itself, this incredibly expensive voyage was strikingly improvised. They would have to learn as they sailed.

Shaw kept a detailed journal of the outbound voyage and their four months trading at Canton, further revealing not only the learning curve American merchants faced, but also their remarkable good luck. They sailed from New York to the Cape Verde Islands, then without pause past the Cape of Good Hope to the Straits of Sunda between Java and Sumatra. The crew's experience ended at the edge of Atlantic waters; in the Indian Ocean Captain Green literally relied on instructions from a navigation book, the 1780 fifth edition of Samuel Dunn's *A New Directory for the East-Indies*.[19] At the Malaysian archipelago, the *Empress of China* had the good fortune to fall in convoy with two French ships bound for China.[20] The Americans had allied with the French to win their independence; these French merchants were delighted to learn that the war had finally ended with their British enemy defeated. Shaw recorded in his journal and his post-voyage report that these French allies provided 'constant advice and assistance'.[21] Most helpful were their explanations of the circumscribed nature of trade at Canton: the Co-Hong of state-privileged merchants ran the trade, with one Hong merchant standing responsible for a visiting ship; he 'secured' the crew's behavior and the payment of duties. These merchants were the sole link for foreign ships to the Chinese economy: European merchants were confined to trading factories on a narrow strip of land along the Pearl River. If Mr. Molineaux, the captain's clerk, also helped educate Shaw, the young supercargo revealed no evidence of it to his superiors, preferring to present himself as connected diplomatically to powerful, friendly European actors.[22] Similarly, Shaw showed no sign he recognized their good luck in being assigned in Canton to Phuankequa, a deeply experienced and powerful merchant. Under Phuankequa's unacknowledged guidance, Shaw naively thought 'Commerce [here] appears to be...as simple, as any in the known world.'[23] Shaw's ultimate stroke of good fortune, however, was that the *Empress* and her new French friends were the first trade ships of the season to have sailed up the Pearl River to the anchorage at Whampoa. Their ginseng and silver dollars initially faced little competition in the trade seeking tea.

It took Shaw four months to exchange their cargo in Canton. Thomas Randall, Shaw's old army comrade and fellow supercargo on the voyage, later crowed 'we had a larger quantity of genseng than ever had been brought to the Chinese market,' and calculated their ship carried more than the British and Portuguese ships combined had brought to Canton in 1784.[24] The ship's purser was more realistic: 'Our cargoe turned out but so so. We brought too much Ginseng... Our [ship] is much too small.'[25] Nevertheless, the *Empress of China* returned to New York in May 1785 laden with 400,000 pounds of tea, exotic cloth, and porcelain. Shaw's partner Randall followed in a separate vessel carrying an additional cargo of tea. All told, the voyage posted a comparably modest profit of 25% percent, which naturally the partners immediately entangled in lawsuits. But no matter. American newspapers celebrated, praising the voyage: 'it presages a future happy period in our being able to dispense with that burdensome and unnecessary traffick, which hitherto we have carried on with Europe – to the great prejudice of our rising empire, and future happy prospects of solid greatness.'[26]

Lobbying for Preferment and Monopoly

The story of the voyage of the *Empress of China* is well known to history, and its scholars have drawn a straight line drawn from this lucky beginning to the wealth Americans gained from the China trade in the first generations of the 19th century. Continuing the examination of Shaw into his subsequent voyages, however, shows a more complicated picture of these first American forays. The first voyage was defined by the struggle to obtain needed expertise, but Shaw's further efforts reveal his attempts to deploy that experience to satisfy his private ambitions. Above all, Shaw's work reveals expectations that American trade with China should develop under the aegis of a government-chartered company with a sanctioned monopoly.

Shaw had already begun to seek independence from Robert Morris and his partners before the return to New York; Randall and Shaw had ham-handedly chartered a second ship, the *Pallas,* to carry additional tea and merchandise to America. Only delicate negotiation averted the two cargoes from competing in the marketplace.[27] Shaw and Randall also planned to establish a sideline business as consultants in Canton for inexperienced American merchants, promising to 'do the business of any vessel in the most advantageous manner'.[28] Perhaps most promising was the fruits of some technological espionage: while in Canton, Shaw had acquired the design for a proper English Indiaman, *The New Triumph.* Before he sailed out on his second voyage to China, Shaw forwarded the plans to New England for immediate building of a ship 'expressly for the Canton trade'.[29] This vessel, *Massachusetts,* was the largest ship yet constructed in America; Shaw made his third voyage to Canton with the ship, where it was praised at Batavia and Canton 'as perfect as the then state of the art would permit'.[30] Shaw told Randall, regarding the *Massachusetts,* 'I anticipated a moderate fortune to ourselves in the course of three voyages more.'[31] In reality, the ship had been improperly constructed with unseasoned wood, had its hold inadequately ventilated, and suffered catastrophic rot in its inaugural voyage through the tropics. Shaw sold it to Danish merchants in Canton for a loss; 'but I saved my credit,' he explained to his patron Henry Knox, 'and by this well timed piece of disinterestedness, I have laid the cornerstone of a moderate fortune.'[32]

In these efforts, Shaw worked as a normal merchant; but he also had plans to reinforce his ambitions in China with his connections with the nascent national elite, rooted in his Continental Army days. Shaw was a founding member of the Society of the Cincinnati, a fraternal society of army officers, the national elite, who quickly embedded themselves in the Congress, state governments, elite merchant firms, and frontier land speculation schemes.[33] In his first stint in Canton, Shaw commissioned porcelain dishes painted with emblems of the Cincinnati. He gave a Chinese painter two engravings of Minerva, an elegant figure of a military man, and a medal of the order as models for a design, but found the artist 'unable to combine the figures with the least propriety; though there was not one of them which singly he could not copy with the greatest exactness'.[34] (On subsequent voyages Shaw commissioned more attractive porcelain pieces with the logo of the order, specially personalized and monogrammed for well-placed Cincinnati members.) With these exotic goods, Shaw marketed the China trade – and himself – to the military-governmental elite upon his return to the United States. George Washington bought this first set of this chinaware, albeit only when it went on sale.[35] In Canton, Shaw had learned that it was customary to subtly bribe the Hoppo, the Mandarin who controlled access to the trade, with gifts of ingenious clockwork gadgets.[36] This Cincinnati porcelain showed the other side of the coin: Shaw tried to use glamorous, politically-freighted objects to curry favor and access with American government officials.

Shaw used words as well as objects to advance himself. His first voyage with the *Empress of China* yielded two key texts: a detailed travelogue journal and a formal report to John Jay, the Foreign Secretary for the United States. The journal has long been mined for quotations and descriptions; Shaw clearly wrote it with an eye for circulation among friends and patrons. His 1785 letter to Jay, however, points to his emerging strategy that he would deploy across his subsequent voyages to Canton. As soon as the he was returned to New York, Shaw drafted his letter to Jay 'to communicate... for the information for the fathers of the country, an account of the reception its citizens have met with, and the respect with which its flag has been treated'.[37] Captain Green had been dispatched with copies of the Declaration of Independence and the subsequent treaties that ended America's war with Great Britain; Shaw's first report to the officers of the US government spoke to those diplomatic concerns. He was glad to have done his patriotic duty 'to attract the attention of the Chinese towards a people of whom they had but very confused ideas... [and] to place the Americans in a more conspicuous point of view'.[38] Though it took a bit of time for the Chinese to 'fully comprehend the distinctions between us and Englishmen,' the merchants and mandarins 'styled us the *New People'*. The first Americans in China, Shaw proudly reported, were ultimately treated 'in all respects as citizens of a free and independent nation'.[39] Shaw conveyed to Jay two pieces of silk, diplomatic presents from the *Fuen* in Canton, 'a mark of his good disposition towards the American nation'. 'To every lover of his country, as well as to those...immediately concerned in commerce, it must be a pleasing reflection, that a communication is thus happily opened between us and the eastern extremity of the globe.'[40]

With this letter, Shaw was angling for preferment and a government position. It worked. (He needed it; accounting irregularities and lawsuits had frozen his expected commission from the profits of the voyage of the *Empress of China*.) After a half-year ashore, Shaw contracted to sail again for China. 'I am now certain that this undertaking will answer my most sanguine expectations,' he explained to his brothers.[41] On this voyage, which departed New York in February 1786, he would also go with a commission as Consul at Canton for the United States. Given no salary, he would conduct his own business for profit alongside his diplomatic responsibilities as representative of his nation's interests. When the Dutch governor-general at Batavia later met Shaw, another American sailor recalled how the European official 'seemed surprised to see a consul doing business like a common speculating merchant'. The official snidely remarked, 'if the American government was not able to pay consuls, it would do better not to send them abroad. It... would only hurt the commercial interest of other commanders and supercargoes, as they naturally had every advantage over the man who had not the confidence of his own government to help him.'[42] Nevertheless Shaw and his superiors knew what they were doing, mixing government power with private enterprise. Shaw would attempt to capitalize on this new access to the corridors of political power.

Shaw's subsequent communications to his superiors in government took a different tone from his first report to Foreign Minister Jay in 1785. Before he secured a government position, Shaw was careful to hold back details about the mechanics of the China trade: 'The situation of the Europeans at Canton is so well known,' he had demurred, 'as to render a detail unnecessary.'[43] His two official communications from Canton in January and December 1787 to Jay, by contrast, not only carefully explained the procedures of the trade system, but also sought to reshape the policies of the American government. These letters precisely described America's European competitors in China. Shaw recounted each country's methods and advantages, repeatedly distinguishing between trade conducted by 'companies' or 'national establishments' and trade undertaken by individuals or 'private persons'.[44] The Dutch, for example, were well-positioned in the China trade because of their southeast Asian colonies. The Portuguese, by contrast, though possessed of a base at Macao, were 'scarcely the shadow of their former consequence' and were reduced to trading voyages sponsored by individual merchants. He suspected the Spanish were forming a new company based in Manila. The French at present had no organized national company, but 'experiments' by their king would 'very probably' result in 'the forming of a new company'.[45] Most striking of all was the example of the British East India Company, the holder of a government-sanctioned monopoly. Shaw explained to Jay how internal accounting within the company allowed arbitrage of different commodities in different ports, which ideally 'render[ed] it unnecessary for the company to export... any specie for carrying on their commerce with the Chinese'. The agents of the British East India Company 'seem to be not only aiming at a monopoly of the tea trade for Europe, but appear to have in view the exclusive commerce of this division of the globe'.[46] Shaw was hopeful for his country's position: America's access to ginseng could be 'as beneficial... as her mines of silver and gold' had been for the Spanish empire, but purposeful organization and regulation was necessary. Shaw proposed a legislative inquiry 'whether it would be for the interest of America to prevent the export of it in any but American bottoms'.[47] Looking at the highly-controlled, government-chartered European companies, Shaw suggested the young United States act in kind in their trade with China.

In these 1787 letters, Shaw was implicitly making the case for the chartering of an 'American East India Company' – an idea floating among the governmental elite of the young republic.[48] After the Americans reorganized their national government with the ratification of a strongly centralizing constitution in 1788, Shaw continued to push the point. Reappointed as consul by the newly inaugurated President George Washington, Shaw suggested his patron, Secretary of War Henry Knox, lobby to expand the commission to 'Consul for China and the Countries Eastward of the Cape of Good Hope or something to that extent'. [49] Shaw desired a diplomatic portfolio that mirrored the monopoly of the British East India Company. (Shaw would also undertake exploration of British trade ports in India). While Knox failed to deliver for his protégé, Shaw continued to make the case for government intervention in the China trade, pushing in his 1790 report to President Washington that 'the commerce of a nation [is] one of the principal objects of the attention of its rulers...'. He particularly pointed to an episode at Batavia where 'to my no small astonishment, I was informed... that all commerce with the Americans was prohibited'.[50] He warned President Washington that the American merchants in the Indies lacked collective clout, and as a result were 'classed with smugglers' and 'evil reports' of poor behavior by one merchant had closed the trade to all. 'I feel for the honor of my country,' he lamented.[51] Only a government organized trade, he implied, would have the strength to counter these anti-competitive machinations by European competitors.

These suggestions about reorganizing America's trade with the east reached their maximum amplification in a 1791 letter addressed to Secretary of the Treasury Alexander Hamilton from Shaw's trading partner and vice consul Thomas Randall. The product of many conversations between Shaw and Randall in slow, hot days in Canton and on long ship voyages, this letter answered Hamilton's request for a report on the China trade. Hamilton had set in motion a massive reorganization of America's tax and financial system; Randall and Shaw knew the Secretary's goal was 'to promote and secure a happy and prosperous trade to your country'.[52] This text detailed their experiences at Canton since 1784, reporting how American merchants had 'suffer[ed] under the operation of the Chinese government' and suggested 'regulating the mode of conducting trade... to that port'.[53] Randall detailed how he and Shaw had realized the Chinese merchants 'possessed more power than we were aware of;' they systematically suppressed payments for imported commodities by playing foreign merchants against each other.[54] The situation was deteriorating: in their second voyage, Randall and Shaw had found 'the extortion of the mandarins for fees of every kind had rather increased,' and found themselves 'defrauded in the quality of the teas' they were allowed to purchase.[55] Looking forward, Randall saw promise in America's supply of ginseng, which could sell 'for nearly its weight in gold,' but only if properly sorted and processed. 'But this I think cannot take place at present... in the hands of individuals.' It required 'too much capital for a single merchant to risque,' and partnerships 'made up in too hasty a manner' were inadequate to the task.[56] 'I must confess it is my opinion,' Randall concluded, echoing Shaw's earlier suggestions, 'that the trade to Canton will never flourish, but in the hands of a well-regulated company, which will not be easily formed without the protection of government.'[57]

Shaw's official letters as Consul in Canton, amplified by his partner Randall's report to Hamilton, point to a sustained project: convincing the powers-that-be in the United States that a nationally chartered company, something like an American East India Company, was necessary. As Shaw had explained to Jay in 1787, even the 'co-hoang' of Chinese merchants at Canton was 'expressing our idea of a trading company.'[58] Just as Shaw had copied the design of the *Massachusetts* from a British ship, he had British corporate forms in mind. Randall spoke in concert with Shaw, explaining to Hamilton 'the necessity of a nation showing an active protection of their subjects trading to China'. State-backed companies possessed leverage to counter the 'frauds and impositions...practiced by the Chinese on every trading nation there'. A proper monopoly company would control bad actors and defend the collective interests of American merchants.[59] When Shaw encountered the challenges of this complicated trade system in Canton, the mental sea chest into which he reached held the standard 18th-century idea of state-sponsored monopoly companies. This was an era of constitutional flux in America: the powers of the state and federal government were unsettled. In one plain reading of the new Constitution, chartering corporations had been a privilege reserved to the state governments and a power not expressly given to the federal Congress. But Treasury Secretary Alexander Hamilton had successfully pushed past opposition to a federally incorporated bank that could organize the national money supply and channel lending and investment in ways that strengthened the new central government. Shaw and Randall, as veteran members of the Society of the Cincinnati and good federalists, wanted Hamilton to incorporate America's burgeoning China trade into his controversial financial and industrial plans for building the United States a respectable fiscal-military state.

***

It never came to pass. A centralized American East India Company was never chartered. Instead, individual American firms would profitably compete against each other and less agile European state-backed companies for shares of the lucrative Canton trade. Samuel Shaw never saw this outcome. He fell grievously ill in China and died at sea in 1794, returning from his fourth voyage to the Indies. Shaw's death was not remarkable: his younger brother Nathaniel had died at sea on the ill-fated voyage of the *Massachusetts*. Amasa Delano, one of that ship's officers, later charted in his autobiography that 18 of that ship's 62 man crew died in the east, along with five of her eleven officers.[60] But even if Shaw had survived, this vision for a government-sponsored company to monopolize America's trade with China was doomed. Alexander Hamilton's ambitious financial plans had run into unexpected opposition in the Congress; he got his central bank and his tax policy, but expended practically all his political capital provoking this first crisis of constitutional interpretation in the United States. Thereafter, the centralizing Federalist Party could not even establish a national university, despite President Washington's call for this otherwise uncontroversial institution in his first annual address to Congress in 1790.[61] Reorganizing and centralizing trade with China, despite Shaw's insistence on its necessity, would be orders of magnitude more politically challenging. It was not attempted.

Historians point out how the profitable half-century of free-wheeling competition among American mercantile firms helped prove Adam Smith's denunciation in *The Wealth of Nations* of the inefficiencies and corruption of state-chartered monopoly companies.[62] In his own moment, however, Samuel Shaw envisioned no such thing. He expected the chartering of an American East India Company. He lobbied for it. For who would be better than the indispensable, well-connected Major Shaw to stand close to its helm?

The 'China trade' the newly independent Americans sought to join in 1784 was remarkably different in mechanics from the ancient overland routes of the Silk Roads: tall ships undertook hemisphere-spanning voyages in contrast to the short-distance, hand-to-hand exchange that shaped the trade across Central Asia to the Indian subcontinent and the Mediterranean. Nevertheless, Shaw's story connects to the broader history of the Silk Roads and its myriad stories of individual ambition and the cultivation of patronage with exotic objects from distant lands. Shaw's story shows a set of elite individuals trying to figure out and access this source of power, and then attempting to reorganize their own side of the mercantile exchange. This process certainly iterates back through the centuries. While Shaw's ideas about monopoly trade were rooted in early modern mercantile practices, and the constitutional questions at the heart of Shaw's advice to his government were particular to the early American republic, his ambitions, experiences, and the exotic goods he carried were as old as the Silk Roads themselves.

Notes

1 Adams 1853, 8:343–44; Dudden 1992, 4.
2 For studies patriotically emphasizing 'firstness', see Dulles 1930; Tamarin and Glubok 1976; Smith 1984. On tea, material culture, and economic evolution of the trade, see Christman 1984; Dudden 1992; Downs 1997; Fichter 2010; Haddad 2013; Merritt 2017. On race and identity, see Morrison 2014. For an important inquiry into the finances of both American and Chinese merchants in this Canton tea trade, see Du 2018.
3 Quincy 1847, 5.
4 Ruddiman 2014, 131.
5 Quincy 1847, 110; Dulles 1930, 7.
6 Ver Steeg 1953, 2.
7 Faragher 1993, 54; Haddad 2013, 9–11.
8 Smith 1984, 25.
9 Christman 1984, 45, 50.
10 Smith 1984, 58.
11 Smith 1984, 70.
12 Smith 1984, 75.
13 Smith 1984, 104.
14 Ver Steeg 1953, 9; Christman 1984, 44.
15 Transformations of the economy over two centuries make value comparisons complicated: in addition to comparing as a portion of GDP, this 1784 capital outlay compared to labor value today would equal perhaps only 150 million dollars. The overall size of the economy has grown exponentially faster than the cost of unskilled labor. Calculators at Williamson 2020.
16 Morris 1973, 9:64
17 Shaw 1775–1887, 24 December 1783; Quincy 1847, 112.
18 Morris 1973, 9:66.
19 Smith 1984, 104.
20 Quincy 1847, 154; Dulles 1930, 9.
21 Quincy 1847, 164.
22 Quincy 1847, 337–341.
23 Quincy 1847, 342; Haddad 2013, 18.
24 Syrett 1965, 9:41.
25 Swift 1886, 485.
26 Smith 1984, 224–225.
27 Smith 1984, 226–227.

28 May and Fairbank 1986, 260; Merritt 2017, 134.
29 Quincy 1847, 117; Delano 1970, 25.
30 Quincy 1847, 117.
31 Christman 1984, 66.
32 Christman 1984, 67.
33 Hünemörder 2006.
34 Quincy 1847, 198–199.
35 Christman 1984, 61. Surviving pieces from Cincinnati porcelain sets are held in the collections the Museum of Fine Arts in Boston and at the Society of the Cincinnati headquarters in Washington, D.C.
36 Quincy 1847, 176.
37 Quincy 1847, 337.
38 Quincy 1847, 337.
39 Qunicy 1847, 338.
40 Quincy 1847, 341.
41 Quincy 1847, 114.
42 Delano 1970, 38.
43 Quincy 1847, 339.
44 Quincy 1847, 346–347.
45 Quincy 1847, 344–345.
46 Quincy 1847, 343, 353.
47 Quincy 1847, 351.
48 Fichter 2010, 39–45; Haddad 2013, 28.
49 Christman 1984, 67.
50 Quincy 1847, 359.
51 Quincy 1847, 118, 357–358.
52 Syrett 1965, 38.
53 Syrett 1965, 39.
54 Syrett 1965, 40.
55 Syrett 1965, 42.
56 Syrett 1965, 44.
57 Syrett 1965, 54.
58 Quincy 1847, 347.
59 Syrett 1965, 50, 53.
60 Delano 1970, roughly 26–29.
61 Thomas 2015, 64–65.
62 Fichter 2010, 23–25.

Bibliography

Adams, C.F. 1853. *Works of John Adams*, Boston.

Christman, M.C.C. 1984. *Adventurous Pursuits: Americans the China Trade, 1784–1844*, Washington, D.C.

Delano, A. 1970. *Narrative of Voyages and Travels in the Northern and Southern Hemispheres*, Upper Saddle River, N.J.

Downs, J.M. 1997. *The Golden Ghetto: The American Commercial Community at Canton and the Shaping of American China Policy, 1784–1844*, Bethlehem, P.A.

Du, D. 2018. Green Gold and Paper Gold: Seeking Independence through the Chinese-American Tea Trade, 1784–1815, *Early American Studies* 16, 151–191.

Dudden, A.P. 1992. *The American Pacific: From the Old China Trade to the Present*, New York.

Dulles, F.D. 1930. *The Old China Trade*, Boston.
Faragher, J.M. 1993. *Daniel Boone: The Life and Legend of an American Pioneer*, New York.
Fichter, J.R. 2010. *So Great a Proffit: How the East Indies Trade Transformed Anglo-American Capitalism*, Cambridge.
Haddad, J.R. 2013. *America's First Adventure in China: Trade, Treaties, Opium, and Salvation*, Philadelphia.
Hünemörder, M. 2006. *The Society of the Cincinnati: Conspiracy and Distrust in Early America*, New York.
May, E.R. and Fairbank, J.K. 1986. *America's China Trade in Historical Perspective*, Cambridge.
Merrit, J.T. 2017. *The Trouble with Tea: The Politics of Consumption in the Eighteenth-Century Global Economy*, Baltimore.
Morris, R. 1973. *The Papers of Robert Morris, 1781–1784*, Pittsburgh.
Morrison, D.A. 2014. *True Yankees: The South Seas and the Discovery of American Identity*, Baltimore.
Quincy, J. 1847. *The Journals of Major Samuel Shaw, the First American Consul at Canton with a Life of the Author*, Boston.
Ruddiman, J. 2014. *Becoming Men of Some Consequence: Youth and Military Service in the Revolutionary War*, Charlottesville.
Shaw, S. 1775–1887. *Samuel Shaw Papers*, Ms.N-49.47, Massachusetts Historical Society.
Smith, P.C.F. 1984. *The Empress of China*, Philadelphia.
Swift, J.W. 1886. Notes and Queries, *Pennsylvania Magazine of History and Biography* 9, 485–86.
Syrett, H.C. 1965. *Papers of Alexander Hamilton*, New York.
Tamarin, A. and Glubok, S. 1976. *Voyaging to Cathay: Americans in the China Trade*, New York.
Thomas, G. 2015. *The Founders and the Idea of a National University*, New York.
Ver Steeg, C.L. 1953. Financing and Outfitting the First United States Ship to China, *Pacific Historical Review* 22, 1–12.
Williamson, S.H. 2020. Seven Ways to Compute the Relative Value of a U.S. Dollar Amount, 1790 to present, *MeasuringWorth.com*. www.measuringworth.com/calculators/uscompare/index.php (accessed 20 March 2019).